WJEC
GCSE Science and GCSE Additional Science

Philip Barratt, Morton Jenkins, George Snape
Editor: Morton Jenkins

EDUCATION
AN HACHETTE UK COMPANY

Acknowledgements

The Publishers would like to thank the following for permissions to reproduce copyright material:

p. 3 *t* Mary Evans Picture Library, *b* Courtesy BFI; **p. 4** *t* NHPA / John Shaw, *cl* NHPA / T.Kitchin & V.Hurst, *cl* Corbis / Alan & Sandy Carey / zefa, *cr* NHPA / Manfred Danegger, *bl* FLPA / Nigel Cattlin; **p. 5** *t* NHPA / Mike Lane, *c* Alamy / Danita Delimont, *b* Corbis / George D. Lepp, *bl* FLPA / Nigel Cattlin, *br* NHPA / Martin Harvey; **p. 6** *t* FLPA / Nigel Cattlin, *c* NHPA / Daniel Heuclin; **p. 10** Becca Law; **p. 12** *t* Science Photo Library / Dr Jeremy Burgess, *c* Empics / PA; **p. 13** National Library of Medicine / Science Photo Library; **p. 14** Science Photo Library / George Bernard; **p. 16** *c* Science Photo Library / Alfred Pasieka, *b* Ria Novosti / Science Photo Library; **p. 17** *tl* Science Photo Library / Peter Menzel, *cl* Science Photo Library / CNRI, *bl* Science Photo Library / Professor P Motta, Department of Anatomy, University La Sapienza; **p. 18** *all* Science Photo Library / Dr Bernard Lunard; **p. 19** Biophoto Associates; **p. 23** Science Photo Library; **p. 26** Science Photo Library / Hattie Young; **p. 28** Science Photo Library / Professor K Sneddon & Dr T Evans; **p. 29** Science Photo Library / A Barrington Brown; **p. 38** Biophoto Associates / Science Photo Library ; **p. 39** Science Photo Library / Astrid & Hanns-Frieder Michler; **p. 40** Science Photo Library / Manfred Kage; **p. 53** NHPA / George Bernard; **p. 54** Science Photo Library / Eye of Science; **p. 58** Science Photo Library / Will McIntyre; **p. 64** *t* Life File Photographic Library Ltd / Emma Lee, *c* Science Photo Library / Sheila Terry, *b* Life File Photographic Library Ltd / Emma Lee; **p. 72** Primrose Peacock / Holt Studios / FLPA; **p. 88** Science Photo Library / Simon Fraser; **p. 91** *t* Science Photo Library / Vanessa Vick, *b* Life File Photographic Library Ltd / Barry Mayes; **p. 92** *t* Life File Photographic Library Ltd / Jan Suttle, *b* Life File Photographic Library Ltd / Jeremy Hoare; **p. 93** *t* NHPA / Mark Bowler, *c* NHPA / John Shaw; **p. 94** *t* NHPA / Bill Coster, *c* NHPA / Manfred Danegger, *b* FLPA / D P Wilson; **p. 95** Still Pictures / Martin Wendler; **p. 97** Mary Evans Picture Library; **p. 98** Science Photo Library / NASA / ESA / STScI; **p. 99** Empics / Deutsche Press-Agentura / DPA; **p. 102** Getty Images / Rischgitz / Hulton; **p. 107** *t* Science Photo Library / Andrew Lambert, *b* Science Photo Library / Charles D Winters; **p. 108** Science Photo Library / Andrew Lambert; **p. 109** *t* Science Photo Library / Andrew Lambert, *c* Alamy / The Photolibrary Wales; **p. 117** Science Photo Library / David Leah; **p. 127** Harald Finster; **p. 128** Alamy / Jeff Morgan; **p. 132** *l* Still Pictures / Ray Pfortner, *r* Science Photo Library / Simon Fraser; **p. 133** Getty Images / John Lamb; **p. 134** *cl* Empics / AP, *bl* Empics / AP; **p. 139** Rex Features / AGB Photo Library; **p. 146** Alamy / Robert Harding Picture Library; **p. 150** Arnold Fisher / Science Photo Library; **p. 153** John Jaszczak; **p. 158** Science Photo Library / Biophoto Associates; **p. 159** Northwest Railway Museum, www.trainmuseum.org; **p. 160** Alamy / Jeff Morgan; **p. 162** Alamy / Jeff Morgan; **p. 163** *t* Courtesy Audi, *b* Alamy / Jeff Morgan; **p. 164** Getty Images / Hulton Archive; **p. 166** *t* Courtesy Comalco, *b* Science Photo Library / Astrid & Hanns-Frieder Michler; **p. 167** National Museums & Galleries of Wales; **p. 168** *t* Alamy / South West Images Scotland, *c* Picture reproduced by courtesy © Synthes GmbH 2006, Switzerland, *b* Corbis / Peter Johnson; **p. 169** *c* Alamy / Warren Kovach, *b* Getty Images / Chris Graythen; **p. 174** *t* Science Photo Library / Planetary Visions Ltd., *c* Alamy / Mark Boulton, *b* Alamy / blickwinkel; **p. 175** Science Photo Library / Sheila Terry; **p. 176** *t* Alamy / John Smaller, *c* Lorna Ainger; **p. 179** Science Photo Library; **p. 180** *t* FLPA / Nigel Cattlin, *b* Photo courtesy of Kellogg Brown & Root LLC and Caribbean Nitrogen Company Limited; **p. 183** *t* Science Photo Library / Martin Bond, *b* Alamy / Jeff Morgan; **p. 185** *t* Courtesy Swish Building Products, *b* Science Photo Library / Charles D Winters; **p. 186** *t* Corbis / Alen MacWeeney, *b* Science & Society Picture Library / Science Museum; **p. 187** NHPA / Andy Rouse; **p. 191** Data from figure 10.2: United Nations Statistics Division; data from figure 10.3: National Resources Defense Council; **p. 195** *bl* North Hoyle, North Wales©onEdition2004, *br* Courtesy Lego; **p. 196** Data from figure 10.12: The British Wind Energy Association; **p. 202** Lorna Ainger; **p. 205** Alamy / D Hurst; **p. 206** *t* Alamy / Jeff Morgan, *c* Courtesy Honda; **p. 207** Renewable Devices Ltd; **p. 213** Getty Images / Marco Di Lauro; **p. 223** U.S. Navy photo by Photographer's Mate 2nd Class Floyd Grimm, *b* Alamy / Ashley Cooper; **p. 224** *t* Science Photo Library / Dr P Marazzi, *c* Science Photo Library / Maximilian Stock Ltd, *b* Alamy / PCL; **p. 228** Science Photo Library / Steve Horrell; **p. 230** *t* Science Photo Library / Tom Van Sant / Geosphere Project, Santa Monica, *b* Alamy / Robert Haines; **p. 233** *t* courtesy Carphone Warehouse, *b* Rex Features; **p. 238** NASA Jet Propulsion Laboratory; **p. 239** NASA; **p. 240** *t* Lowell Observatory Archives, *c* Science Photo Library / Dan Schechter, *br* NASA; **p. 241** *tl* NASA, *tc* Photodisc; **p. 244** *t* NASA, *b* Science Photo Library / NOAO; **p. 245** both NASA; **p. 246** NASA; **p. 247** NASA; **p. 248** Anglo-Australian Observatory; **p. 249** *c* Science Photo Library / David A. Hardy, *b* Science Photo Library / Tony Hallas; **p. 260** Science Photo Library / Blair Seitz; **p. 263** Science Photo Library / CC Studio; **p. 267** Science Photo Library / CC Studio; **p. 272** Science Photo Library / Andrew Lambert; **p. 274** Science Photo Library / Andrew Lambert; **p. 276** George Snape; **p. 279** George Snape; **p. 280** *t* Corbis, *c* Courtesy www.screwfix.com, *b* Science Photo Library / Sheila Terry; **p. 285** George Snape; **p. 297** Getty Images / Joe McBride; **p. 299** Science Photo Library / Takeshi Takahara.

Every effort has been made to trace all copyright holders, but if any have been inadvertently overlooked the Publishers will be pleased to make the necessary arrangements at the first opportunity.

Although every effort has been made to ensure that website addresses are correct at time of going to press, Hodder Education cannot be held responsible for the content of any website mentioned in this book. It is sometimes possible to find a relocated web page by typing in the address of the home page for a website in the URL window of your browser.

Risk Assessment

As a service to users, a risk assessment for this text has been carried out by CLEAPSS and is available on request to the publishers. However, the publishers accept no legal responsibility on any issue arising from this risk assessment: whilst every effort has been made to check the instructions of practical work in this book, it is still the duty and legal obligation of schools to carry out their own risk assessments.

Hachette's policy is to use papers that are natural, renewable and recyclable products and made from wood grown in sustainable forests. The logging and manufacturing processes are expected to conform to the environmental regulations of the country of origin.

Orders: please contact Bookpoint Ltd, 130 Milton Park, Abingdon, Oxon OX14 4SB. Telephone: (44) 01235 827720. Fax: (44) 01235 400454. Lines are open 9.00 am–5.00 pm, Monday to Saturday, with a 24-hour message answering service. Visit our website at www.hoddereducation.co.uk

First published in 2006 by Hodder Education,
an Hachette Uk company,
338 Euston Road
London NW1 3BH

Impression number 10 9 8 7 6 5 4
Year 2011 2010 2009

Cover photos: brain, Sovereign, ISM/Science Photo Library; wind turbine, The Photolibrary Wales/Alamy; woodpecker, © Niall Benvie/Corbis
Illustrations by Barking Dog Art
Typeset in New Baskerville 10/12pt by Pantek Arts Ltd, Maidstone, Kent
Printed in Italy

A catalogue record for this title is available from the British Library

ISBN-13: 978 0 340 88582 6

Contents

Introduction

What makes science different?

A scientific method is a logical, orderly way of trying to solve a problem. It is this logic and order that make scientific methods different from ordinary hit-and-miss approaches.

Remember, though, that scientific methods are not magic. Even the best planned investigations can fail to produce meaningful data. On the other hand, failure can lead to final success. By carefully studying each result, the scientist may find a new direction to take. Often this new direction leads to an even more important discovery than the one first expected.

Several methods are used in science, depending on the nature of the problem. Perhaps the most important is the investigative method. It is by investigating that new knowledge is gained and new concepts are developed.

There are many opportunities to investigate in your study of science. The steps in this investigative method are logical and orderly.

Decide on what you want to investigate

You cannot understand why something is happening unless you observe it in the first place. Science calls for the kind of mind that makes observations and asks questions about them. For example, why do you breathe faster when you run? Why do some metals melt at a lower temperature than others? Why are rainbows created?

Well planned investigations can answer each of these questions. However, in science, every answer raises new questions. Successful research leads to new research and new knowledge that sometimes solves problems.

Collect information

Scientists must build on the work of other scientists. Otherwise science could not advance beyond what one person could learn in one lifetime. Before beginning an investigation, the scientist needs to study all the important information that has to do with the observations.

Often it turns out that people have already answered some of the questions. For this reason, access to a library of scientific books and journals is an important part of any investigation. As a special kind of library, the internet is of help too, as are CD-ROMs and software packages. This textbook will also assist in your background reading.

Make a prediction

The information available may not fully explain the scientist's observations. The researcher must then begin to investigate. At this point, a hypothesis is needed. The hypothesis is a sort of working explanation or prediction. For example, you might predict that the rate of a chemical reaction will increase as the temperature increases.

This gives the researcher a point at which to aim. However, no matter how reasonable the hypothesis seems, it cannot be accepted until it has been supported by a large number of tests. A research worker must be open-minded enough to change or drop the original hypothesis if the evidence does not support it.

Test the hypothesis

The scientist must plan an investigation that will either support or disprove the hypothesis. This means that the investigation must test only the idea or condition involved in the hypothesis. All other factors must be removed or otherwise accounted for. The one factor to be tested is called the single variable.

A second investigation is often called a control, carried out along with the first. In the control, all the factors except for the one being tested are the same as in the first investigation. In this way, the control shows the importance of the missing variable.

Observe and record data from the investigation

Everything should be recorded accurately. Correct international standard units must be used. The record of the observations may include notes, drawings, graphs and tables. From time to time, sensors (devices that plug into a computer) can be of help in making measurements in investigations.

Observations and measurements are called data, and in modern research, data are often processed using information technology. For example computer spreadsheets can help the researcher to draw graphs and analyse data. A word processor can help a scientist to prepare experiment reports.

Draw conclusions and evaluate

Data have meaning only when valid conclusions are drawn from them, and validity can be assessed only when the investigation has been evaluated critically. Such conclusions and evaluations must be based entirely on observations made in the investigation. If other investigations continue to support the hypothesis, it may become a theory.

Improving your grade

Some people worry about examinations and tests. Others are more relaxed. We hope this advice will help everyone. Certainly examinations need not be horror stories, provided that you (a) *want* to do well, and (b) are *well prepared.*

The first thing to do if you want to improve your grade is to recognise if and when you are going wrong. Students often seem interested in marks or grades given for work. The danger is that the marks or grades are remembered but the mistakes are forgotten.

Follow the steps below to help you do better in answering test and examination questions in the future.

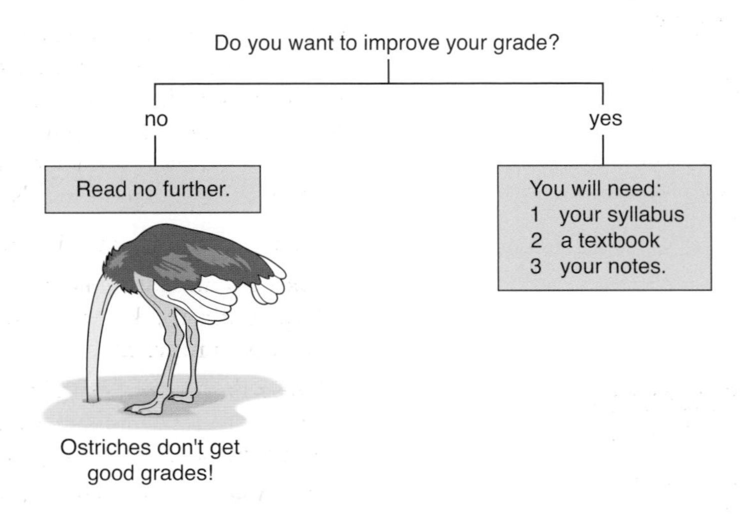

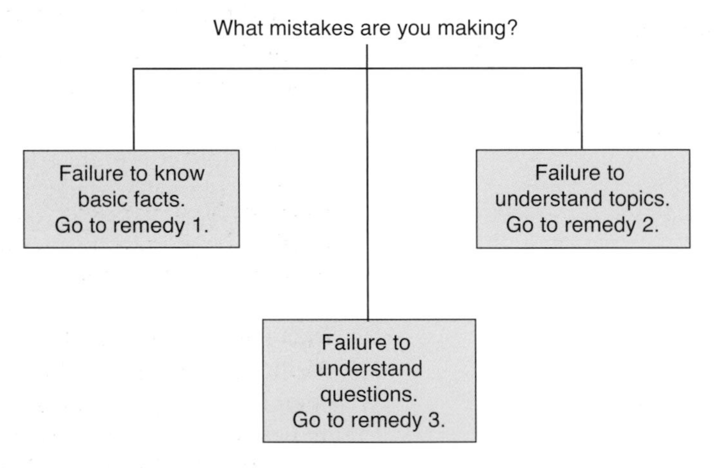

Remedy 1

DAILY DOS AND DON'TS

DO *spend a sensible amount of time learning facts.*

DON'T *leave learning facts until the last minute.*

Learning facts is the most time-consuming task of all. You do not really know a fact until it is more difficult to forget it than to remember it. Try these.

- Repeatedly read a concise set of notes. If you think that is boring, so are low grades.
- Draw simple flow diagrams, using key words.
- Summarise topics, using key words as headings and subheadings.
- Divide your learning time into 30 min periods separated by 10 min breaks.

Remedy 2

- Ask for *explanations* of topics. If you are invited to ask questions, then do ask. Do not be shy.
- If you do not understand a topic, do not try to learn it like a parrot. Go back to the beginning of the topic, and build up your understanding step by step, or even better, get someone else to explain it to you.
- Use past examination questions and answers. You can find them in textbooks, study aids, and computer software. Study the topic first, and then look at the types of question set on the topic. Try to answer the questions *before* looking at the answers.

DAILY DOS AND DON'TS

DO *ask your teacher for help.*

DON'T *learn like a parrot if you don't understand what you are learning.*

Remedy 3

Terms that introduce questions are sometimes misunderstood. This can lose you marks because examiners have a rigid mark scheme for each question. Even though you may have written a lot of correct information, it will not gain you marks if it does not fit the exact question you are answering. Take care with questions beginning with the following common words:

- '**State**' means present in the form of a *concise* statement.
- '**Explain**' means make known in detail and make understood. Explanations take up more time and space than statements. They are worth more marks.
- '**Describe**' means explain, in words or with the help of diagrams, the form or function of something. It can also mean explain how a process works or how you carried out an experiment.
- '**Summarise**' means give a brief statement of …
- '**List**' means write, one after the other, in the style of a catalogue.
- '**Distinguish between**' means state essential features that make something different from something else.

DAILY DOS AND DON'TS

DO *read the question carefully.*

DON'T *write answers that are irrelevant.*

Here's how to do it!

Question: State the meaning of the term 'respiration'.
Answer: Respiration is the way in which living things release energy from glucose or other substances.

Question: Explain clearly the term 'respiration'.
Answer: Respiration is the way in which living things release the chemical energy in complex carbon compounds for use in their activities. They usually use oxygen. and produce carbon dioxide and water. Respiration takes place in the mitochondria of all cells.

Question: Describe the process of respiration briefly.
Answer: Respiration involves the breakdown of complex carbon compounds, such as glucose, and the release of energy. Oxygen is usually used in this process, which involves a complicated series of reactions. The chemical products of respiration are usually carbon dioxide and water.

Safety in the laboratory

Questions

1 Look up the meanings of the following words:

a corrosive
b irritant
c flammable
d explosive
e toxic.

2 Explain what **each** of the following symbols means.

a 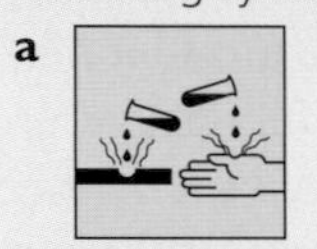

b 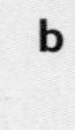 

c 

d  

e 

f 

- You must not enter a laboratory unless instructed to do so by a teacher.
- You must not do anything with equipment or materials unless told to do so by a teacher. Instructions should be followed exactly.
- You must wear eye protection when told to do so, and keep it on until told to take it off when all practical work, including clearing away, is finished.
- When instructed to use a Bunsen burner, make sure that hair, scarves, ties, etc. are tied back or tucked in to keep them well away from the flame.
- When working with liquids, you should normally stand up; then you can move out of the way quickly if there is a spill.
- Never taste anything or put anything in your mouth when in the laboratory unless your teacher tells you to do so. This includes sweets, fingers and pencils, which may have picked up dangerous chemicals from the bench.
- If small amounts of chemicals or microbiological cultures get on your hands or any other part of your body, wash them off. Wash your hands after working with chemicals or with animal or vegetable matter.
- Report any accident to the teacher. This includes burns, cuts or chemicals in the mouth and eyes or on the skin.
- Keep your bench clean and tidy, with bags put in a place where people will not trip over them. Wipe up small splashes with a damp cloth and report bigger ones to the teacher. Put waste solids in the correct bin, never in the sink.
- Be aware of other students working around you.
- Always read the labels on bottles carefully. Chemicals should be labelled with appropriate hazard symbols. All chemicals are potentially dangerous. Even water can be a hazard.

Radioactive

Biohazard

Harmful

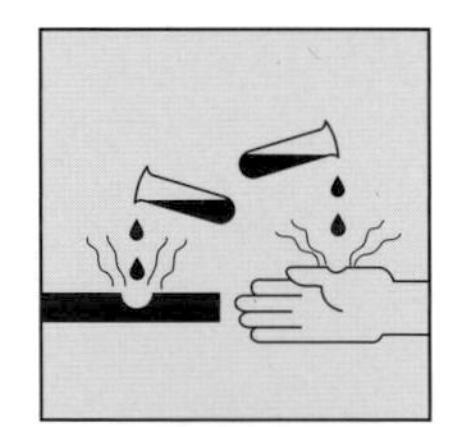

Corrosive

Oxidising

Irritant

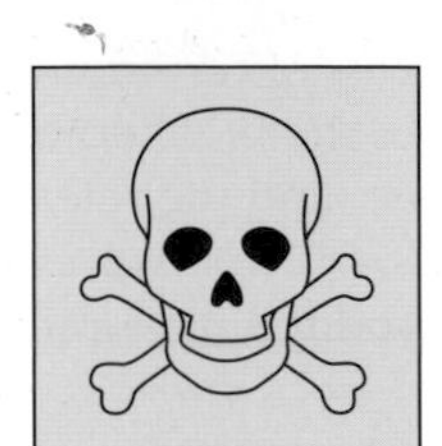

Toxic

Explosive

Flammable

Some hazard warning labels

Chapter 1 Genes and variety

By the end of this chapter you should:

- understand why and how organisms are classified;
- know that the distribution and numbers of organisms in a habitat can be explained in terms of adaptation, competition, predation and pollution;
- know that sexual reproduction leads to genetic variation, and that asexual reproduction produces clones;
- know that a mutation may be harmful or be a benefit, and is a source of genetic variation;
- understand that mutations may have several causes;
- know that variation and selection lead to evolution or extinction;
- know that there is a fossil record that is evidence of evolution;
- know that the nucleus of a cell contains chromosomes that carry genes;
- know that the gene is a section of DNA;
- understand that cells divide by mitosis so that growth takes place, and that cells divide by meiosis to produce gametes;
- know that there is a mechanism of monohybrid inheritance when there are dominant and recessive alleles;
- know that some diseases are inherited;
- understand how gender is determined in humans by sex chromosomes;
- understand the basic principles of genetic engineering;
- understand the potential benefits and ethical dilemmas created by advances in cloning and genetic engineering.

Adaptation and competition

Variety

In any small area of land, you will find many different kinds of living things. In a garden, for example, you may come across earthworms, slugs, snails, centipedes, insects, spiders, birds, and even moles and voles. There are very obvious differences between these creatures. Biologists use these differences to classify animals into groups, the members of which have certain features in common. The same is true of plants. Your garden may provide a home for mosses, ferns and plants with flowers, for example.

Millions of types of living things are already known to science, and many others are being discovered each year. In fact, there are so many that it would be impossible to study every one of them in a lifetime.

Biologists put living things into groups. In each group there are organisms that are similar in various ways. The similarities may be based on features of the organisms that you can see easily. However, scientists now usually use the structure of an organism's **DNA** (deoxyribonucleic acid) to determine whether it is closely related to another organism.

The group or kingdom called *animals* includes creatures as different as house flies and elephants. In what way are they similar? One likeness is that they move in search of their food, like all animals at some stage in their lives. The group or kingdom called *plants* includes such organisms as mosses and pine trees. In what way are they similar? One similarity is that they can make their own food by photosynthesis (see Chapter 3). This is the basis of a logical way of classifying organisms. Although the word 'classification' has a dull and complicated sound, it is essential for the understanding of the variety of life on Earth.

The principle of classification

The system upon which the classification is based is shown in Figure 1.1. It is the basis of the classification of animals.

We shall look at the stoat to illustrate the main principle of the logical system of classification. You can see that each of the arches in Figure 1.1 ends with the stoat.

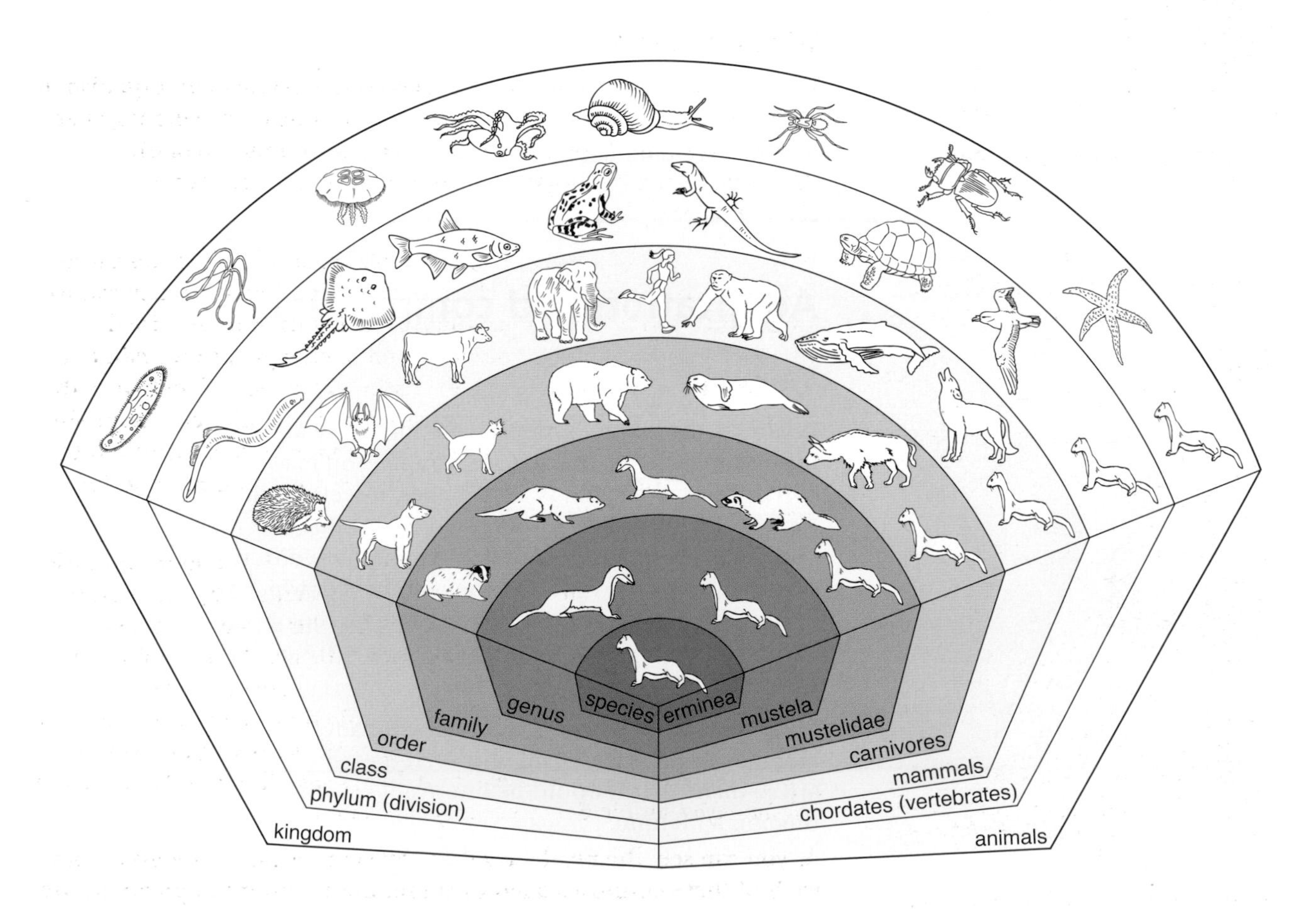

Figure 1.1 The principle of classification of animals

All living creatures are included in the animal kingdom, and so examples of many different kinds of animal (that is, biodiversity) are included in the top arch. It starts with a type of animal that has only a single cell, but also includes worms, molluscs and insects.

The first division of the complete animal kingdom to which the stoat belongs is the **chordates**, which includes all animals with backbones (**vertebrates**). The **division** (which used to be called the **phylum**) is then divided into **classes**. The stoat belongs to the **mammals** class. Mammals also include hedgehogs, cows, dolphins, bats and humans.

What do all these have in common? Classes are divided into **orders**. The stoat belongs to the **carnivores** order. Orders are divided into **families**. The stoat belongs to the family called the **Mustelidae**, often called the weasel family. This family includes the polecat and the badger. Note that members of the same family have many more similarities than members of the same class.

The family is divided into **genera** (the plural of genus). The stoat belongs to the genus *Mustela*. Other genera of the same family include *Meles*, the badger, and *Lutra*, the otter.

Finally we arrive at the stoat considered as its own **species**. All living organisms that have been named by scientists have been given two names, a genus and a species. This so-called binomial system is used in all the countries in the world. The stoat is of the genus *Mustela* and the species *erminea*, so it is called *Mustela erminea*. Biologists in China, Egypt and Wales would all use the same scientific name for the stoat.

Figure 1.2 Carolus Linnaeus, the father of classification

What's in a name?

It is very important that the correct scientific name of an organism is used for international communication, because every language has its own word for a particular organism (the common name). Scientific names for organisms are usually based on Latin or classical Greek.

Why were those languages chosen? It all began with Carl von Linne (1707–78) (see Figure 1.2). His great contribution to science was to name and classify living organisms in the logical way described above. His love of Latin influenced him to change his own name to the Latin form, Carolus Linnaeus. Latin and ancient Greek were the languages used at that time in universities where science was taught to students. Today, we are left with the remains of this custom, with Latin and classical Greek words being used for organisms and for many parts of the body.

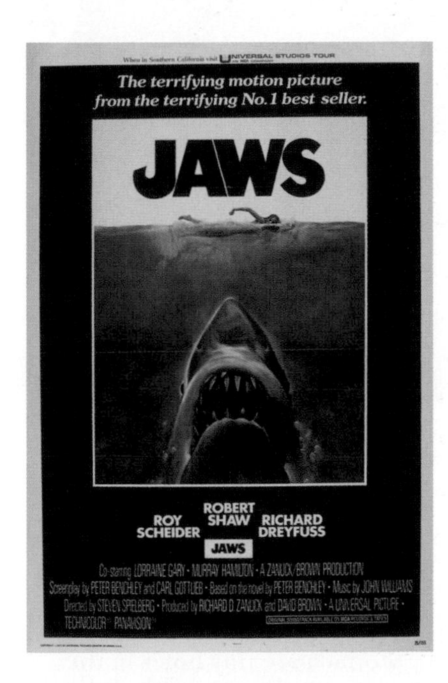

Figure 1.3 Great white shark, as seen by millions of people

The film *Jaws* has been seen throughout the world, and its soundtrack has been translated into most languages (see Figure 1.3). In English, we know that *Jaws* is a great white shark, but this name is not used universally. The great white shark is one of the best known animals in the world because of the film, but its common name in some European languages would only be recognised by those who speak that language. For example, in German, the name is *weisshal*, in Spanish, *jaqueton*, in French, *requin blanc*, in Italian, *squalo bianco*, in Maltese, *kelb il-bahar abjad*, and in Serbo-Croat, *pas ljudozder*.

As you can see, the result could be total confusion if biologists from each of these countries tried to communicate information about the species using their own native language. Thanks to Linnaeus, they

Figure 1.4 Desert fox

Figure 1.5 Arctic fox

Figure 1.8 This cactus has adapted to be able to survive in desert conditions.

all use the same scientific Latin name for the great white shark. This is *Carcharodon carcharias*, first used by Linnaeus in the eighteenth century.

Every time a new type of organism is discovered, it is given two universally understood scientific names, a genus and a species. An international organisation called the International Commission on Zoological Nomenclature (ICZN) decides on the names of newly discovered animals.

Adaptation

All organisms living today have become **adapted** to their environment in many specialised ways. Each part of the Earth has its own special animals and plants.

For instance the arctic fox (Figure 1.5) is closely related to the desert fox (Figure 1.4), but they have both developed adaptations to help them survive. The large ears of the desert fox help to radiate heat away from the animal. In contrast, the very small ears of the arctic fox allow it to retain as much heat as possible. The red fox is found in Wales and other European countries. It has ears that are not obviously very large or very small because it does not have to survive at extreme temperatures.

The arctic hare (Figure 1.6) has a white coat in winter, whereas the European hare (Figure 1.7) blends into the background with its greyish-brown coat. The body fat and insulating fur coat of arctic mammals are thicker than those of desert-living types.

Figure 1.6 Arctic hare

Figure 1.7 European hare

Plants too show many adaptations to their environment. The ability to survive in dry conditions depends on how good a plant is at retaining water. The reduction of leaves to spines, succulent stems, thick **cuticles** and lack of **stomata** are all used to conserve water, in a cactus for example (see Figure 1.8). Stomata are tiny holes in the surface of leaves and stems. They allow water to be lost by evaporation. Cuticles are waterproof layers on leaves.

Figure 1.9 These red deer stags are displaying competition within a species.

Figure 1.10 Here, wolves and magpies are competing for the same food.

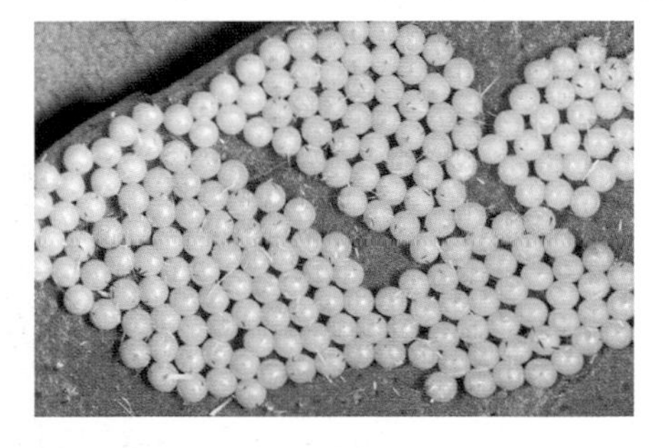

Figure 1.11 Some insects lay hundreds of eggs.

Competition

Competition is a relationship that organisms have with one another if they live in a natural environment and share a limited resource. It may be between the many members of the same species (see Figure 1.9), or between members of different species (Figure 1.10). There is a struggle between animals when too many mouths reach for too little food. In plants, competition is seen when one slowly dies because its neighbour is taller and shades it from sunlight.

Members of the same species

Competition arises between members of the same species when far more offspring are produced in each generation than can possibly survive. However, if a species can produce large numbers of offspring which vary slightly, then there is a good chance that some may be able to adapt to natural changes in the environment.

If a species produces large numbers of offspring that are spread over a wide area, then the chances of the next generation surviving are greatly improved. Therefore, although it leads to competition, the formation of large numbers of offspring can be an advantage to the species. Some insects lay hundreds of eggs for example (see Figure 1.11), and many plants produce more seeds than most of us would want to count (Figure 1.12).

Wherever you look, you can find examples of this general rule of organisms producing very large numbers of offspring. Mammals (particularly humans) and birds are an exception to this rule. More parental care is given, and there is not as much competition between individuals as there is in other groups (see Figure 1.13).

Members of different species

A different sort of competition exists when several species rely on the same sources of food. What happens when you throw a handful of breadcrumbs outside on a frosty morning? Birds such as blue tits, sparrows, starlings, blackbirds and perhaps a robin all jostle for position. The larger birds drive off the small ones, but, through cunning and cheek, they never seem to go away with less than their share.

Figure 1.12 A dandelion clock dispersing its seeds

Figure 1.13 This chimpanzee is caring for its young.

Figure 1.14 The trees and plants in this wood are competing for light, minerals and water.

This type of competition for resources also occurs with plants. For instance, when an acorn falls to the ground and starts to grow, it has to compete with, and dominate, a succession of different plants at various stages of development. The forest floor may be closely covered by a variety of woodland grasses and herbs. The oak seedling must compete successfully for space and light with all these in turn as it grows through them. Having done this, it may have to face competition from other young saplings, or from plants such as bramble. When a fully grown tree falls in the forest and opens up a gap in the canopy of leaves, it is a common sight to see 50–100 young saplings all reaching towards the light.

Oak, ash, beech, sycamore and other species are commonly found competing with one another to replace the fallen tree until one dominates the rest. The competition, with many individuals pushing down roots towards the same sources of water and minerals, or stretching up towards the same patch of light, is severe (Figure 1.14). The plant that can outgrow its neighbour will have a valuable advantage. The animal that can reach the feeding ground first will not starve.

Figure 1.15 This leaf insect matches its background perfectly.

Population size, predators and prey

It is not just the availability of resources that affects the size of a population. The numbers of **predators** and **prey** affect each other. If the population of a species increases, there is more food for predators so predator numbers increase. That means more prey are eaten, so prey numbers go down – so the numbers of predator and prey follow each other in a pattern.

Animals are able to protect themselves in many ways against the threat of an attack from another animal. Some fight back, while others are able to conceal themselves because their colouring matches their background (see Figure 1.15 for example).

Indicator species as a sign of pollution

These types of population change take place in natural environments and provide stable conditions in which living and non-living things interact and materials are recycled. In other words, there is a natural balance. Sometimes, though, this balance is upset by the introduction of harmful materials into the environment, which results in pollution. Organisms that live in areas where pollution is suspected can be compared with the types of organisms that live in unpolluted areas to see the extent of the damage. For example, you can determine the extent of pollution of fresh water by taking samples from the water and analysing them.

Practical work

Using indicator species to determine the pollution in a river

Plants and animals that live permanently in a river 'sample' the water continuously throughout their lives, so you can estimate the amount of pollution in a river by recording the presence or absence of certain **indicator species**. The major groups of freshwater indicator species are shown in Figure 1.16, but you may well find other animals that are not illustrated in this book. Ask your teacher for help and use identification guides. It is usually not necessary to name individual species; major groups of animals are sufficient.

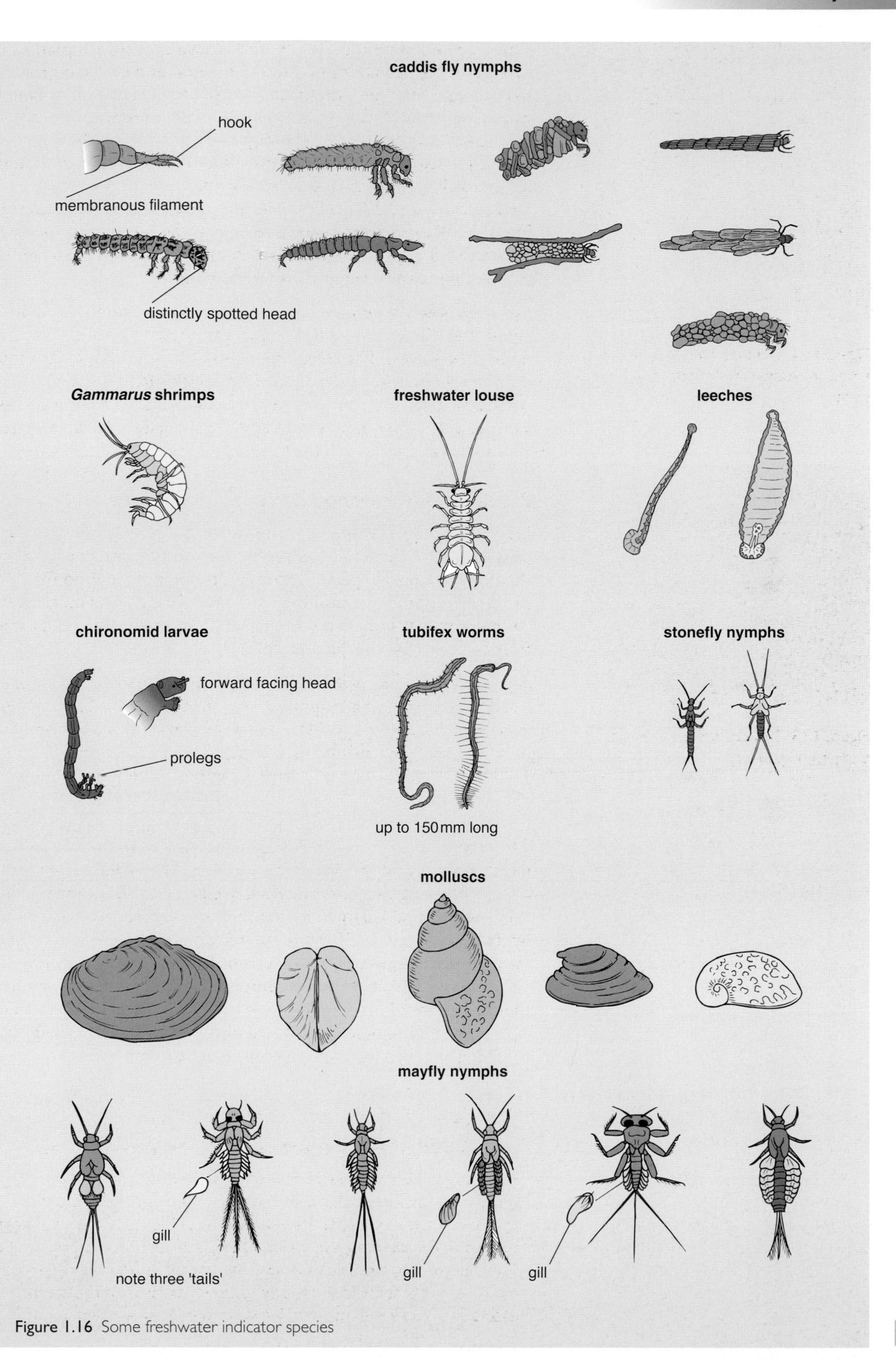

Figure 1.16 Some freshwater indicator species

Practical work *continued*

People with an interest in river management and the purity of water often need a standard method of measuring pollution, and one of the simplest is the Trent Biotic Index, which is based on the fact that animals are adapted to withstand various degrees of pollution. The basic units of the index are the groups listed in Table 1.1.

Before you proceed with this activity, the area *must* have been assessed and considered safe by your teacher. The river should be shallow and not fast-flowing, and you should be supervised *at all times* by a teacher, who will tell you where you can *safely* take samples.

You will need:

- a pond net
- sample containers
- sorting trays
- specimen tubes
- a pH meter or test papers
- a thermometer
- identification guides.

Method

1 Use your net to sweep samples of bottom-living animals from sites at various distances downstream.

2 Measure and record the date and time, water temperature and pH at each site.

3 When you are back at school, identify the animals, using Figure 1.16 and other resources if necessary.

Table 1.1 Trent Biotic Index data

1		2	3	4	5	6	7
Indicators present		**Other species**	**Total number of groups**				
			0–1	**2–5**	**6–10**	**11–15**	**16+**
			Biotic index				
1	Stoneflies	More than one	–	VII	VIII	IX	X
2		Only one	–	VI	VII	VII	IX
3	Mayflies	More than one	–	VI	VII	VIII	IX
4		Only one	–	V	VI	VII	VIII
5	Caddis flies	More than one	–	V	VI	VII	VIII
6		Only one	IV	IV	V	VI	VII
7	*Gammarus* shrimp	Above groups absent	III	IV	V	VI	VII
8	Freshwater louse	Above groups absent	II	III	IV	V	VI
9	Tubifex worms and/or red larvae of flies (chironomids)	Above groups absent	I	II	III	IV	–
10	Above groups absent	Some groups that do not need oxygen may be present	0	I	II	–	–

4 Sort the animals using Table 1.1 and the Trent Biotic Index. You calculate the index by moving down column 1 until you reach the appropriate line. For example, if there are no stoneflies or mayflies, but there are caddis flies, either line 5 or line 6 is correct, depending on the choice in column 2. You can then find the Trent Biotic Index in columns 3–7, depending on the number of animal groups present. The groups

at the top of column 1 can withstand the least pollution, and those at the bottom of the column the most. Table 1.2 explains what the Trent Biotic Index tells you about the river.

Table 1.2 Interpretation of the Trent Biotic Index

Index	Condition	Characteristic animals
XI–X	Very clean	Trout, salmon, stoneflies, mayflies, caddis flies, freshwater shrimps
VII–X	Clean	Assorted fish plus the above arthropods
VI–VIII	Clean	Fewer species of the above groups
V–VI	Fairly clean	Few fish, freshwater louse, leeches and molluscs
III–V	Doubtful	Fewer species of the above
II–IV	Doubtful	As above, but no fish
I–III	Bad	Only red chironomids (midge larvae) and worms
0–1	Bad	Only anerobic organisms

5 Plot your data on a graph, showing the relationship between the Trent Biotic Index for your various sites and (a) the pH, and (b) the temperature.

Questions

- Discuss the state of this river with respect to pollution.
- How might natural changes in the river complicate your pollution investigation? List any factors that could make your results inaccurate.
- List the factors that may affect the size of a population in a natural environment.

Other indicators of pollution

A high presence of tubifex worms or chironomid midge larvae in a river is an indicator of low oxygen levels. These organisms are adapted to live in low oxygen levels – they are red because they have plenty of haemoglobin to pick up oxygen in poorly oxygenated water. Changes in oxygen levels or pH that could indicate pollution could also be measured directly, for example with a datalogger.

Lichen can be used as indicators of air pollution. Different species of lichen have different sensitivities to sulphur dioxide. Some are so sensitive that a trace of the gas in the air will kill them. This means that by looking at the lichens still growing, we can use them as indicators of the level of sulphur dioxide pollution in the air.

Variation

What makes us different?

No two living things are exactly alike, even when they have the same genes. For instance if you looked quickly at a group of the same species of penguin, you might think that they were identical. However, anyone who has to care for animals soon learns to distinguish between them. Plants can also vary, just as much as animals.

A boy tends to look like his brother or his father more than distant relatives; and he looks like his relatives more than his unrelated friends. The similarities are mainly due to heredity. The differences that enable us to tell one identical twin from another are partly due to their environment.

Practical work

Observing variation due to heredity

You can observe **variation** due to heredity by recording certain characteristics of students at your school. In this context, a characteristic is a single human feature, such as hair colour or eye colour. Variation is the distribution of these features throughout the group. The larger the group, the more valid your conclusions will be, and you can get a complete picture of variation only by collecting all of the results for the whole group. Variation can be either **discontinuous** (each member of the group either has the characteristic or they do not) or **continuous** (where everyone has the characteristic but it varies for each member).

Your teacher will explain that more than one gene controls the hair or eye colour of a baby. For instance, children of brown-eyed parents can have blue eyes. Simple heredity rules are often complicated by other factors.

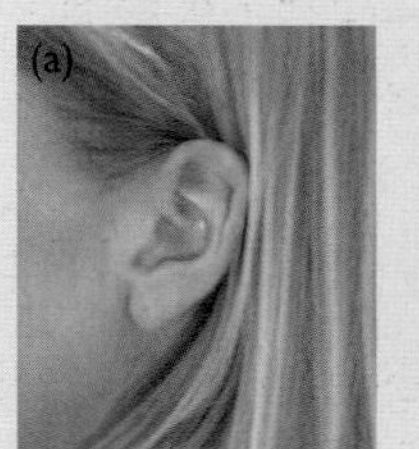

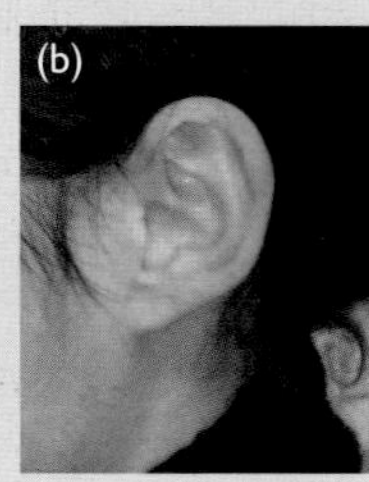

Figure 1.17 Ears with free lobes (a) and attached lobes (b).

Free and attached ear lobes (discontinuous variation)

1. Find the percentage of your class with free ear lobes like Figure 1.17(a).
2. Plot a histogram of percentage of the class against those with free ear lobes and those with attached ear lobes.
3. Look at the section on the work of Gregor Mendel (see page 23) then explain your observations in terms of dominant and recessive alleles.

Eye colour (continuous variation)

1. Work in pairs. Look at your partner's eyes in good light.
2. Record their colour as blue or brown if they are definitely that colour. If they are not, describe the colour as precisely as you can.
3. Collect the data for blue and brown eyes for the class.
4. Now decide on descriptions for the other colours. Can some of these colours be grouped together, or is there a continuous eye colour gradation across the students that does not allow for grouping? Does anyone in your class have eyes of two different colours? What could be the explanation for this?

Hair texture and colour (continuous variation)

1. Work in pairs. Classify your partner's hair as being straight or curly.
2. Find the percentages of straight or curly for the class. Is there a distinct grouping, or is there gradation?
3. Hair colour is not at all easy to describe. It can be black, dark brown, medium brown, light brown, blond, auburn, etc. Look at the hair colour of the members of your class, and modify your hair colour code if necessary.
4. Record the number of students in each of your hair colour groups.
5. Are the differences within the colour groups as great as those between groups? If so, what does this tell you about the variation of hair colour in your class compared with the variation in tongue-rolling ability?

Environmental factors can cause variation

The outward appearance of an organism is called its **phenotype**, and it depends on both the organism's genes and its environment.

Consider a litter of German shepherd puppies for example. They all have genes (units of heredity) from the same parents. If we compared the growth of these puppies, we might see that one of the litter had become weaker and smaller than the rest. This could be because it had not been given the correct diet and exercise. In this case, the effects of the environment would have influenced its phenotype. In contrast, a litter of Pembrokeshire corgis will grow up to be small dogs no matter how well fed they are. If a group of animals have the same environment, their **genes** determine their phenotypes. It is impossible to change genes unless they are altered as a result of **mutation**.

Genetic variation

Offspring become genetically different from their parents as a result of **sexual reproduction**, in which during the process of **fertilisation**, an **egg** fuses with a **sperm**. The genes in the egg are mixed with different genes in the sperm. The cell formed as a result of fertilisation (the **zygote**) has a set of genes from the father and a set from the mother (see page 20).

Those organisms that reproduce by **asexual reproduction** do not mix their genes because fertilisation does not take place. One individual produces offspring that are genetically identical to each other and to the parent. These are called clones.

In the 1990s, mass production of economically important organisms by **cloning** became common. Hundreds of genetically identical offspring can be obtained from a single cell at an early stage in the embryonic development of sheep and cattle. Egg cells are removed from ewes or cows and are fertilised in containers with sperm from rams or bulls, respectively. This is called ***in vitro fertilisation*** (IVF). Cells from the embryos formed in this way are separated. They are grown in special culture dishes containing nutrients, and then transplanted into the wombs of adult animals. In 1997, Australian researchers produced 500 cloned cattle embryos in this way. In the twenty-first century, this method of producing domestic animals is widespread throughout the world.

Plants can be cloned using a **micropropagation** method. Small pieces of plant are cut from healthy stock. The pieces may be flower buds, leaves or parts of stems. They are sterilised in an **antimicrobial chemical** and cultured on a medium containing nutrients and **plant hormones**. After the plants have been grown at a controlled temperature for 3–9 weeks in 10–14 hours of light per day, they develop new roots. These are removed and placed on a new culture medium. This process is repeated every few weeks. In this way, a few plants can give rise to thousands. The method is used with oil palms, bananas, orchids and any other cultivated plant that has potential for sales on a worldwide scale. If these clones are grown in identical environments, the only way in which they can become different from one another is through gene mutation.

What is a mutation?

Mutation is a change in one or more of the following:

- the chemical structure of a gene
- the arrangement of genes on **chromosomes**
- the number of chromosomes found in a cell.

If the mutation is an unfavourable change for the organism, it will eventually disappear, because the organism with the mutation will be

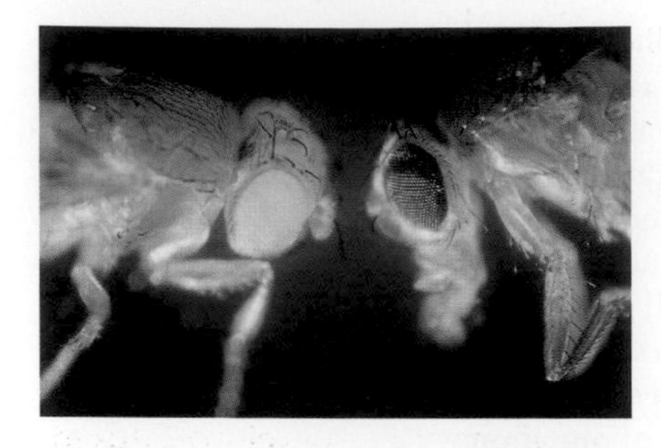

Figure 1.18 The fruit fly on the right-hand side is a normal fly and the one on the left is a white-eyed mutant.

unlikely to have offspring that will inherit the mutation. However, sooner or later the mutation will occur again. An example of a mutation is shown in Figure 1.18.

Although we speak of mutant genes and normal genes, the genes we now call normal were once mutants. Because they were favourable, they have been passed on from generation to generation and have become part of the normal collection of genes in a population. Mutation is all part of the process of evolution.

How are mutations in genes caused?

All living things are exposed to a certain amount of **radiation**. The three natural sources of this radiation are:

- cosmic rays from space
- radioactive materials in the Earth's crust
- ultraviolet radiation from the Sun.

This background radiation is, in part, responsible for the mutations that occur in organisms.

Consider radiation as a series of rapidly fired mini bullets. Think of these bullets hitting the molecules that make up the genes of organisms. These molecules are then damaged, and this affects the genes that are inherited by the organism's offspring. It is the energy carried by the radiation that causes the damage. Any biological molecule that is hit by radiation is destroyed. This is enough in itself to cause considerable damage to a living cell. It has been known for a long time that an increase in radiation causes an increase in mutation rate. In other words, there is an increase in the rate at which genes are altered.

Figure 1.19 The nuclear test site at Mururoa Atoll in the Pacific Ocean

This knowledge has been used in genetic experiments to produce mutations in experimental organisms. The treatment has produced some useful new plant varieties. It is also known that even very low levels of radiation can cause mutations. These findings have real significance for us in our nuclear and technological age.

The human race has increased radiation levels on Earth in several ways.

- There has been an increase in the use of **X-rays** for medical purposes.
- We now use energy generated in **nuclear power stations**. Unfortunately, there have been accidents that have caused pollution by radioactive materials. The disposal of radioactive waste from nuclear power stations is a problem that has not yet been completely solved with modern technology.
- The use of nuclear weapons began towards the end of the Second World War, leading to **nuclear fallout**. Since then, even more powerful nuclear weapons have been tested (see Figure 1.19).

Questions

1. What is the meaning of the term 'mutation'?
2. List the major causes of mutation.
3. Why are mutations that occur in body cells not important for a species?
4. Explain the difference between continuous and discontinuous variation. Give examples of each in humans.

This increase in man-made radiation is only a fraction of the naturally occurring background radiation. However, any increase in radiation increases the gene mutation rate. From a biological point of view, the most important damaging radiations are X-rays and **gamma rays**. X-rays and gamma rays have great penetrating power and are a massive hazard when their source is outside the body. Great care must be taken to minimise exposure when X-rays are used. The most active tissue, that in developing embryos for instance, is most sensitive to radiation. X-raying pregnant women is risky for the **embryo**, and it is rarely, if ever, carried out.

Evolution

Figure 1.20 Charles Darwin, the great British naturalist and father of evolution

Natural selection

With a knowledge of variation, we can now understand the idea of **natural selection**. Charles Darwin (see Figure 1.20) developed this in his famous book *The Origin of Species*, published in 1859, which was written after his voyage of discovery on HMS *Beagle* in 1834. He based his theory on three observed facts and two deductions from them.

The first fact is that animals and plants have a tendency to multiply at an ever-increasing rate: this is in multiples of 2 and then 4 and then 8 and then 16 and then 32, and so on. Offspring usually tend to be more numerous than parents. We have only to look at the human population explosion in certain parts of the world to see this principle in action.

The second fact is that while all living organisms can multiply at an increasing rate, they seldom do so for ever. Few species, apart from humans and some animals and plants dependent on us, have been observed to increase rapidly for very long. The species that have done so have often been presented with new opportunities by humans. For example, gulls feed on our refuse dumps and rats breed in our sewers. Humans introduced the rabbit to Australia, where they had no natural enemies and became a national nuisance within a few years.

From these two facts, Darwin deduced that there was a 'struggle for existence', or a competition for the chance to reproduce. Almost everywhere in nature, organisms produce more young than can possibly survive to the age at which they reproduce. They must compete for food and for all their other needs. For example we are not completely covered with flies, despite their enormous breeding rate.

The third fact is that all living things vary. We have seen this in our studies of **genetics**. Darwin hence deduced that there is a mechanism of natural selection. The principle of natural selection states that there is competition for existence within groups of individuals that vary. Some individuals are better adapted to their environment than others. Those with favourable variations are more likely to survive and reproduce than those with unfavourable variations.

Many variations are inherited. Favourable inheritable variations have a better chance of being passed on than unfavourable variations. Natural selection is the principal cause of **evolution**. However, although Darwin was convinced that variations within species could be passed on to future generations, he had no knowledge of the causes of variations through mutation.

Problems with predictions

H

Charles Darwin was aware of the importance of population size in determining the survival of individuals competing for limited essentials such as food. In this process, predation plays an important part.

Figure 1.21 The varieties of the snail *Cepaea*

H

One of the few valid criticisms of Darwin's work was that he could not test his theory with experiments. Today, however, much evidence for Darwin's theory has been gained from studies in genetics and by experiments on natural selection. There have been many studies of predators selecting the least well camouflaged prey, most notably the land snails of the genus *Cepaea* (see Figure 1.21).

Case study: natural selection in snails

Only two species of *Cepaea* occur in Britain: the brown-lipped snail *Cepaea nemoralis* and the white-lipped snail *Cepaea hortensis*. The brown-lipped type is the more variable of the two, the colour of its shell being one of three distinct shades: brown, pink and yellow. The markings on the shell vary from unmarked forms to those with 1–5 dark bands. The white-lipped type is usually yellow, and there are either no markings or five distinct bands; intermediates are relatively rare. In both species, shell colour and pattern have been shown to be under genetic control.

Studies of the brown-lipped types in a variety of habitats such as grassland and woods showed that there is a high correlation between yellow shells and the green backgrounds of open short grass. In darker places, ranging from shaded hedgerows to deepest beech woods, the proportion of yellow shells gradually declines almost to zero. The banding on the shells shows a clear correlation with the environment: the more varied and dark the environment (for example in woodland), the greater the proportion of banding.

The main predator of the brown-lipped snail is the song thrush. In mid-April, the vegetation is mainly brown, and yellow shells are at a selective disadvantage. By late April, yellow is of neutral survival value relative to the other colours, and by mid-May it is at an advantage.

There is clear visual selection by predators; certain shell patterns seem to be camouflaged against particular backgrounds. However, the white-lipped type of snail also has variable shell patterning and may live in the same locality as the brown-lipped snail. They are hunted by the same predators, but their gene frequencies may be very different. In beech woods, almost all the brown-lipped snails are unbanded browns, whereas the white-lipped ones are mostly banded yellows. Studies of natural selection through predation often involve computer simulations or other procedures that provide **models** to show the effects of camouflage and predation. In this way, predictions of what could happen over a long period of time can be made by carrying out short-term investigations. Care must be taken when gaining information from such models because in a natural environment, several variables are involved. These include the following:

- The size of the population being studied is a significant factor.
- Each individual of reproductive age may or may not have an equal chance of finding a mate.
- Immigration or emigration into or out of the population may take place.
- Random mutations occur.
- Differences in colour may lead to other problems, for example black shells absorb heat more readily and may influence the snail's survival.

Antibiotic resistance

H

One example of how natural selection is taking place today is the development of resistance to antibiotics by some bacteria. **Antibiotics** are substances produced by microbes that kill bacteria. Antibiotic resistance in certain disease-causing bacteria has led to serious health problems on a worldwide scale.

The genes involved in resistance to antibiotics vary between species and within a species. The overuse of antibiotics has led to those types of bacteria that have mutated being selected to become resistant to antibiotics. The resistant types survive to breed and pass the mutated gene on to their descendants. Because of the rapid rate of reproduction of bacteria, millions of clones are soon formed. **Penicillin** resistance has been recognised since the 1940s, and it began because of the action of a bacterial **enzyme** that can break down penicillin. Resistance to antibiotics has already destroyed or limited the usefulness of several valuable antibiotics that were once used to treat a variety of infectious diseases. Recently there has been a build-up of antibiotic resistance in several strains of bacteria that cause tuberculosis. The so-called super-bug **MRSA** (methicillin-resistant *Staphylococcus aureus*) has evolved resistance to all commonly used antibiotics and it is now responsible for many deaths in hospitals throughout the UK.

Super-rats in Wales

H

Rats have competed with humans for food ever since we began to cultivate grain and store it. They have also been responsible for the spread of diseases such as the plague, and they are the indirect cause of the deaths of thousands of people throughout the world every year. In an attempt to control them, humans have used several types of poison. One, called **warfarin**, was first introduced in 1950. It became a popular rat poison because it was not very poisonous to farm animals. It acts by preventing blood from clotting. When rats eat it, their blood capillaries become very fragile and the blood fails to clot. The capillaries burst and the rats bleed to death.

Warfarin-resistant strains of rats appeared in Welshpool in central Wales in 1959. They spread out from their original site at a rate of three miles per year, and by 1972 these super-rats were breeding in 12 other areas of Britain. The basis of the evolution of the new strain can be explained in terms of a single gene mutation in which resistance is caused by a **dominant allele** (dominant form of the gene). The blood-clotting process of the resistant strain of rats does not react to warfarin, so the blood clots normally even if the rat eats warfarin.

A few mutants carrying the dominant resistance allele are favoured by natural selection. They breed and pass on the allele to future generations. This proves the rule that, given a sufficiently large population with a sufficiently high breeding rate, natural selection leads to evolution. In this case, natural selection led to the evolution of a new type of rat in less than ten years!

An alternative view

H

Before Charles Darwin's time, most people believed that all natural species had been created as they appeared to them, and had never

H

Did you know?

The word 'biology' was first invented by Jean-Baptiste Lamarck. He suggested a theory of evolution of animals without backbones (invertebrates) before Charles Darwin. However, his ideas were fundamentally flawed because he believed that evolution occurred because characteristics that were developed during the lifetime of an organism were inherited by its offspring. For example, if Lamarck had been correct, children of well developed trained weightlifters would always inherit strength and large muscles.

changed. This view was called **creationism**. In 1859, when Darwin's ideas about evolution were first published, the majority of people still believed in creationism, and many argued against Charles Darwin's theory of evolution. Then some people began to think that species could change gradually into other species by evolution, and this belief was called **Darwinism**.

Today, almost all biologists believe in Darwin's ideas about natural selection. The combination of genetical studies of inheritance, fossil evidence, and an increasing amount of experimental data has made the theory of natural selection a proven fact.

Hard evidence for evolution

Charles Darwin built up a large collection of **fossils** to provide evidence that species change over time. He found that more primitive animals and plants occurred as fossils in the oldest rocks. The word 'fossil' comes from the Latin *fossilis*, meaning something dug up from the ground. Today the term is applied to any organism preserved in the Earth's crust, and traces of such organisms.

In past ages, occasionally a plant or animal, when it died, was caught in some liquid that cut off air and so prevented the dead body's decay by bacterial action. The organism would then stay virtually unchanged. An example is the ancient insects found preserved in amber, which is a fossilised resin once produced by trees (Figure 1.22). Sometimes the liquid would destroy the flesh but leave the bones intact, as in the animal skeletons found in the tar pits of California, USA.

Figure 1.22 This insect has been preserved in amber.

However, fossils of whole bodies or skeletons are extremely rare. Most animals and plants are eaten or decay when they die. Whether you are fossilised depends on when and where you die, and the chances of someone finding the fossil are remote. Usually the finds are just fragments, a few teeth or bones. A fossil may be a shell, the tunnel of a worm in sand, a footprint, an impression of an animal or plant in mud, etc. Coal is probably the best-known fossil of all. It is the fossilised remains of ancient tree ferns that lived many millions of years ago.

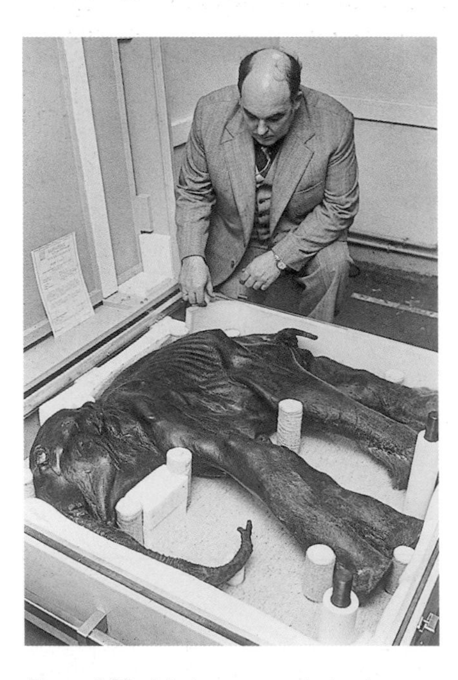

Figure 1.23 A baby mammoth that has been preserved in ice

Some parts of the Earth's crust have proved to be rich in fossil specimens. Ice in parts of the Arctic has preserved organisms, often in near complete condition. There are well preserved remains of mammoths and other large creatures, for example (Figures 1.23 and 1.24).

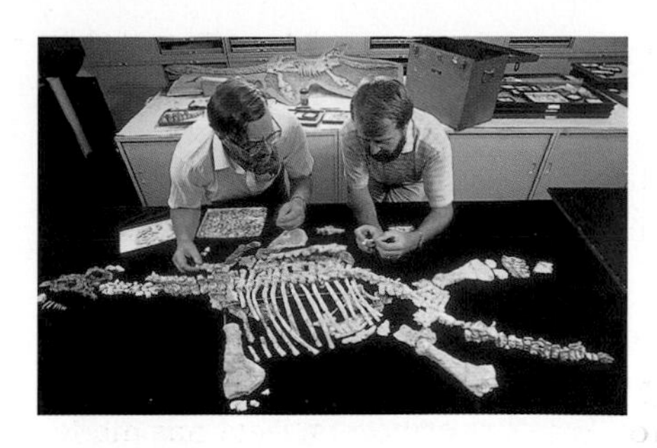

Figure 1.24 A fossil dinosaur being examined by palaeontologists

The best sources of fossils are the floors of seas past and present. Dead organisms sink to the bottom unless they are eaten first. Their shells or bony parts are converted to hard limestone by chemical action. Sand settles on top of them over millions of years, and although the soft parts of the organism decay, the hard parts remain. The fossils may not be found until long after the sand has become solid rock and the waters of the sea have receded.

Questions

5 What is the main difference between Darwin's and Lamarck's ideas about evolution?

6 How can fossils be used as evidence for evolution?

7 Describe examples of evidence to show that evolution is taking place today.

Inheritance

Where do you keep your genes?

In any study of biology we recognise variation and can explain it. However, we need to consider its basic cause in more detail. Variations are handed down from one generation to the next through genetic material. Those cells that form a link between generations – the sex cells, or **gametes** – contain genetic material, and they are therefore the best starting point for our study.

Figure 1.25 Human sperms

Figure 1.25 shows male gametes (sperms). In common with other cells, these have a **nucleus** and a **membrane**. Unlike the female gamete (egg) (Figure 1.26), there is very little **cytoplasm**. All the genetic material of male gametes is contained in their nucleus, which is the so-called head of the sperm. Sperms can move through liquid using a swimming action. When fertilisation takes place, the sperm's nucleus enters the female gamete and fuses with the egg's nucleus (see Figure 1.27). The nucleus contains all the chemicals that are used to control the development of the individual.

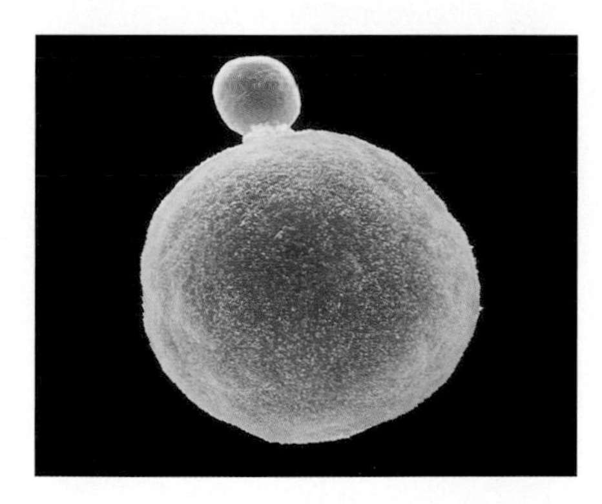

Figure 1.26 A human egg just before ovulation

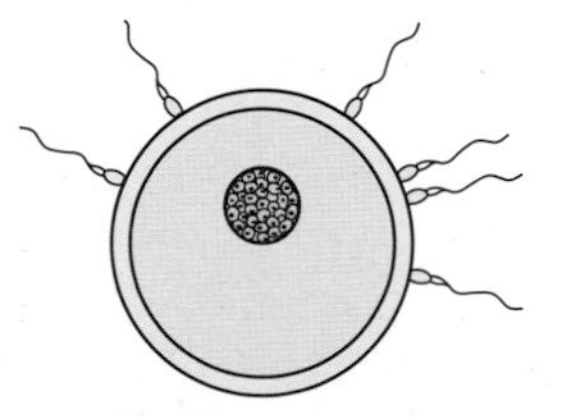

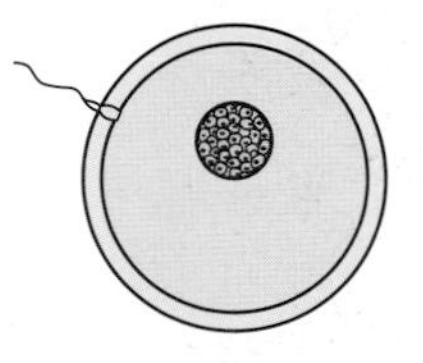

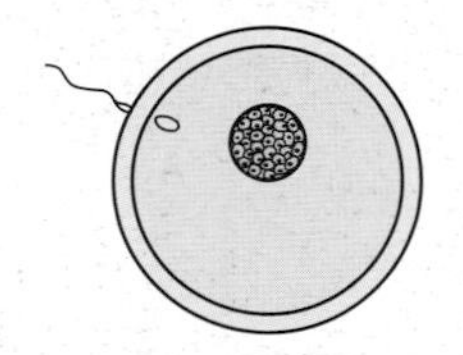

Figure 1.27 Fertilisation

A closer look at the nucleus

Practical work

Looking at the nucleus

You will need the growing region of a root tip of garlic. This plant is available throughout the year and is easy to grow. The garlic bulb should be placed on the top of a test tube containing water. Roots appear in a few days. Your teacher will give you a root tip that has been stained with Feulgen's stain (this reacts with chemicals in the nucleus to produce a pink colour).

You will also need ethanoic acid solution, which is corrosive. Your teacher will have stored this in a safe place in the laboratory, and may give you a microscope slide with a drop of ethanoic acid solution already on it.

Method

1 Wear eye protection.
2 Carefully cut 2 mm off the end of the root tip using a sharp scalpel.
3 Place the tip on a microscope slide in a drop of 45% ethanoic acid.
4 Use two mounted needles to pull the root tip apart. Try to break down the tissue into very small parts.
5 Place a cover slip on the material.
6 Rest the slide on some filter paper, and put a few layers of filter paper on top of it. Carefully press your thumb straight down on the region of the cover slip without using any sideways pressure. This should flatten the cells and separate the genetic material (**chromosomes**).
7 After about 5 seconds, peel off the filter paper. The ethanoic acid solution should just fill the space under the cover slip.
8 After removing your eye protection examine your slide under a microscope. Use the high-power objective after focusing under low power. You should be able to see the chromosomes. Chromosomes contain **genes**, the carriers of genetic information. Draw the cells.

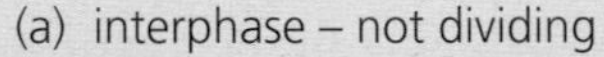

(a) interphase – not dividing

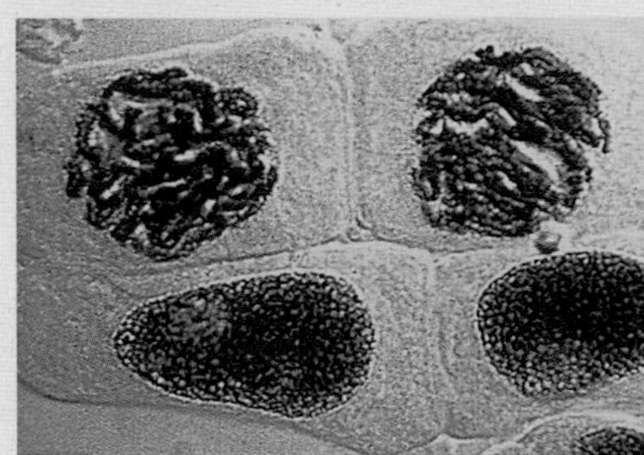

(b) prophase – stage 1

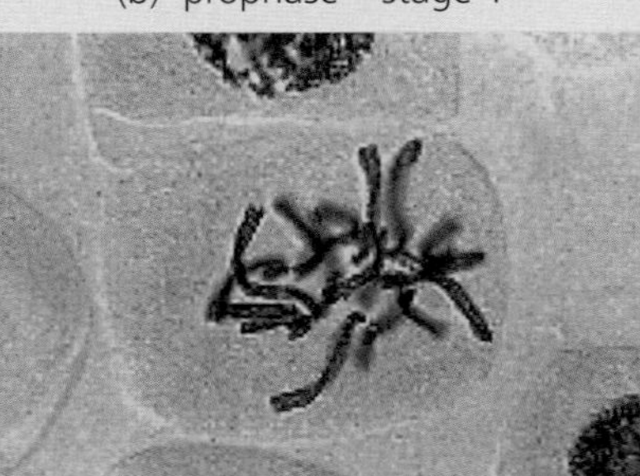

(c) metaphase – stage 1

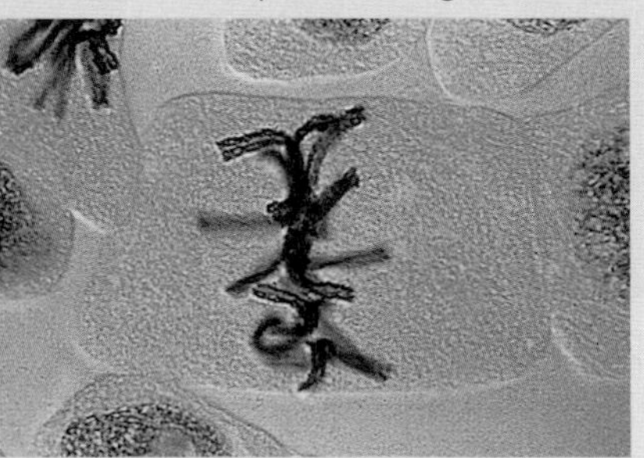

(d) anaphase – stage 3

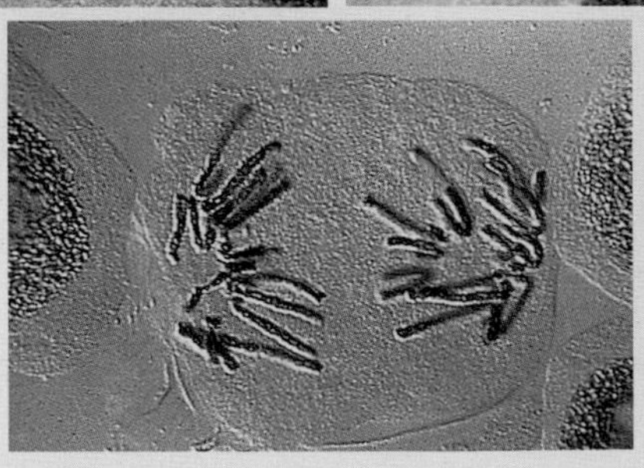

(e) telophase – stage 4

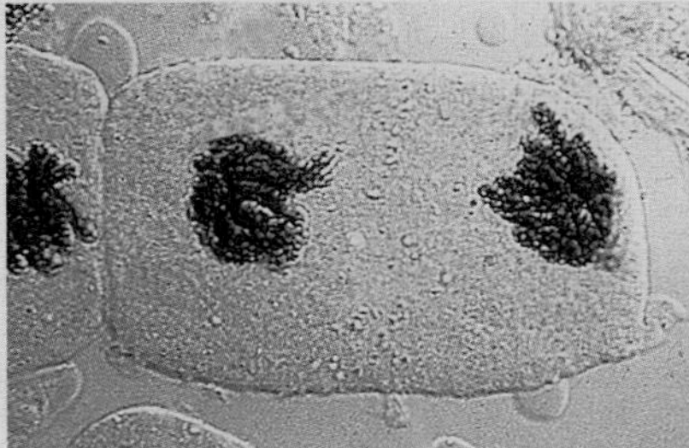

Figure 1.28 Cells from a bluebell root showing stages of mitosis in the sequence in which they occur during growth

9 Look at several cells and compare them to the series of photographs in Figure 1.28. Try to find the earliest stage at which the chromosomes are visible as double structures.

You should have observed the following:

- The nucleus is the most likely site of genetic material.
- The nucleus can contain chromosomes.
- Chromosomes are visible only in cells that are actively dividing.

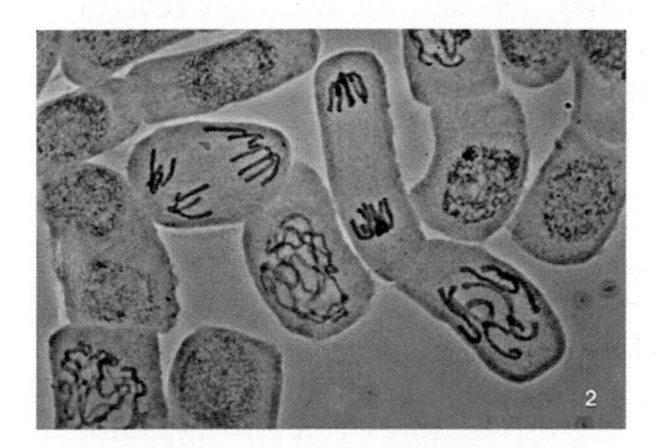

Figure 1.29 Cells from a root tip at different stages of mitosis

Cell division

The type of cell division that took place in the above practical work is called **mitosis**. During this process, a cell produces two daughter cells that are genetically identical. This type of cell division continues through our lifetime and is involved in, for example, growth and repair of skin and blood cells. Apart from the special case of gamete formation, all cells undergo mitotic division of chromosomes during development. Figure 1.29 illustrates the four different stages of mitosis.

Before mitosis begins, each chromosome replicates. At the start of mitosis, the original and the copy separate and migrate to opposite ends of the cell. Two daughter nuclei are thus formed, each surrounded by a nuclear membrane. In plant cells, a cell wall separates each cell. In animal cells, the cytoplasm invaginates to form two daughter cells (see Figure 1.30).

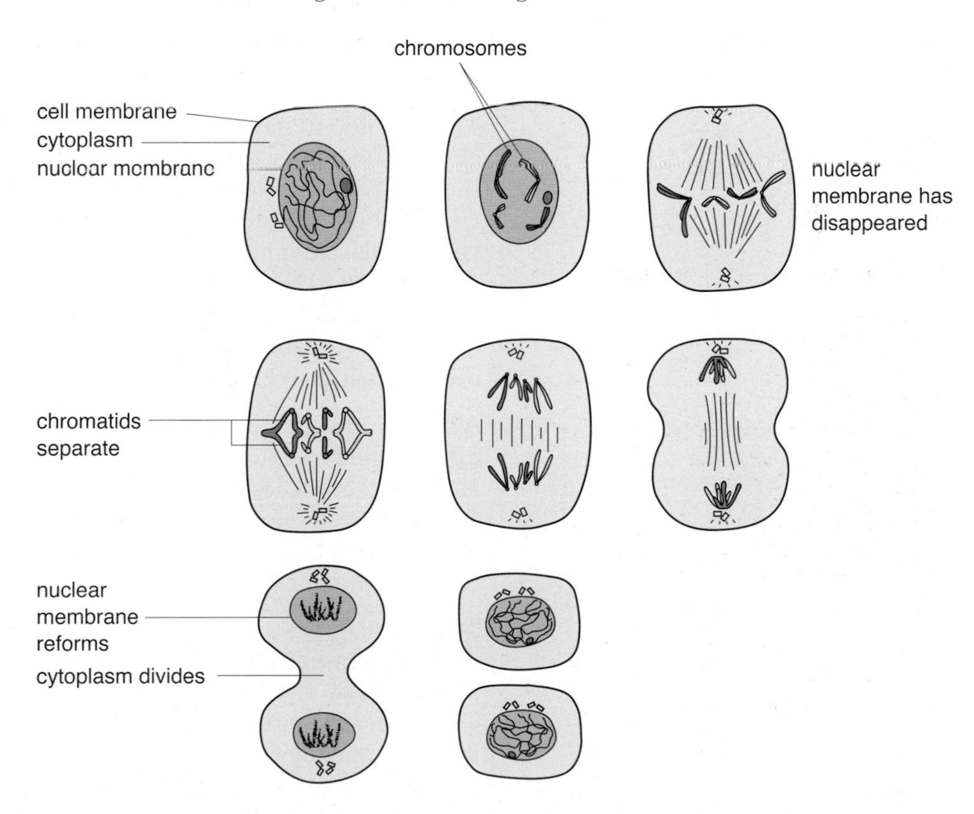

Figure 1.30 Stages in mitosis

Mitosis enables each new cell (daughter cell) to have a set of chromosomes identical to the original cell. It is the means by which cells in an embryo develop from a fertilised egg. It also allows organisms such as bacteria to reproduce asexually. Thus all asexually produced offspring from one organism are genetically identical to their parent.

Sexually produced offspring arise from the fusion of male and female gametes (sex cells).

Fertilisation and chromosome number

When a sperm swims to an egg, its nucleus merges with the egg nucleus (this is fertilisation). The nuclei of both types of gamete have fewer chromosomes than other cells in the body of the same organism. What would happen if they did not? Cells in the human body have 46 chromosomes (23 pairs). If each matured sex cell, or gamete, also had 46, then a baby's cell would each have 92, and its children's cell would have 184. Yet all normal cells in the human body have 46. Human cells normally contain 23 pairs of chromosomes. One set comes from the mother's egg and the other set comes from the father's sperm. Mitosis ensures that each new cell gets a full set of chromosome pairs.

Microscopic studies of developing human sperms and eggs show that they have only 23 chromosomes, one from each pair. How does an organism produce a cell with half the usual number of chromosomes?

Cell division to form gametes

Of all the countless millions of cells in our bodies, only egg-producing and sperm-producing cells divide in such a manner as to split up the chromosome pairs. Logically enough, the process is called reduction division or, more commonly, **meiosis**. To see how it works, we can follow meiosis as it occurs in an animal that has only two pairs of chromosomes.

The first step in meiosis is similar in some ways to mitosis. Each chromosome lines up with its opposite number across the equator of the cell. Each chromosome has duplicated itself. There is a double chromosome for each original, the halves of which are called **chromatids**. Now, just as in mitosis, the **nuclear membrane** disappears and the cell divides, but unlike in mitosis, one member of each double pair goes to the new cells. At this point we have two cells, each containing two double chromosomes, one from each original pair (see Figure 1.31).

A brief resting period follows, and then a new wave of activity begins, during which the double chromosomes break apart and each chromatid becomes a separate chromosome. Then the cells divide again. There are now four cells, each with two chromosomes, one of each pair of chromosomes in the original. This is why gametes have only half the normal number of chromosomes.

Meiosis takes place during the formation of sperm and egg. When the sperm fuses with the egg, each provides half the chromosomes for the new individual. Since it is random which chromosome a gamete gets from the original pair during meoisos, it is incredibly unlikely that any two gametes will have exactly the same 23 chromosomes. Which sperm fuses with which egg is also random. This means each fertilised egg contains a unique set of chromosomes. The advantage of sexual reproduction lies in the fertilised egg, which makes each new living organism genetically different from either of its parents. This variation among offspring may produce one that can adapt to changing conditions in the environment.

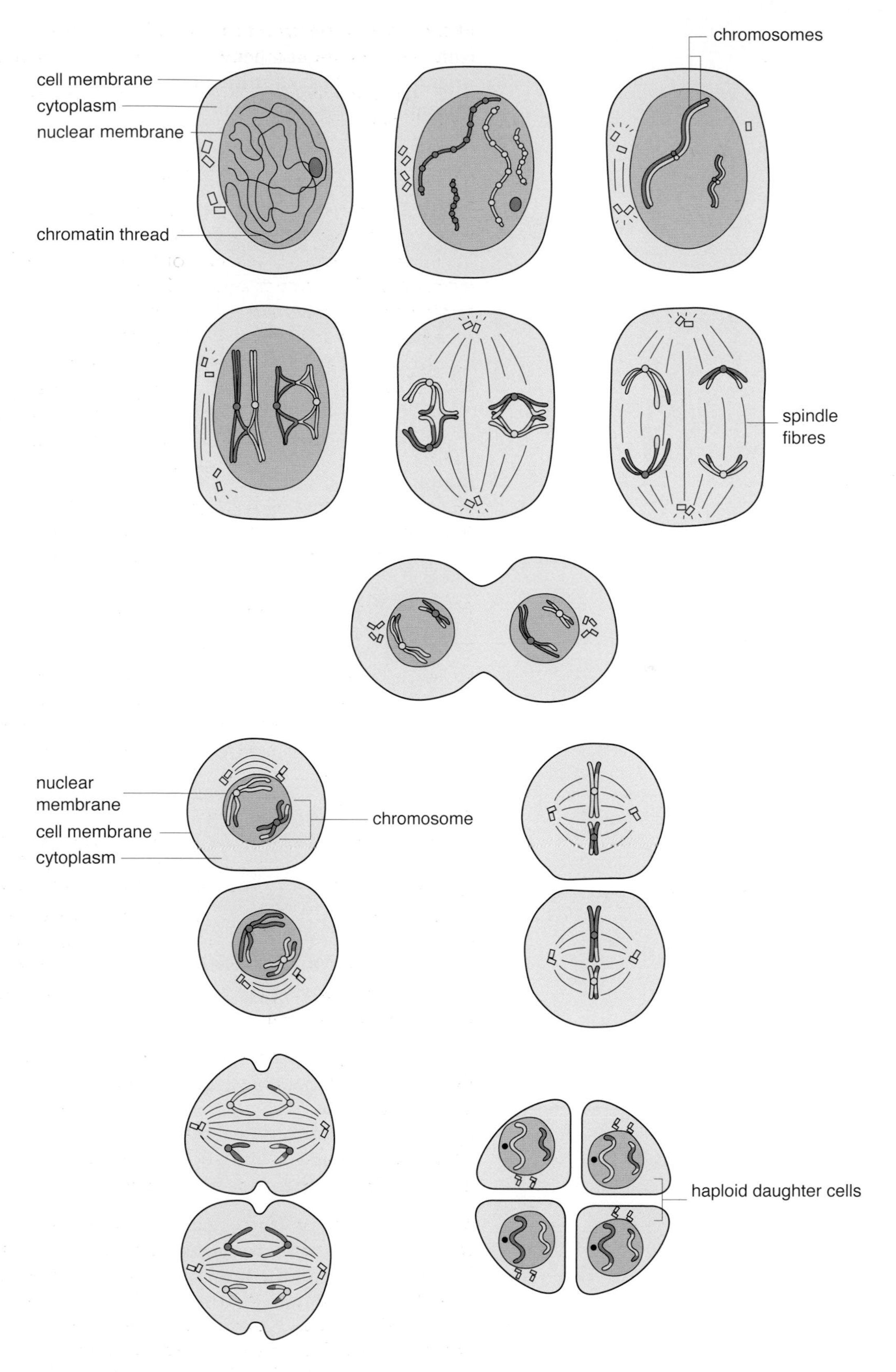

Figure 1.31 Stages in meiosis, for a cell with just two pairs of chromosomes

Once the egg has been fertilised, all further cell divisions produce cells with the full number of chromosomes. Eventually the new organism reaches maturity, and its time comes to reproduce. Its reproductive organs then produce sperm or eggs, and the cycle of life comes full circle.

Practical work

Making a model showing a stage of meiosis

You will need:

- four pipe-cleaners of one colour and four pipe-cleaners of another colour
- two dustbin-liner wire ties
- some adhesive tape and four small adhesive labels
- a piece of plain A4 paper.

Method

1. Draw a horizontal line through the middle of the A4 paper. This represents the equator of the spindle (see Figure 1.31). Mark with a cross, 6 cm above the line, the position of a pole of the spindle. Do the same below the line to show the opposite pole.
2. With the pipe-cleaners, make a model that shows the structure of a pair of chromosomes as they appear on the equator of the spindle. Use pipe-cleaners of one colour to show the chromosome from the mother and pipe-cleaners of the other colour to show the chromosome of the father. Use the dustbin-liner ties to represent the points of contact between the chromosomes.
3. Assume that the chromosome carries the gene A, and that your model represents a heterozygote (an individual with different alleles) for this gene. Using your adhesive labels, add the appropriate labels to your model to show the positions of the genes.
4. Use adhesive tape to attach your model to the piece of paper to show the correct positions.
5. Indicate, by drawing arrows on the paper, the direction of movement of the chromosomes in the next stage of meiosis.

Meiosis tutorial:
www.biology.arizona.edu

Meiosis animation:
www.csuchico.edu

Photomicrographs of mitosis in onion root tips:
micro.magnet.fsu.edu

Passing on information

The link that connects one generation to the next is microscopic. It consists of an egg cell and a sperm cell. Within these tiny bits of living matter are the plans for the next generation.

The sperm fertilises the egg and then one of the most remarkable series of changes known to science begins. The instructions within the human cell ensure that the embryo develops into a human, rather than an elephant or a mouse, and there are similar instructions for every other organism that reproduces sexually.

The instructions are in the form of chemical messages coded in genes. These messages are passed from one generation to the next by **inheritance**. The study of inheritance has become a branch of science called genetics. The terms 'genetic engineering', 'genetic

> **Did you know?**
>
> The science of genetics began when a monk experimented with pea seeds in 1865. Using this unlikely material and a great deal of patience, he set out the first principles of inheritance. Because of his unique contribution to science, he is often called the father of genetics. His name was Gregor Mendel.

counselling' and 'genetically inherited disease' are often used in newspapers and on radio and television.

We are probably most interested in the medical aspects of genetics. This is not surprising, because we hear of so many disorders that can be passed from parents to their children through genetic inheritance. If you were a farmer or a gardener, you would also be interested in knowing about methods of passing useful characteristics from one generation to the next via **selective breeding**.

The work of Gregor Mendel

Figure 1.32 Gregor Mendel, the father of genetics

Gregor Mendel (see Figure 1.32) used garden peas for his experiments for the following reasons:

- They had characteristics that he could see. For example, the colour of the seeds was always yellow or green. The plants grew to be tall or short, and never an intermediate size.
- The plants were easy to cultivate and cross, and the time they took to grow was reasonably short.

Mendel made sure that the plants he started with all had the same alleles for each characteristic he was studying (that is, they were pure-bred). He did this by not letting the plants cross-fertilise with other plants for a number of generations. The offspring of each generation were studied to make sure they were all like one another and like the parent plants.

Then Mendel crossed plants showing one characteristic with those showing the contrasting characteristic. For example, he crossed plants from a type that only grew to be tall with those that only grew to be short. In every case he found that all the offspring resembled one of the parents and showed no sign of the characteristic of the other parent. Thus all the crosses between tall plants and short plants produced seeds that always grew into tall plants. One characteristic seemed to 'dominate' the other. Mendel called this the **dominant character**.

Inheritance of a single characteristic

Mendel made his greatest contribution to genetics by explaining his observations on the inheritance of a single characteristic (**monohybrid inheritance**). He began by using symbols to represent the characteristics he was dealing with. He assumed that the characteristic for tall plants was caused by a dominant factor. He used a capital **T** to symbolise this factor. The characteristic for short plants, the only alternative to tall plants, was caused by a **recessive** factor, **t**.

Next, Mendel assumed that every plant had a pair of factors for each characteristic. These pairs of factors are now called **alleles**. An allele is one of two alternative forms of a gene. For example, if **T** represents tall and **t** represents short, the pairs of alleles can be **TT** or **Tt** or **tt**.

Mendel was convinced of the existence of alleles because some parent plants with the dominant factor produced some offspring with the recessive factor. Therefore, every one of the first-generation offspring (**F1**) plants had to have had each type of allele. The **F1** plant could be represented by **Tt**.

A plant from parents that always produced tall plants was therefore **TT**. A plant from parents that bred true for short plants was **tt**. These paired symbols representing the genes of an organism are its **genotype**. When an organism has identical alleles in its genotype (for example **TT** or **tt**), it is called **homozygous**; if it has different alleles (e.g. **Tt**), it is called **heterozygous**.

The outward expression of the genes is known as the phenotype. For example, the phenotype of **Tt** or **TT** is tall. The phenotype of **tt** is short. Using this logic, Mendel was able to test his ideas about inheritance. If he knew the genotype of each parent, he could predict the kinds and proportions of gametes each parent could produce. From this, he could predict the kinds and proportions of the offspring.

If every plant had a pair of alleles for each characteristic, was there any rule about how these were passed on to the next generation? Mendel thought about the short plants that appeared in the **F2** second generation. These could not be carrying the dominant gene **T**, so they must have received their recessive **t** from their **F1** parents. (The second generation was produced by self-fertilisation of the **F1** generation.)

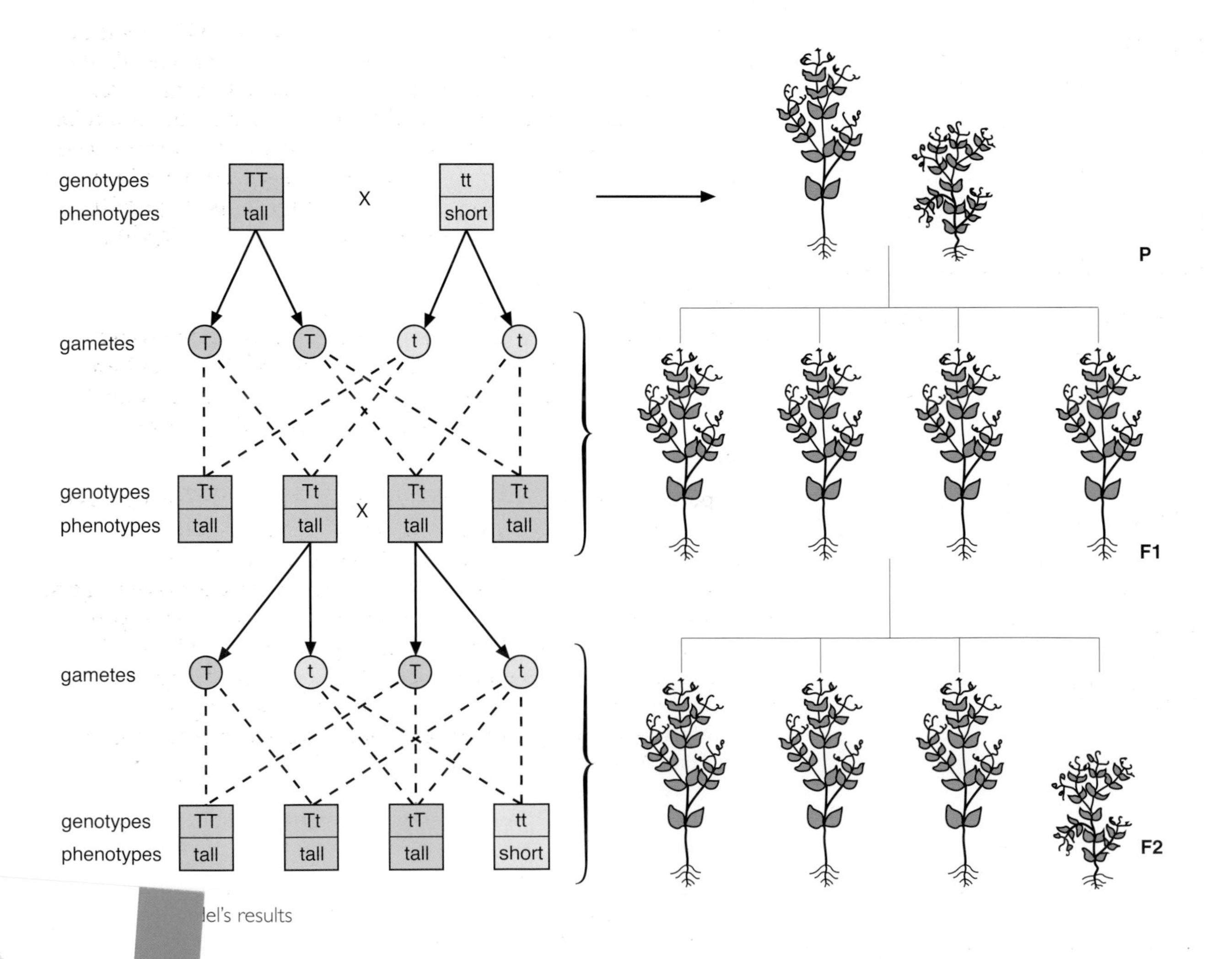

del's results

Figure 1.33 shows Mendel's classic experiments with peas. If a pure-breeding tall variety was crossed with a pure-breeding short variety, all the offspring were tall. When two of these offspring were crossed, three tall offspring were produced for every short one. Mendel saw that these results could be explained if the characteristics were inherited as 'particles'. Each plant had two of these particles, and one was dominant to the other. In this case the tall characteristic (**T**) was dominant to the short one (**t**).

Mendel thus arrived at a general rule, the law of segregation: 'The two members of each pair of germinal units must separate when gametes are formed, and only one of each pair can go to a gamete.' The significance of Mendel's work was not recognised for many years, partly becuase no one had seen chromosomes inside cells. In 1910, many years after Mendel carried out his work, the factor that he called a germinal unit was given the name 'gene'.

If a parent is **TT**, its gametes all inherit one or other of its **T** genes, but not both. If the parent is **tt**, its gametes all inherit one of its **t** genes. If the parent is **Tt**, half the gametes inherit its **T** gene, and the other half inherit its **t** gene.

Punnett squares

gametes	T	t
T	TT	Tt
t	Tt	tt

Figure 1.34 A monohybrid cross in a Punnett square

A **Punnett square** is a grid named after its inventor R C Punnett in 1905. It is used to predict the results of genetic crosses. The alleles that could be present in female gametes are placed down the left-hand side of the grid (see Figure 1.34) and the alleles that could be present in the male appear across the top of the grid (or vice versa). The alleles from both are combined in the relevant squares in the grid. The grid then shows all the possible different pairings of the alleles and therefore all the possible genotypes of the offspring.

Family trees and pedigrees

Pedigrees of family trees have been used to trace characteristics through family histories for hundreds of years. Such studies have shown that some human genes behave like the ones that Mendel studied and have simple patterns of dominant and recessive inheritance.

A simplified version of one of the first known recorded pedigrees is shown in Figure 1.35. The pedigree showed the 'short fingers' condition running down through a Norwegian family. The condition is caused by a dominant allele. If you have average fingers, then you must be homozygous recessive for this character. If your partner also has average fingers, then any children you have will have similar average fingers. If your partner has short fingers, then what will your children's chances of having average digits be? It depends on whether your partner is homozygous or heterozygous for the condition.

Did you know?

The word 'pedigree' comes from the French *pied de grue,* which means crane's foot, the shape of which resembles the linked lines of a family tree. (A crane is a large stork-like bird.)

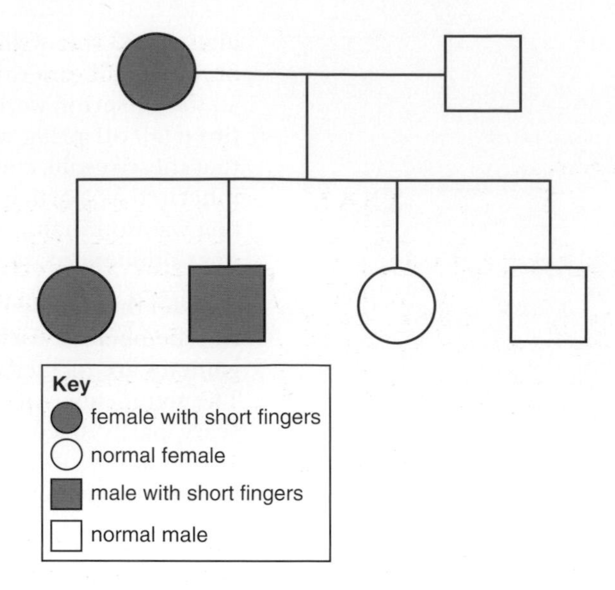

Figure 1.35 One of the earliest known pedigrees, showing a condition called 'short fingers' in humans

Genetic science learning centre:
gslc.genetics.utah.edu

Gregor Mendel:
www.mendelweb.org

When things go wrong in humans

There are more than 4000 disorders in humans that can be inherited because of faulty genes. Probably the most common genetic disorder is **cystic fibrosis**. This is a condition that affects the **pancreas** and the **bronchioles** of the lungs. People with this problem may suffer from failure of the pancreas to function properly, intestinal obstruction, and an accumulation of mucus in the lungs (see Figure 1.36).

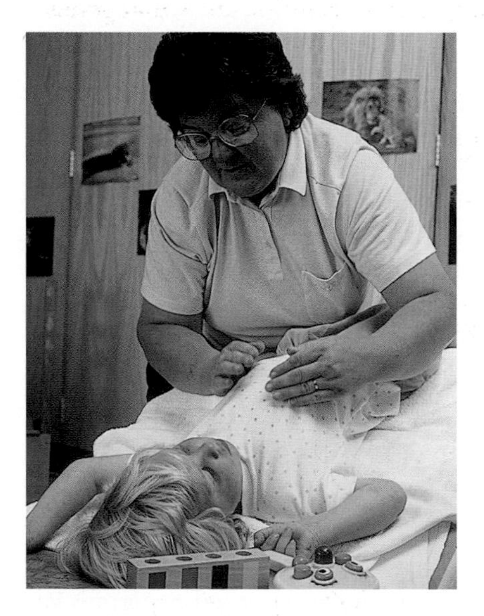

Figure 1.36 A patient receiving therapy for cystic fibrosis

Cystic fibrosis is still one of the most common serious illnesses. It is inherited as a recessive allele in the following way. Let **N** be the allele for normal pancreas and bronchioles. Let **n** be the recessive allele for cystic fibrosis. A person suffers from the disease only if he or she has two recessive alleles for cystic fibrosis, i.e. the genotype is **nn**. A person with the genotype **Nn** is a **carrier** of the disorder but does not suffer from the problem. There is a one in four chance of a child suffering from the disorder if two carriers have a child (see the Punnett square in Table 1.3).

Table 1.3 Punnett square for cystic fibrosis

gametes	N	n
N	NN	Nn
n	nN	nn

The disorder occurs about once in every 2000 births, and it accounts for 1–2% of admissions to children's hospitals.

There is no cure for cystic fibrosis but treatments are improving, and people who suffer from it are living longer into adult life. **Gene therapy**, in which normal copies of the dominant gene are

Did you know?

It is estimated that the recessive gene that causes cystic fibrosis has survived for more than 2500 generations in the human race.

Does this suggest that in the heterozygous condition, it may be an advantage to some people? Otherwise, why has it not become extinct?

introduced into the lungs of sufferers by means of aerosols, holds out possibilities for the future.

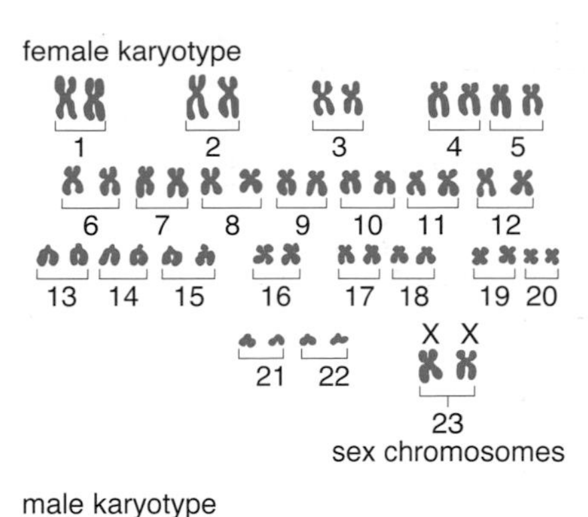

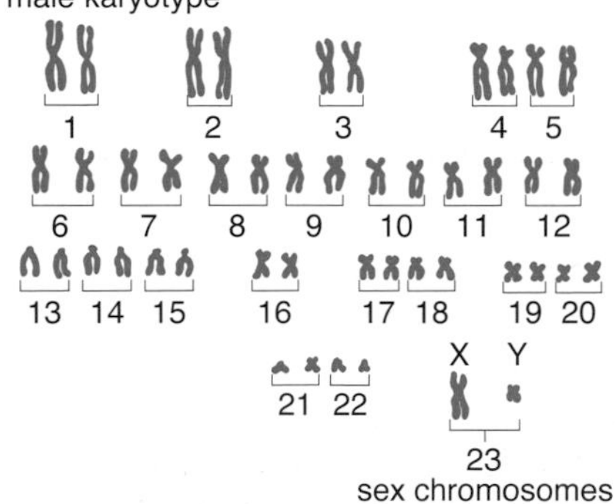

Figure 1.37 The complete set of chromosomes of a man and a woman

Boy or girl?

In all of our body cells that contain nuclei, one of our 23 pairs of chromosomes carries genes that determine our sex (see Figure 1.37). These are our **sex chromosomes**. In females the chromosomes are the same, **XX**, and in males they are different, **XY**. The **Y** chromosome is very small by comparison with the **X** chromosome.

We can explain sex determination in humans as shown in Figure 1.38. There is a 50% chance of a child being male or female.

Genes and DNA

An organism's characteristics are passed down from generation to generation by genes. What are genes, where can they be found, and what do they actually do?

Life is a series of chemical reactions. It is not surprising, therefore, that quite early in the history of genetics, scientists knew that genes must be chemicals.

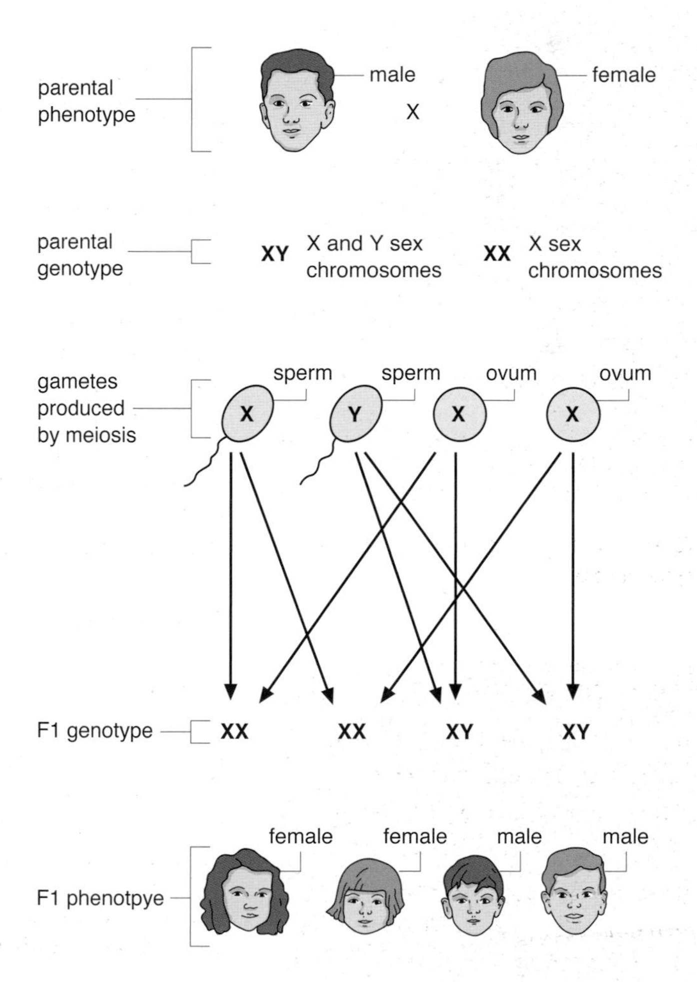

Figure 1.38 Human sex chromosomes

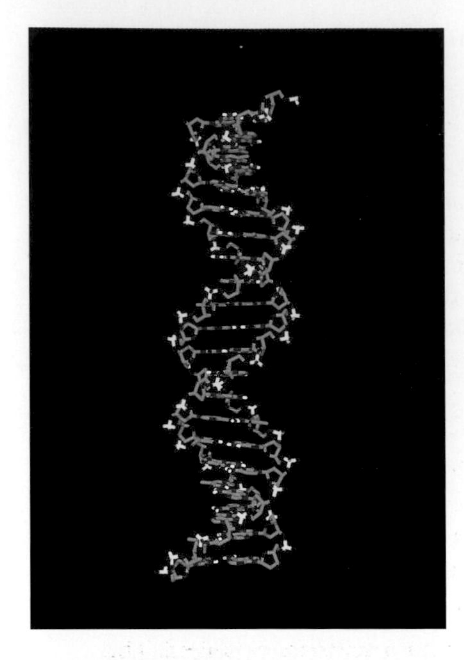

Figure 1.39 A model of DNA.

Proteins are the most essential chemicals found in living cells. Hence the early geneticists guessed that genes probably existed as protein contained in the chromosomes in the nuclei of cells.

By 1950, it had became clear that it was not the protein in chromosomes that passed on the code of life from generation to generation; it was another component of cells: **nucleic acids**. These are some of the largest and by far the most fascinating of all life's molecules.

Two forms are known: **deoxyribonucleic acid** (**DNA**), which is found in all chromosomes, and **ribonucleic acid** (**RNA**), which is found in the cytoplasm and nuclei of cells (see Figure 1.39). A chromosome is a single long molecule of DNA.

It is DNA that carries the **genetic code**. The structure of the DNA molecule was discovered in 1953 by an American, James Watson, and the English scientist Francis Crick, working at the Cavendish Laboratory in Cambridge, UK.

Nucleic acids such as proteins are made up of many units strung together. DNA has a ladder-like structure of two long sugar–phosphate chains joined together by connecting bases (rungs). The ladder is twisted to form a three-dimensional **double helix** (see Figure 1.40, which shows a model of a strand of DNA).

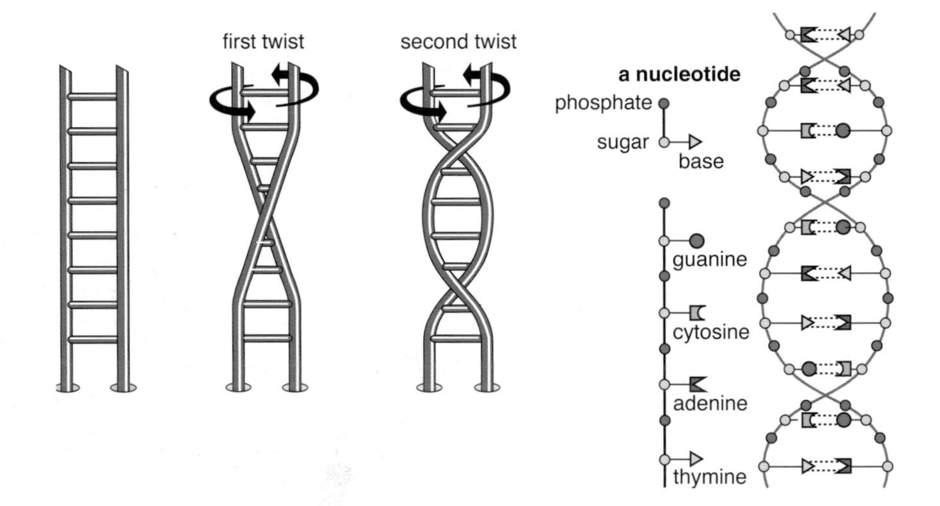

Figure 1.40 The structure of DNA

Watson and Crick discovered that the rungs of the ladder are made of four different types of base: guanine, cytosine, thymine and adenine. The bases fit together as shown in Figure 1.41. Guanine pairs only with cytosine, and thymine pairs only with adenine.

The differences between one DNA molecule and another, or one gene and another, depend on the pattern of these base pairs. This is how genes produce certain effects in organisms. For example, the order of base pairs that produces blue eyes is different from the order of base pairs that produces brown eyes.

There is an almost unlimited number of possible arrangements of base pairs, when you consider that a strand of DNA can be over 10 000 base units long. Therefore there is an almost unlimited number of possible genes in plants and animals.

How do genes work?

How do genes control the development of an organism? How can such microscopic particles have such an enormous effect on life?

Genes consist of DNA. DNA expresses a code that determines which chemical reactions take place in a cell and at what speed. It does this by determining which proteins are made (synthesised) in the cell. The growth and development of a cell is determined by the type and speed of the chemical reactions taking place within it. Hence, by controlling **protein synthesis**, DNA controls the life of the cell, and hence the development of the organism.

How does DNA control protein synthesis?

Proteins are made of building blocks called **amino acids** (see the section on proteins and their uses in Chapter 3). The amino acids are linked together in chains. The different ways in which various amino acids are linked together determines the type of protein synthesised.

DNA is able to regulate how the amino acids are arranged. The types and arrangement of the bases in the DNA molecule act as a code that determines which amino acids are linked together. By determining the form and arrangement of these basic building blocks, DNA controls protein synthesis.

The following is a very simplified account of how DNA is used for making proteins:

1 The long molecule of DNA (remember that it is like a twisted ladder) unwinds and splits along its length between the bases.

Figure 1.41 James Watson and Francis Crick discovered the structure of DNA in 1953.

2 One half of the molecule now acts as a pattern for the formation of a 'copy' molecule. These strings line up opposite their partners on the original half of the DNA. In this way, they form a single strand. The result is that the code originally present in the DNA is now also on the copied molecule.
3 The molecule then passes through the nuclear membrane to structures in the cell called **ribosomes**.
4 The code on the 'copy' molecule then controls which amino acids from the cytoplasm of the cell become linked together, and therefore what type of protein is synthesised.

The discovery that DNA carries the code of life was an enormous step forward in molecular biology. The next step was the discovery of how the molecule replicates itself.

DNA replication

Remember that before a cell divides, the chromosomes duplicate themselves. As we have just seen, the chromosome actually consists of a long chain of DNA. It is this that duplicates itself.

The two long strands of DNA separate, and then the free bases present in the cell nucleus align themselves with the separated strands (guanine with cytosine and thymine with adenine). This results in the creation of two new double helices (see Figure 1.42).

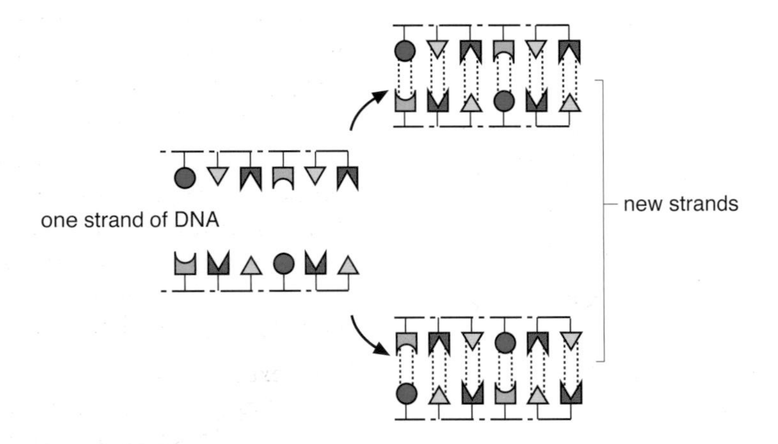

Figure 1.42 DNA replication

Questions

8 What does the word 'genetics' mean?

9 What were the advantages of Gregor Mendel using peas for his experiments?

10 State Mendel's principle of dominance.

11 What do you understand by the term 'genotype'?

12 In humans, the allele for free ear lobes, **R**, is dominant to the allele for attached ear lobes, **r**.

- **a** Using these symbols, give the genotypes of children that could be born to parents where the father is heterozygous and the mother has attached ear lobes.
- **b** State whether the children would have free ear lobes or not.

13 Gregor Mendel's first law of genetics states: 'Of a pair of contrasted characters, only one can be represented in a gamete by its germinal unit.'

- **a** Give the modern name for 'germinal unit'.
- **b** State where germinal units are found in gametes.
- **c** A red-haired woman marries a brown-haired man. All their six children have brown hair. Explain this genetically.

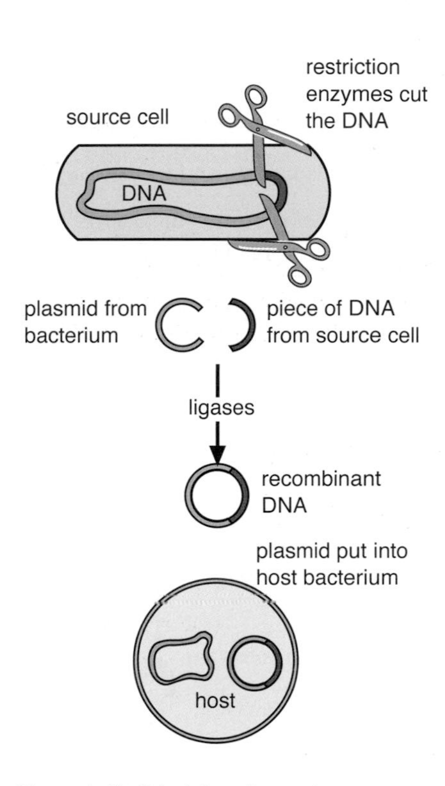

Figure 1.43 Principles of genetic engineering

Gene technology

Genetic engineering

Genetic engineering is based on isolating a gene from one organism and putting it into another gene of a different species. Why would you want to do this? The gene might be responsible for carrying out a very useful function. For example, scientists often isolate genes from human chromosomes that control the production of proteins such as hormones. They insert these useful genes into bacteria or yeast cells. The idea of transferring human genes to the bacteria or yeast cells is to increase production of the useful substance. The microbes multiply very rapidly and can be cultured relatively cheaply. In fact, they can provide an almost unlimited amount of substances that are practically unobtainable in bulk in any other way.

The process starts with an enzyme that acts like a pair of biological scissors (see Figure 1.43). The enzyme is added to chromosomes so that it can cut out chunks of DNA at points where useful genes are known to exist. The enzyme therefore enables the scientist to cut out very precisely the gene that is needed. It may be one out of hundreds on a particular chromosome.

The next stage is to insert the gene into a bacterium (the host cell). It is put directly into a circular piece of bacterial DNA. The circular piece of DNA, like chromosomes, carries bacterial genes that control all the activities taking place inside the bacterium. The circular piece of DNA is cut open with enzymes and the foreign gene is inserted. The break is then sealed with another enzyme.

The altered circular pieces of DNA are then mixed in a test tube with bacteria that do not have any such pieces. Some altered circular pieces of DNA move inside these bacteria. They now have the foreign gene in them, and can instruct the bacteria to make the required protein (see Figure 1.43). This means that the host cell obeys the instructions carried by the foreign gene and makes human proteins.

Life-saving genetics

There is a **hormone** called **insulin** that is made by certain cells in the pancreas (see Chapter 2). Its function is to keep the concentration of **glucose** in the blood at a constant level (0.1 g per 100 cm^3 of blood). If the glucose level falls much below this, the body does not have enough fuel to function properly. A glucose level above the normal concentration also disturbs body functions. In particular, the kidneys fail to cope and glucose is lost in the urine. The body is eventually drained of fuel.

Insulin plays a vital role in getting rid of excess glucose from the blood. It helps to do this in a number of ways, one of which is to change glucose into an insoluble form of **carbohydrate** called **glycogen**. This is then stored in the liver until a new supply of glucose is needed. Without enough insulin, people cannot control their glucose level and suffer from **diabetes mellitus**. Fortunately, it is possible to control this condition by injecting insulin into a diabetic person.

Until recently, the insulin used for this treatment was taken from the pancreases of various slaughtered animals. It was difficult to produce enough insulin in this way to meet the high demand from sufferers of diabetes. Also there are several cultures and religions that do not allow the use of products from various animals, for example pigs and cattle. Strict vegetarians sometimes object to animals being used in this way.

Genetic engineers have solved these problems by genetically modifying bacteria. The method is summarised in Figure 1.44. The main advantage of this technique is the mass production of many otherwise scarce and expensive proteins that help save lives. Bacteria can be used as living factories to produce these proteins on demand. Theoretically, any protein can be mass produced using the same principle.

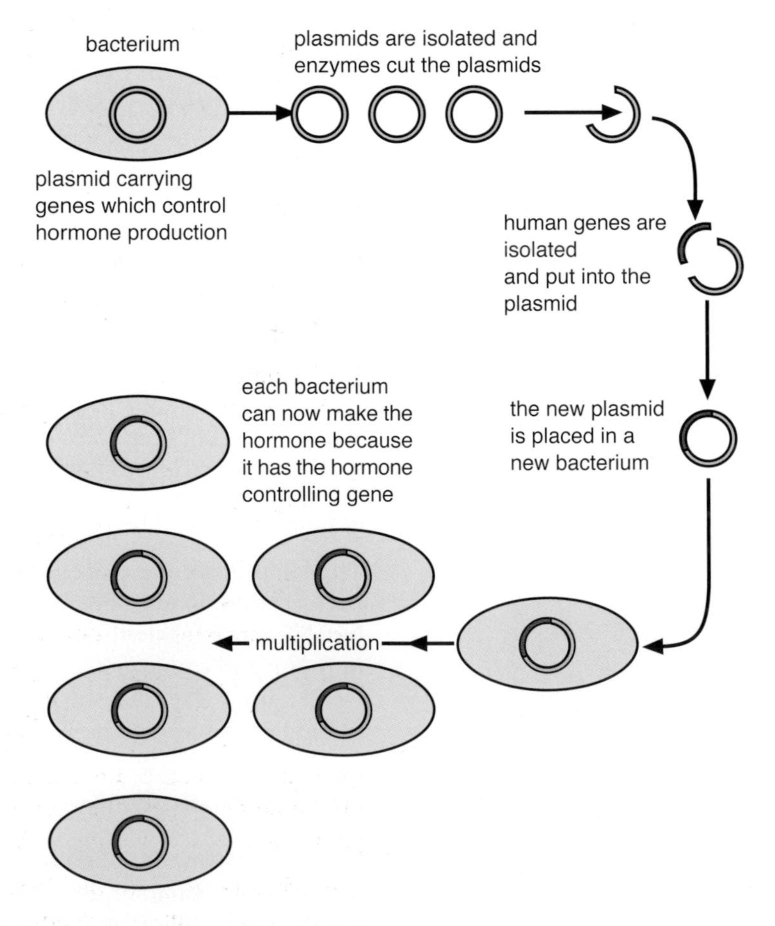

Figure 1.44 The production of a human hormone

Another example of life-saving gene technology began to be used in 1993. Sheep were genetically modified to produce protein that helped blood to clot. Human genes that normally control the production of a blood-clotting factor were introduced into early stages of female sheep embryos. When these grew into adults, they produced milk containing the blood-clotting factor.

This technology is used today to treat people suffering from a condition called **haemophilia**. Haemophilia is an inherited blood disorder in which an essential blood-clotting factor is either partly or completely missing. Because their blood does not clot properly, people suffering from haemophilia are in danger of massive blood loss if they cut themselves accidentally.

Genetically modified crops

In the 1980s, the first commercial **genetically modified** (**GM**) crop was modified so that it was resistant to insects and pests. It was the potato, and it was modified so that it made its own built-in **insecticide**. The insecticide was an insect poison normally produced by a type of bacterium that lives in the soil. The gene for the poison's production was transferred to potato plants, which then made the plant resistant to insect pests.

Since then, many types of crop have been modified using exactly the same principle as the one used for making potatoes resistant to insects. Resistance to **herbicide** is now common in genetically modified crops. In 1996, the European Commission approved the sale of maize that was genetically modified for resistance to herbicide as well as being resistant to insect pests. By 1999, 30% of maize and 50% of soya harvested in the USA was genetically modified for herbicide and insect resistance.

Weeds compete with crops if left unchecked. For a great many years, farmers have attempted to get rid of weeds by using chemicals called herbicides. However, selective herbicides that kill only weeds and not plants that are wanted are difficult to produce. A herbicide-controlling gene can be taken from a bacterium that normally grows in soil and transferred to a plant such as soya.

Unfortunately, there have been some potential problems with this technology to which some people have objected. For example, during the process of introducing the gene into the plant, a **marker** gene is used to mark the cells that contain the modified chromosomes. The marker gene produces resistance to antibiotics. Objectors suggest that in the case of maize, there is a small chance that if cattle eat the maize, the gene could pass into bacteria that normally live in the cattle's digestive system. This would help to spread antibiotic resistance.

Other concerns include the possibility of herbicide-resistant plants escaping into the environment and flourishing. How could they be destroyed if herbicides cannot kill them? The answer is to ensure that the plants are sterile and can only reproduce asexually. Another unwanted side effect found in herbicide-resistant soya was that many plants split their stems in hot climates and could not support themselves.

The advantages of herbicide-resistant and insect-resistant plants are that it is necessary to introduce far fewer chemicals into the environment to kill insects and weeds. Theoretically, high yields from crops can be maintained without affecting the environment.

Will developing countries benefit?

The case for:

- Crops could be tailor-made to suit the varied farming conditions found throughout the world. In this way, they could provide more nutritional value and a higher income.
- Energy-producing crops could save natural resources and so conserve the environment.

The case against:

- GM crops could reduce the developed countries' reliance on crops from developing countries. This could result in loss of trade and severe economic damage for the developing countries.
- Because of political reasons and mismanagement, there is doubt as to whether the populations of developing countries will actually receive the benefits from genetically modified crops in many cases.

These issues raise important political, ethical and trade questions that are not unique to modern biotechnology. They must be resolved at government and international level to maximise the benefits from gene technology.

Forensic genetics

In 1987, the first case in the UK using **DNA profiling**, or **genetic fingerprinting**, proved that a suspect who had confessed to a rape and a murder was in fact innocent. Two girls had been raped and murdered in the same part of the country, but the crimes were committed three years apart. Investigators suspected a connection between the two, and soon a suspect was being questioned. He admitted to one of the crimes but not the other.

Forensic scientists used DNA profiling that showed clearly that both crimes were committed by the same person, but the suspect was not that person. All men from the area were asked to give blood samples so that their DNA could be analysed. Around 5000 samples were taken. Finally, one sample matched those from both crime scenes, and the person was arrested and convicted.

Genetic fingerprinting can also be used to determine whether a man is the father of a child. The makeup of the DNA of a father and his child is very similar.

The principle of gene profiling is based on the fact that every person's DNA is made up of a particular pattern of chemicals. Unless you are an identical twin, your DNA will be unique to you.

Ethical problems of gene technology

Since the 1980s, advances in the fields of genetics and medicine have led to rapid developments in medical genetics. The complete genetic map of human chromosomes, showing the positions of all the genes, was completed in the early years of the twenty-first century. This was the outcome of the **human genome project**.

By 2002, intensive research was being carried out into the possibility of mapping an individual's genome quickly and cheaply. Our genomes are 99.9% identical. to other people's. Only 0.1% makes us different. An individual's genome is able to tell us whether there is a chance of the person having an inherited disease, as well as having other characteristics that can be inherited.

Many moral and ethical problems are associated with this knowledge. Once imperfect genes have been discovered in someone's genome, will the person's life be made worse? Should other people have access to information about our genes? Insurance companies are certainly interested in a person's genes when pricing life-assurance policies.

It is now possible to create viruses by knowing the pattern of genes and putting their chemicals together in the right order. In 2002, US scientists made the polio virus in this way. The genomes of many viruses are available on the internet, and this raises serious questions about whether it is desirable for the general public to have access to such information. Should there be patents on genomes or newly discovered genes? This would mean that the ownership of such discoveries would be limited.

Who can own a gene?

There are laws called patent laws that protect inventions for a fixed number of years. During this period, no one else can copy or use the invention without permission, which usually has to be bought. There is a controversy about whether this legal patent protection should cover genetically modified organisms.

The case for:

- Patent protection enables inventor companies to get back the money they have spent on research and development. Without this system, far less research funding would be available to scientists.
- The details of any patent must be published and made available to all. Without patent protection, more inventions would be kept secret, which would slow up the development of science.

The case against:

- Genes are not inventions, and should therefore not be subject to patent laws.
- It is doubtful whether GM crops should be controlled by the few who can afford the development costs.
- Patents tend to benefit only the developed countries. It is important that developing countries have access to these new technologies.

In 1998, the European Union (EU) produced a directive on the legal protection of new inventions in biotechnology. This gives patent protection for genes and genetically modified organisms provided that there is a new invention. The function of a gene cannot be patented.

Did you know?

In 2001, a company called the DNA Copyright Institute of San Francisco in the USA was set up to allow pop stars and other celebrities to own the copyright on their DNA to prevent any unwanted cloning.

The human genome:
www.schoolscience.co.uk

Genetically modified plants for food and human health:
www.royalsoc.ac.uk

Genetic engineering – say no?:
www.greenpeace.org

Questions

14 Discuss the advantages and disadvantages of genetically modified crops.

15 Write an account of the commercial applications of cloning animals.

16 Describe how genetic engineering has been able to produce certain proteins in the bodies of cloned animals.

Summary

1 There is a scientific system for the identification and naming of all living organisms, based on similarity of features.

2 The size of a population may be affected by competition for resources, predation, disease and pollution. Species adapt to their environment by evolution or they become extinct.

3 Indicator species can be used as monitors of pollution.

4 Variation may be due to environmental or genetic causes.

5 Sexual reproduction leads to offspring that are genetically different from their parents. Asexual reproduction leads to genetically identical offspring called clones.

6 New genes result from mutation.

7 Individuals with characteristics adapted to their environment are more likely to survive and breed to pass on the genes responsible for these characteristics.

8 Organisms have changed over time, and fossils provide evidence of this change.

9 Cystic fibrosis is the most common genetically inherited disorder.

10 Chromosomes are normally found in pairs in the nucleus of each body cell. Genes are made of DNA.

11 Mitosis is cell division leading to identical cells. Meiosis is cell division that produces gametes with half the normal number of chromosomes.

12 Sex is determined by sex chromosomes. A male has an **X** and a **Y** chromosome in each body cell, and a female has two **X** chromosomes in each body cell.

13 The outcome of monohybrid crosses depends on dominant and recessive alleles.

14 The technique of producing clones is useful in the quick and economical mass production of organisms.

15 Genetic engineering can be used to produce insulin and herbicide-resistant crops.

16 Gene technology can be used to solve crimes and paternity cases.

17 Many ethical and moral dilemmas are associated with gene technology.

Chapter 2 Body maintenance and protection

By the end of this chapter you should:

- know how humans maintain a constant body temperature;
- understand how insulin is used to control the level of blood-sugar;
- know the causes of diabetes and how it is diagnosed;
- know what path is taken by nervous impulses in response to stimuli;
- understand how the reflex arc involves a nerve impulse that is carried via nerve cells and can pass across synapses;
- understand that health is affected by a variety of factors and that it depends on both lifestyle choices and genetic factors;
- know the implications of genetic counselling and gene therapy;
- understand the importance of healthy eating;
- understand the effects of alcohol, nicotine and other drugs on the body of the individual and on society in general.

Homeostasis

Temperature regulation by the skin

Animals must regulate the conditions inside their bodies to keep them at a constant level. This is so that the complex chemical reactions that take place in the body can occur in a controlled, favourable internal environment.

How sweat glands work:
www.howstuffworks.com

The skin (see Figures 2.1 and 2.2) excretes water, minerals and some urea when we produce sweat. This fluid has a useful role because it helps to regulate body temperature (see Figure 2.3).

Questions

1 Why is an increased salt intake recommended for humans in hot weather?

When liquid water is changed to water vapour by evaporation, heat is used for this change of matter from one state to another. As water in sweat evaporates from the body surface, heat is withdrawn from the outer tissues. The skin acts like an automatic radiator. It is richly supplied with warm blood, and so heat is carried to the surface in this way. At the same time, the production of sweat increases and it bathes the skin. This increases the rate of evaporation and the amount of heat lost.

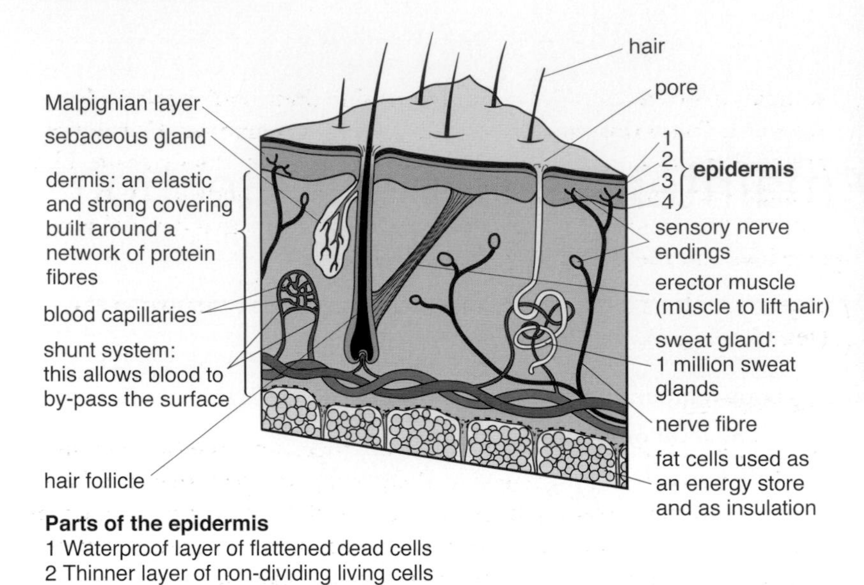

Parts of the epidermis
1 Waterproof layer of flattened dead cells
2 Thinner layer of non-dividing living cells
3 Pigmented layer: protective, absorbs harmful ultraviolet light
4 Inner layer of dividing cells to replace those worn away

Figure 2.1 A section through human skin

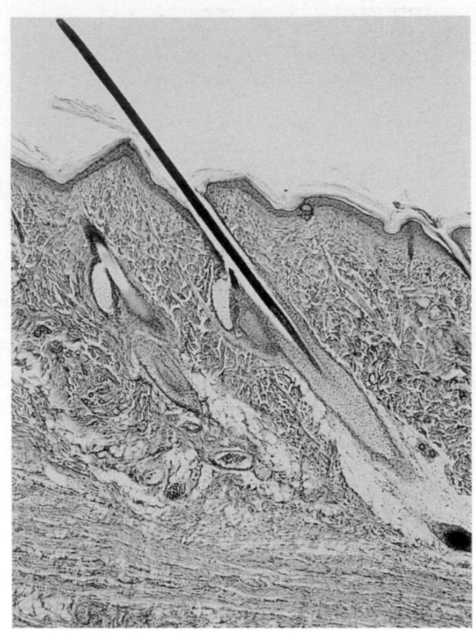

Figure 2.2 A section through human skin showing its internal structure

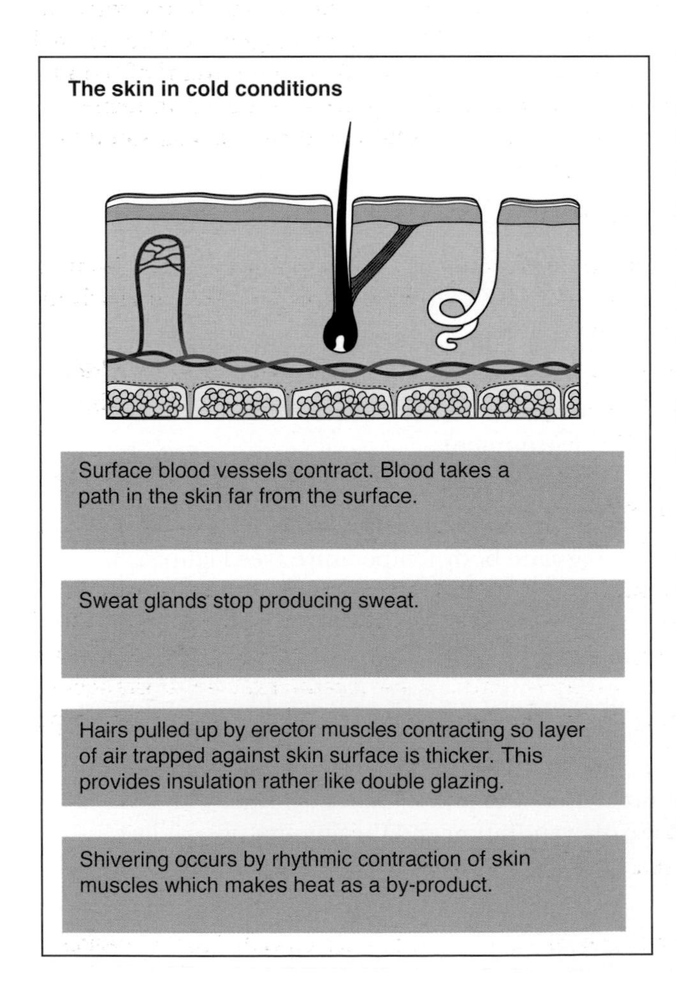

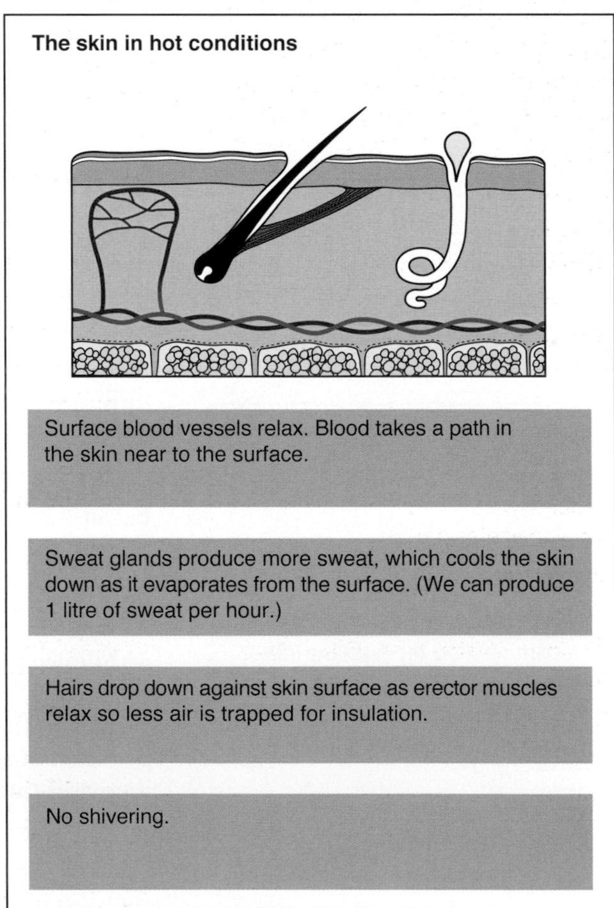

Figure 2.3 Temperature regulation

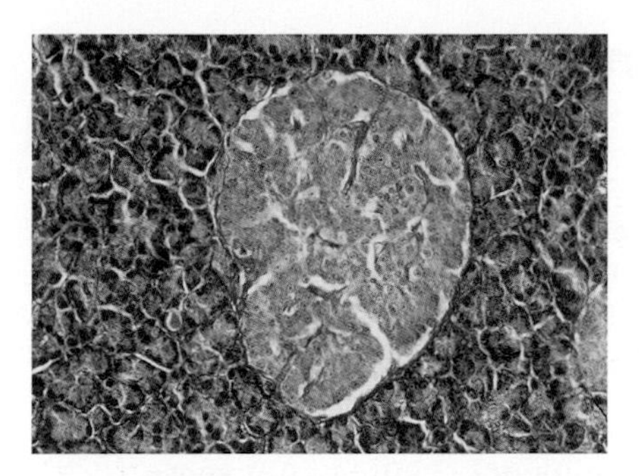

Figure 2.4 A section through a human pancreas showing an islet of Langerhans

The regulation of glucose

Animals need to keep the conditions inside their bodies relatively constant. To do this they use chemicals called **hormones**. Hormones are made in glands that pass them directly into the bloodstream. The bloodstream then carries them to the parts of the body that they control. Some hormones are chemical messengers that control some activities and the development of the body.

One example of a gland that produces hormones is the pancreas (see Figure 2.4).

The pancreas

In addition to producing enzymes to aid digestion (see the section on the pancreas and digestion in Chapter 3), the pancreas also has special cells that produce hormones to regulate the concentration of glucose in the blood. The hormone that reduces the concentration of glucose when it becomes too high is called **insulin**.

Insulin enables the liver to store glucose in the form of glycogen and speed up the use of glucose to release energy in cells during respiration. A person who does not have insulin is unable to store or use carbohydrate properly. As a result, carbohydrate, in the form of glucose, builds up in the blood. This can be dangerous. The normal concentration of glucose in the blood is 0.1 g per 100 cm^3 of blood. If it rises above this, the kidneys allow glucose to be lost in urine. This condition is called **diabetes mellitus** and can be diagnosed by excess glucose in the urine.

There are three main types of diabetes:

- In juvenile diabetes, either insulin is not produced or the body does not react to insulin in the normal way. It is detected early in the life of a person.
- Maturity-onset diabetes is confined mainly to overweight, older people, and occurs when there is too little insulin to meet the demands of the body.
- Stress diabetes may occur during pregnancy or when a person is under physical or mental stress. The body fails to respond to insulin at its normal levels in the blood.

Treatment by regular injection of the correct amount of insulin normally allows people with diabetes mellitus to lead a normal life if they pay attention to carbohydrate intake in their diet. Medication in tablet form is also used regularly for treatment. The substances taken in this way do not include insulin, but, depending on the type of diabetes, they regulate the blood sugar level.

Did you know?

In 2005, part of the pancreas of a Japanese woman was successfully transplanted into her daughter, who was suffering from diabetes. The transplant cured the condition. More recently, cells from a healthy pancreas have been cultured and then transplanted to people suffering from diabetes. Many of these operations have been successful.

Diabetes UK:
www.diabetes.org.uk

International Diabetes Federation:
www.idf.org

Excellent general information site:
www.howstuffworks.com

Questions

2 How does the blood help in the work of the hormone-producing glands?

3 Why does sugar appear in the urine of a person with diabetes mellitus?

The nervous system

Nervous control

Animals use their **nervous system** to make them aware of changes in their surroundings (stimuli) and to help coordinate a reaction to stimuli in a way that is to their advantage. The main part of our nervous system is called the **central nervous system**, and it consists of a **brain** and a **spinal cord**.

We react to changes in our environment because of our **sense organs**. These contain groups of **receptor cells** which respond to the stimuli of light, sound, touch, temperature and chemicals. For example, when you stub your toe, the pressure received by the cells in your skin causes an electrical signal to travel to your central nervous system. This **nerve impulse** is relayed very quickly back to muscles in your foot. These muscles then contract to move your foot away. In the mean time, your brain has registered the feeling of pain and you are aware of the problem.

Nerves consist of bundles of **nerve cells**. They are something like an electrical cable made up of insulated wires bound together. Nerves that carry impulses to activate muscles are called **motor nerves**. **Sensory nerves** carry impulses away from sense organs towards the central nervous system. Nerve cells never really touch one another because there is always a tiny space between them. These spaces are called **synapses**, and they have to be crossed by impulses before any message can be sent from one motor nerve cell to another or from one sensory nerve cell to another. Impulses bridge the synapse with the help of fast-acting chemical reactions at the endings of nerve cells.

Activity

Use a CD-ROM encyclopaedia or multimedia title about the human body to find out how the nervous system and your senses work.

The human body:
www.bbc.co.uk/science

The spinal cord and spinal nerves

The spinal cord (see Figures 2.5 and 2.6) extends down from the brain, passing through the protection of the bones of the spine, the **vertebrae**. Thirty-one pairs of **spinal nerves** branch off the nerve cord, passing out between the vertebrae and dividing to make up the nerves that supply the organs of the body.

If your spinal cord were cut, serious problems would result. All the parts of your body controlled by nerves that leave the cord below the point of the cut would be totally paralysed. In addition, you would lose all sense of feeling in many areas below the point of damage.

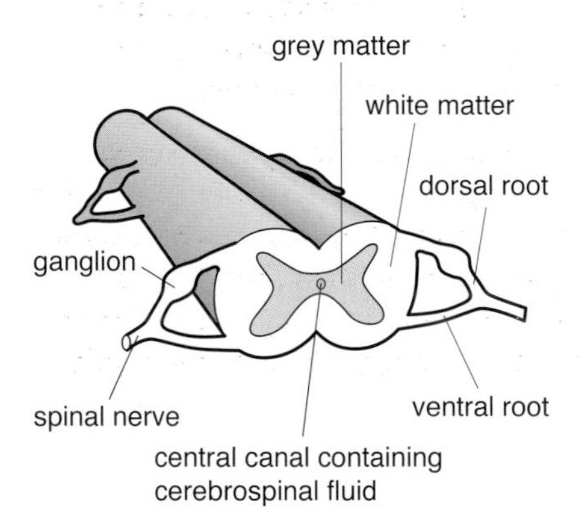

Figure 2.5 The structure of the spinal cord

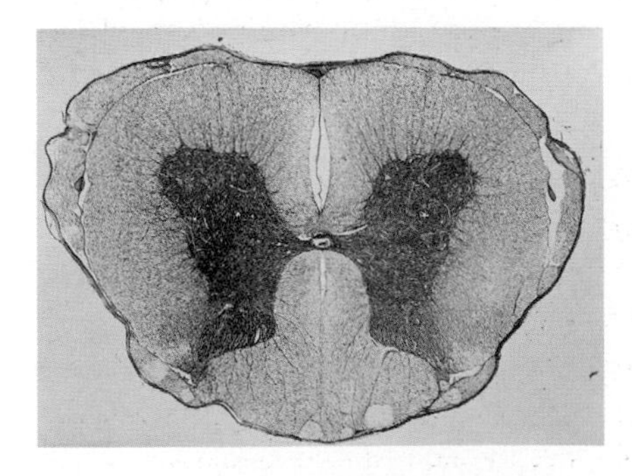

Figure 2.6 Transverse section through the spinal cord

Reflex actions

The simplest kind of nervous reaction in humans is called a **reflex action**. It involves only two or three nerves that link a receptor (sense organ) to an **effector** (muscle or gland) via the central nervous system (the spinal cord or the brain). It is a rapid, automatic reaction that does not involve conscious effort and may not involve the brain.

The knee jerk is a good example of a simple reflex action. Sit on the edge of a table and let your knee swing freely. Then tap your leg just below the kneecap with a narrow object.

Your lower leg will jerk upwards because the tap stimulates receptor cells in the lower leg (see Figure 2.7). An impulse travels along the

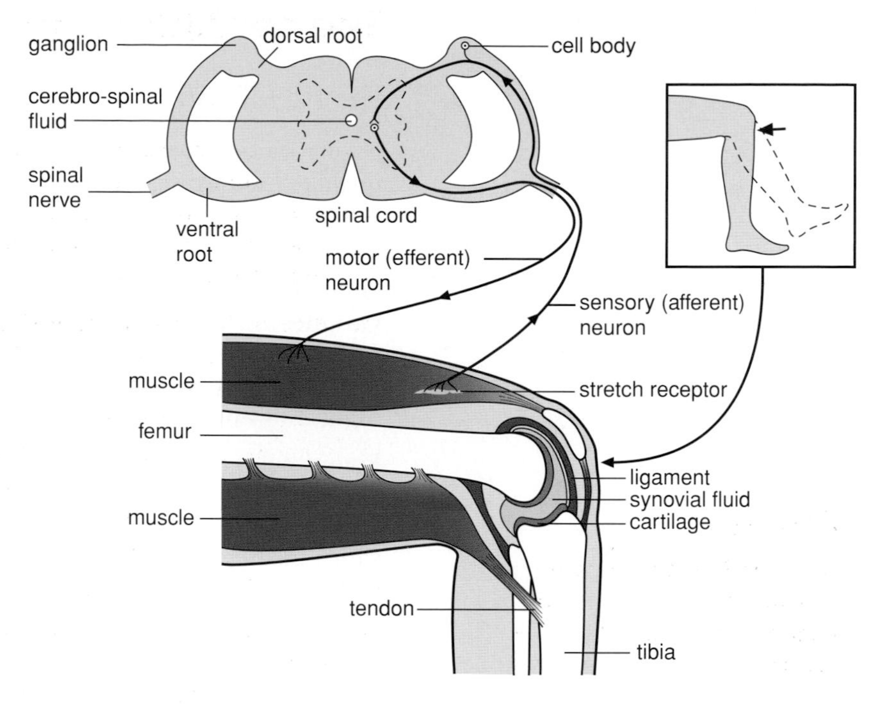

Figure 2.7 The processes involved in the knee jerk reflex, from stimulus to effector

sensory nerve cell and through a synapse to a **connecting nerve cell** in the centre of the spinal cord. The connecting nerve cell stimulates a motor nerve cell through another synapse, and the impulse travels to muscles in the leg that contract, causing the jerking movement. The entire reflex takes a split second.

Reflex actions have a protective function. For example, when you touch a hot object, your hand jerks away almost instantly. The reflex is complete even before your brain registers the pain. If the muscle response were delayed until the pain impulse was complete and interpreted, the effects of the burn would be much greater. Other reflexes include sneezing, coughing, blinking, and the constriction or dilation of the pupil in the eye.

Questions

4 How does the human body check on its surroundings?

5 Why are some of our reactions very fast?

Practical work

Testing your reflexes and reaction time

1 Work in pairs. Hold a 30 cm rule vertically while your partner places a thumb and forefinger on each side of the bottom of the rule without touching it.
2 Let the rule drop without warning. Your partner should catch it as quickly as possible by closing his or her thumb and forefinger.
3 Record to the nearest centimetre by how many centimetres the rule fell before it was caught. Do this by measuring the distance from the end of the rule to the middle of the catcher's thumb and forefinger.
4 Do this three times without practising beforehand, and calculate the average length for each member of the class. The more rapid the reflex action, the shorter will be the distance through which the rule is allowed to fall.

Questions

6 What is a reflex arc?

7 Describe, without the aid of diagrams, the path taken by a nerve impulse during a reflex action.

Health

Healthy living

Health is not just the absence of illness. It is a positive and enjoyable feeling of well-being resulting from efforts to maintain an all-round state of physical and mental fitness.

Although there is no need for all of us to be as fit as Olympic athletes, it is a good aim to be fit enough to live the lifestyle we want. Our lifestyles are probably less active than those of people who lived 100 years ago, but we still have to exert ourselves from time to time, whether in running for a bus or dancing all night. If we can do these things without collapsing, it is obviously to our advantage.

Fitness helps our bodies to function. For example, sustained regular exercise improves our blood circulation and reduces our heart-rate. It also makes the breathing system more efficient. In combination with a healthy **balanced diet** and a non-smoking lifestyle, exercise is certain to benefit us.

Our daily bread

Snails, seaweed and squid all have one thing in common. Each has a place in the human diet, although you might enjoy one of these dishes but be revolted by the thought of others. Humans eat for many reasons. The main one is that food provides enough energy to keep us alive. The foods that we select vary according to availability and our preference.

Eating too much or too little of certain kinds of food can lead to serious physical disorders. Therefore it is important that we have a balanced diet (see the section on food in Chapter 3). The importance of diet has been recognised through the ages. Florence Nightingale was one of the first people to understand the importance of a healthy diet for healing. Since then, the study of diet and health has been the job of **dieticians**. Dieticians use their knowledge to provide or recommend a diet that will keep a person healthy and is also tasty to eat.

Maintaining a balanced diet is not always an easy thing to do. If we look back 100 years, we can see that there has been a marked change in the diet of people living in the UK (see Table 2.1).

Table 2.1 Milestones in the food revolution

1900s	Even though the Edwardian era was a prosperous one, at least one-third of the population in the UK was undernourished. Free school meals and medical inspections were introduced.
1930s	The First World War ended in 1918, but it had shown how unfit people were. Millions failed army medicals. Food production was increased, but people in the UK still suffered from rickets, tuberculosis, anaemia and poor growth.
1940s	In the Second World War, the diet of every individual in the UK was controlled by rationing, and the population was, on the whole, fitter.
1990s	A wide choice of foods was available in the UK from all over the world at any time of year. The use of convenience foods and 'junk' foods led to problems resulting from obesity. It became clear that the use of food additives might be linked to health problems and behavioural problems in children.

In the nineteenth century, most people in the UK ate plenty of fibre in the form of bread and potatoes. However, diseases due to a lack of vitamins or minerals were common. Today, the problems are different. We now tend to eat too much meat, dairy products and sugar, and very little fibre.

Too much of a good thing

We all tend to eat too much at times, but if we keep this up over a long period, some very serious dietary diseases may develop.

- **Heart disease** is linked to eating too much fat.
- **Diabetes** and tooth decay may be linked to eating too much sugar.
- **High blood pressure** and strokes may be linked to eating too much salt.
- **Bowel cancer** may be linked to eating too little fibre.
- **Obesity** may be linked to eating too much fat and carbohydrate.

Our problem nowadays tends to be an excessive intake of food. In the UK, 39% of men and 32% of women are overweight, with 6% of men and 8% of women being classified as obese (excessively heavy). The ideal way to control your diet is to look at your intake of calories.

Carbohydrates and **fats** supply most of our energy. Energy is measured in **calories** or **joules**. Scientifically, one calorie is the quantity of heat that is needed to raise the temperature of 1 g of water by 1 °C, and the joule is the international unit used to measure heat energy. 1 cal = 4.2 J. In the fields of nutrition and dieting, and on food packaging, the term 'calorie' actually means one kilocalorie, or 1000 calories.

The energy requirements for people engaged in various occupations are shown in Figure 2.8.

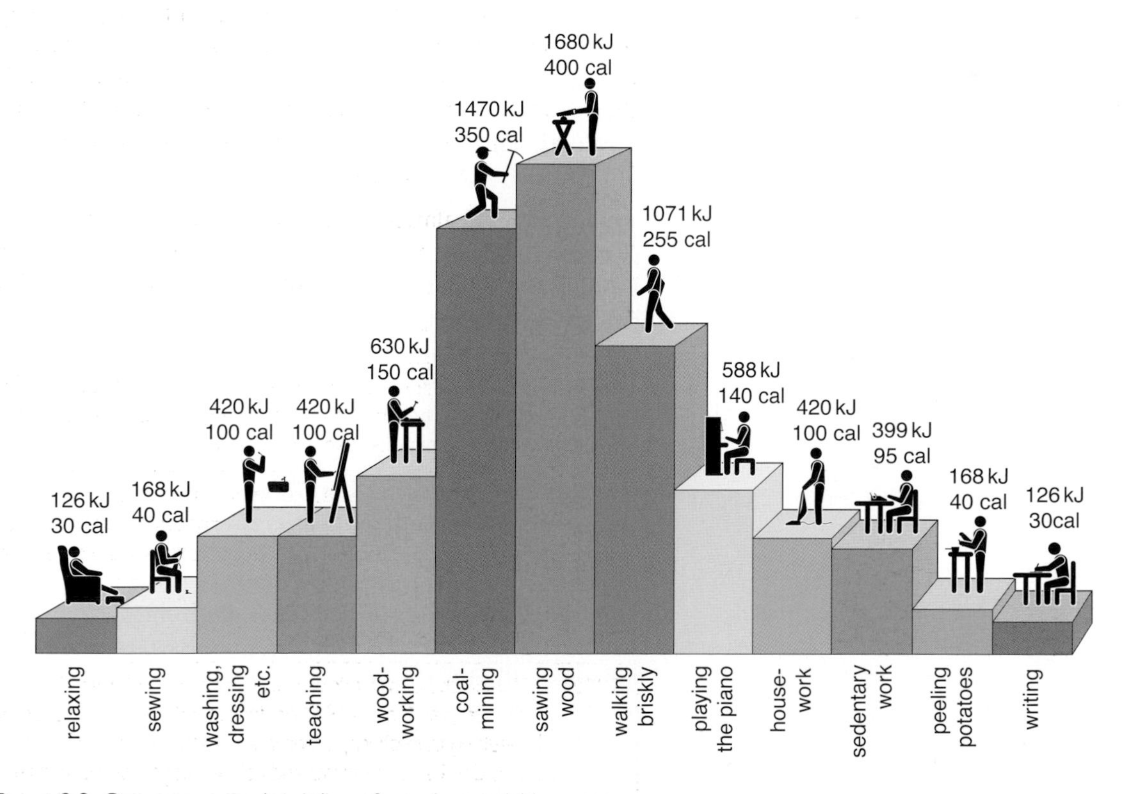

Figure 2.8 Our energy needs per hour for various activities

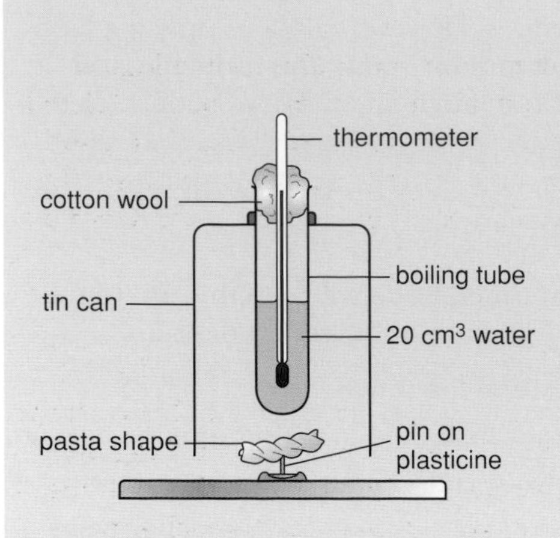

Figure 2.9 Apparatus used to measure the heat energy in a pasta shape

Practical work

Investigating the energy content of a pasta shape

1 Put on eye protection.
2 Weigh a pasta shape and mount it on a pin, as shown in Figure 2.9.
3 Place 20 cm^3 water in a boiling tube.
4 Measure its temperature with a thermometer.
5 Set light to the pasta shape with a Bunsen burner, and place the boiling tube over it to catch as much of the heat as possible.
6 Read the temperature of the water as soon as the pasta has been completely burned.
7 Record the increase in the temperature of the water.
8 Work out the value of the energy in 1g of the pasta shape in joules as follows.

4.2 J raises the temperature of 1g of water by 1°C.

increase in temperature of the water in degrees Celsius = y

mass of water (g) = 20

mass of pasta (g) = x

heat gained by the water in joules = $20 \times y \times 4.2$

heat energy released by 1g of the pasta in joules $= \frac{20 \times y \times 4.2}{x}$

Processed food and additives

Food can be kept in good condition if the growth of microbes, for example bacteria and fungi, is stopped or reduced. Traditional methods used to preserve food include pickling, canning and bottling. The problem with these is that as soon as you open the container, the preservation process is no longer effective. Therefore chemical preservatives are now added to manufactured food products. This allows the products, for example sauces and syrups, to be safe to use for a longer period of time.

The majority of the preserved foods that we buy today contain permitted chemical preservatives. These are identified by their **E-numbers**. The food may also contain colourings, flavour enhancers and extra sugar. These additives are tested fully before food manufacturers are allowed to add them to their products. However, some people are allergic to some additives. They suffer from headaches and feel sick. Young children may react to them by becoming overactive.

Government medical and nutritional experts have set guidelines concerning the amounts of different nutrients that most of us require. These are now called Dietary Reference Values and are similar to Recommended Daily Amounts (RDAs), which are part of the labelling of processed foods and drinks. Besides indicating the RDA for fat (including cholesterol), labels on processed foods also show the RDA for sugar, other carbohydrates and salt. Too much sugar can lead to tooth decay and excess energy intake is stored as fat. Too much salt can result in high blood pressure and heart disease.

Self-inflicted problems

A **drug** is any substance that alters your physical or physiological state. Users of drugs often become dependent on them or addicted to them. They have always caused social and health problems, and they are all harmful when improperly used. The facts about several drugs are given below, and you can use them as a basis for decisions about drug use.

Nicotine addiction, the UK's leading habit

Millions of people in the UK use tobacco in some form. Most smoke cigarettes, and most smokers have developed both a smoking habit and a tobacco habit. The first involves going through the motions of smoking. For example, many smokers automatically reach for a cigarette before finishing the first one. However, smokers eventually develop a physical dependence on the addictive drug in tobacco, **nicotine**, and this is the tobacco habit.

Many young people who smoke feel it makes them seem more mature. However, if they asked people who have smoked for several years, their advice would often be not to start. Unfortunately, the long-term effects of smoking do not worry the beginner, and by the time they begin to appear (see Figure 2.10), he or she has often become addicted.

What's in a cigarette?

Cigarette smoke contains a mixture of tar, nicotine and carbon monoxide. Tar contains over 4000 different chemicals, about 60 of which are known to cause cancer. Nicotine causes addiction because it is a powerful drug that stimulates the nervous system. It also increases blood pressure, which may lead to heart disease. Tar particulates prevent the body's lung-cleaning system of cilia from working. Carbon monoxide reduces the ability of the red blood cells to carry oxygen.

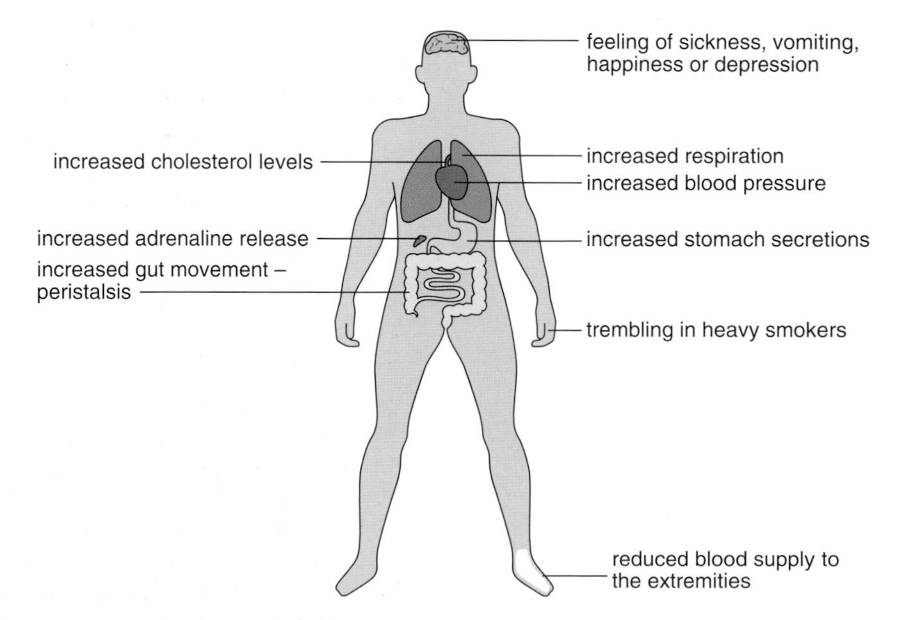

Figure 2.10 How nicotine affects the various parts of the body

> **Did you know?**
>
> Between 1983 and 1992, there were almost twice as many cases of lung cancer among women in the UK aged 25–44 than in men of the same age group.

Changing attitudes

Smoking has decreased since the 1960s as more scientific evidence about its effects was found. However, tobacco is big business, and it is not in the interest of countries that rely on tax on tobacco to ban smoking completely. Hence clever advertising of brands of cigarettes has maintained the sales of tobacco products to young people who continue to pay to die. This is despite the compulsory health warnings on packets of cigarettes and certain restrictions on where advertising should be placed.

Health education attempts by the UK Government have failed to stop young people smoking, perhaps because it spends a small amount on those campaigns in comparison with the millions of pounds spent worldwide on advertising by cigarette manufacturers.

Fortunately, many more non-smoking areas have now been established in public places. This has been accepted by the majority of considerate people. Smoke irritates the eyes and throats of non-smokers, and it can lead to more serious passive-smoker health problems, for example lung cancer.

Alcohol in the body

Alcoholic drinks contain **ethanol**, which is a poison. Ethanol is made by the action of yeast on sugars, and it can be absorbed into your blood system directly through the stomach lining without being altered by digestive enzymes. Within two minutes of ethanol reaching your stomach, it starts entering your bloodstream.

It is carried to your tissues and is rapidly absorbed by the cells. In the cells, most of the ethanol is oxidised very quickly, releasing much heat that then raises your blood temperature. The temperature regulation centre of the brain is consequently stimulated, and it responds by causing increased blood circulation to the skin. Excess heat is radiated away, which gives you a superficial rosy look. The rush of blood to the skin is at the expense of the blood supply to the internal organs, which are deprived of enough blood and heat.

Effects of ethanol on body organs

Only part of the ethanol consumed by the person is oxidised. Part is released into the lungs as vapour, causing the typical alcoholic breath odour. Some goes to the skin and leaves in sweat. Some goes to the kidneys and leaves in urine. All body organs absorb ethanol, and are consequently affected by its presence to some extent.

The oxidation of ethanol in cells produces water, and this is excreted by the skin to help control the body's temperature. Tissues then become dehydrated, and urea is concentrated in the kidneys to such an extent that it can lead to permanent kidney damage. Excessive drinking of ethanol can also affect the stomach. It causes an increase in stomach secretions that can lead to gastritis, a painful swelling of the stomach lining.

Effects of ethanol on nerves

Ethanol is a **depressant**, because it has an anaesthetic or numbing effect on the nervous system. However some people mistake it for a **stimulant**, because the numbing effect on the nerves makes some people less inhibited and less concerned about their behaviour.

The brain shows the first effects of ethanol. Loss of judgement, will-power and self-control are the first signs of drunkenness. When ethanol reaches the vision and speech areas of the brain, vision becomes blurred and speech is slurred. Muscle coordination is affected when ethanol reaches the part of the brain that controls balance, and this causes dizziness and an inability to walk properly. The final stage of drunkenness is unconsciousness, which can be life-threatening because of the possibility of choking on vomit or even dying because of a severely depressed heart-rate.

Alcoholism, the disease

People who suffer from alcoholism depend on ethanol continuously. It may start with occasional social drinking, but this can lead to heavy drinking in an attempt to lessen stress from many causes. About one in ten alcoholics reach a stage where professional treatment and hospitalisation are needed. Symptoms include confusion to the extent that even members of the patient's family are not recognised. Terrifying nightmares occur and uncontrollable trembling takes place.

Health problems may include shortages of certain vitamins. Vitamin deficiency diseases are common among alcoholics. This is because they often eat very little during periods of heavy drinking. As a result, the liver releases its food stores. The liver swells as the carbohydrates are replaced by fats, and, over a long period, a serious liver disorder called **cirrhosis** can occur, where the fatty liver shrinks and hardens as the fats are used.

Drugs and the body

The main reason why people abuse drugs is because it makes them feel good. However, drugs may eventually damage the brain, liver, kidneys and lungs. Another problem is that drugs can be addictive so the user cannot stop taking them easily (see Table 2.2).

Genes and health

Increasing good health, helped by changes in lifestyle and rising standards of living, has reduced infections and other diseases that are caused by environmental factors. However, some diseases are caused by changes in genes. Individuals may inherit a disorder due to a gene mutation (see Chapter 1). At first sight, it seemed that little could be done to overcome these problems, but modern research has created hope that future developments in gene therapy may be able to help.

Tracking genetic disorders: genetic screening and counselling

Genetic counsellors can determine the chances of children being born with a genetic disorder by considering the family histories of both parents. **Genetic screening tests** are also available to find out if you are a carrier for some inherited disorders such as cystic fibrosis. If a faulty gene is known to be present in either partner, then there is a moral responsibility to declare this knowledge before a child is conceived. **Prenatal genetic tests** are also available to test fetuses

Activity

Search the internet using keywords such as 'health education', 'drugs', 'alcohol' and 'smoking'. Visit the websites of the UK health education agencies. You can also look at a CD-ROM encyclopaedia or a CD-ROM about health or drugs.

Drugs:
www.schoolscience.co.uk

Smoking and health:
www.ash.org.uk

Alcoholism:
www.sciencemuseum.org.uk

Alcohol abuse:
www.homeoffice.gov.uk

- Laboratory animals are so different from humans that they do not react to drugs in the same way as humans do.
- Humans do not have the right to subject animals to any form of experimentation.

The laws in the UK that regulate the housing and treatment of animals in laboratories are probably the strictest in the world, but there is still much controversy over whether animals should be used like this at all. The following are reasons why animals are sometimes used for the testing of potentially life-saving drugs:

- Humans should never have their lives threatened by experimental procedures.
- The value of tests on isolated cells or tissue from humans is limited because of the simplicity of the cells or tissues compared with the complexity of live animals.
- No computer simulation is accurate enough to model all the biological processes that take place in a living organism.

Procedures are being developed to enable drugs to be safely tested on some people. This could eliminate some testing on animals, but animal-rights supporters will accept only a total ban on the use of animals. Most medical authorities state that there will always be a need for some form of drug testing using animals. The problem is to balance possible animal suffering against the development of life-saving drugs.

Questions

8 State some of the research findings about death rates among smokers.

9 What are some short-term disadvantages of smoking?

10 Explain the progressive effects of ethanol on the human nervous system.

11 Give examples of addictive drugs that are illegal to take and sell.

12 Why is inhaling smoke from someone else's cigarette harmful to your health?

13 Why is drinking alcoholic drinks on an empty stomach more dangerous than drinking after eating?

14 Why does the presence of ethanol in a person's body give a feeling of warmth?

15 Explain the possible relationship between drug addiction and juvenile delinquency.

Summary

1. Your skin can control your body temperature.
2. The pancreas secretes insulin directly into the bloodstream to help regulate the level of blood sugar in the human body.
3. Diabetes is a condition in which a person's blood sugar level may rise because not enough insulin is produced in the body.
4. Controlling diabetes may consist of injecting insulin, controlling carbohydrate in the diet, and/or transplanting tissue from the pancreas.
5. Humans have the most complex central nervous system of any animal. It consists of a brain and a spinal cord, and it communicates with all parts of the body via nerve cells.
6. Some human responses to stimuli are fast, protective and automatic. These are called reflex actions.
7. Human health is affected by a variety of factors, and medical technology may provide answers to some health problems.
8. Gene therapy has been used to treat some genetic disorders. There have been some successes and some failures.
9. Ethical problems may be associated with genetic counselling.
10. A healthy lifestyle requires attention to diet.
11. The abuse of nicotine, alcohol and other drugs leads to personal and social problems.

Chapter 3 Cells and cell processes

By the end of this chapter you should:

- know that there are some similarities and some differences in the structure of plant and animal cells;
- know that plants and animals have different patterns of growth and development;
- know that stem cells retain the ability to develop into a variety of tissues and can be used to replace damaged or diseased tissue;
- know that chemical reactions in cells are controlled by enzymes that work within limits of pH and temperature;
- know that respiration can occur with or without oxygen;
- know that respiration is the release of energy from glucose and that it takes place in all living cells;
- know that lack of enough oxygen causes humans to release some energy from glucose and make lactic acid, which results in an oxygen debt;
- understand how substances enter and leave cells through the cell membrane;
- know that diffusion is the movement of particles down a concentration gradient;
- know that osmosis is the diffusion of water through a selectively permeable membrane down a concentration gradient;
- know that substances can pass into cells against a concentration gradient by active transport, which requires energy;
- understand the basic structure and functions of carbohydrates, proteins and fats;
- understand the processes involved in digestion, including the role of enzymes;
- understand the role of some of the organs making up the digestive system;
- know that plant cells may contain chloroplasts, which enable photosynthesis to take place;
- understand that the rate of photosynthesis is affected by light, temperature and carbon dioxide concentration.

Cells

The basic units of life

All organisms, apart from bacteria and viruses, are made of one or more cells. Most cells have at least a nucleus, cytoplasm and a cell membrane. The more complex organisms may have thousands, millions, or even billions of cells. An organism's size depends on the number, not the size, of its cells. In general, elephant cells are no bigger than the cells of an ant. The elephant simply has more cells.

A knowledge of cells and their functions is very important to any study of biology. Almost every branch of biology deals with cells in some way.

Did you know?

There are about 50 000 000 000 000 cells in an average person, and about 2000 times as many in a whale. The largest cell produced by any animal alive today is the egg of an ostrich, and red blood cells can be as small as 0.008 mm.

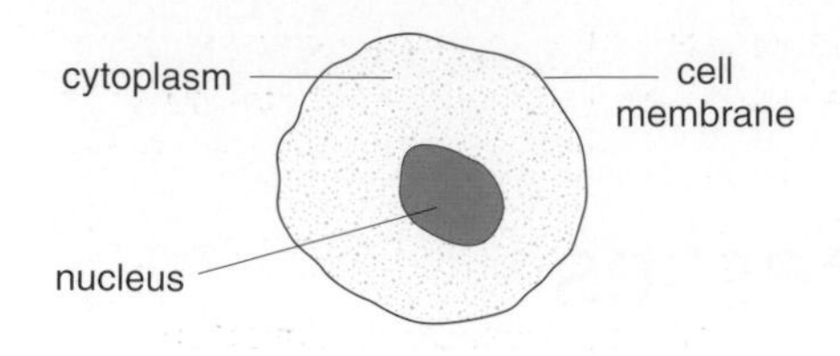

Figure 3.1 A typical animal cell

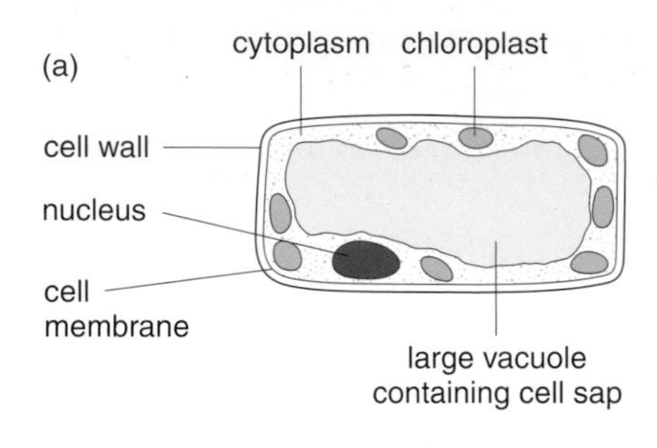

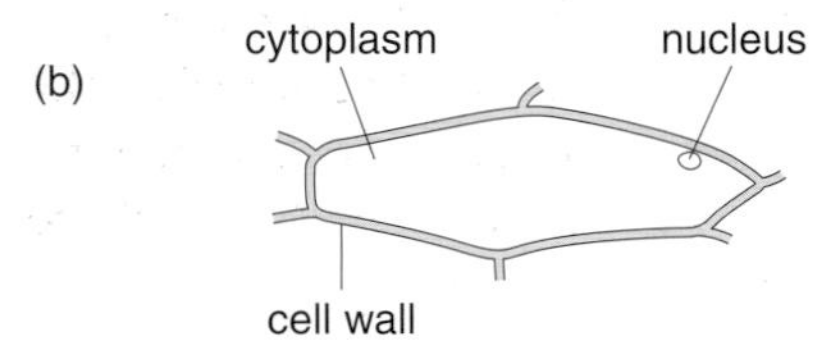

Figure 3.2 (a) A typical plant cell, (b) a cell from an onion epidermis

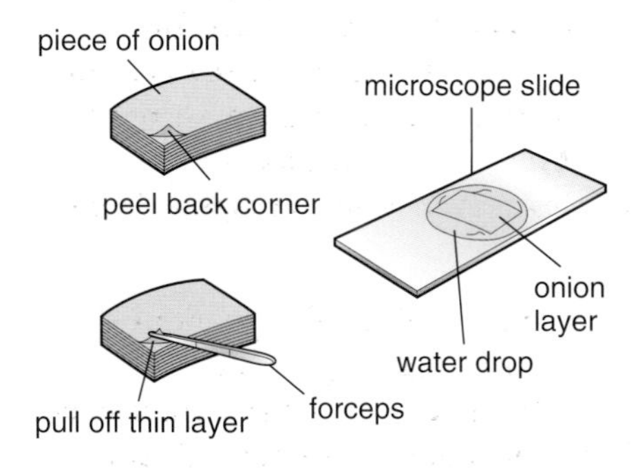

Figure 3.3 Obtaining a layer of cells

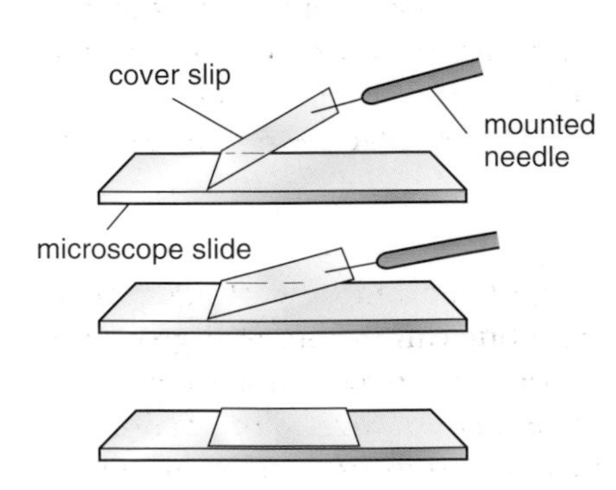

Figure 3.4 Mounting on a microscope slide

The most important parts of all cells

The nucleus is the control centre of all cell activity (see Figure 3.1). Without it the cell dies. It is usually spherical or oval in shape, and often lies near the centre of the cell. It is contained in a thin double-layered membrane that has tiny holes that allow certain substances to pass between the nucleus and the rest of the cell. The nucleus contains the chromosomes that are important in cell division (see the section on cell division in Chapter 1).

Chemicals outside the nucleus make up the cytoplasm. Under a light microscope, cytoplasm can be seen as a clear semi-liquid filling most of the cell. It often moves in the cell, carrying with it the nucleus and changing shape as it moves. At its outer boundary, the cytoplasm is enclosed by a cell membrane that separates the cell from other cells and from surrounding fluids. Like the nuclear membrane, it is not a solid barrier. Molecules can pass through it in a controlled manner. It acts like a gate, allowing some molecules to pass and keeping others out (see the section on how substances enter and leave cells later in this chapter).

Plant cells, specialists in food production

A typical plant cell has a number of structures that are not found in animal cells (see Figure 3.2).

- The cell wall is made of a carbohydrate called **cellulose** (see the section on carbohydrates later in this chapter). This is a non-living, thick, rigid layer around the cell membrane that helps the plant to support itself. The shapes of plant cells vary. They may be angular, rectangular or rounded.
- There is a **vacuole**, which is a fluid-filled space surrounded by a membrane. The fluid contains dissolved food materials, minerals and waste products, and is called **cell sap**. In older plant cells, the vacuoles are large, and their main function is to support the cells through the pressure of the cell sap, rather like a water bed supporting the weight of a person.
- **Chloroplasts** are green discs that contain the green pigment called **chlorophyll**, which is essential for the process of **photosynthesis** (see the section on energy trapping later in this chapter). Chloroplasts are present only in the parts of the plant that are exposed to light.

Practical work

Looking at cells

Method

1. Your teacher may give you a slide with a drop of iodine solution already on it. Otherwise, take a clean microscope slide and place a drop of iodine in potassium iodide on the centre of it.
2. Take a small piece of onion.
3. Using a pair of forceps, start at one corner of the onion piece. Lift up the thin layer that covers the outside of the onion, and peel off as much of this layer as you can.
4. Place the layer onto the iodine on the slide so that it unrolls and lies flat. Carefully lower a cover slip over the onion (see Figures 3.3 and 3.4).

5 Now examine the layer under the microscope. Use a low-power magnification first, focus the microscope, and then examine the slide on high power.

You will not be able to see chloroplasts. Explain why.

One way to see chloroplasts is to look at a whole leaf of the Canadian pond weed *Elodea canadensis* (this is obtainable from most ponds, or can be bought at most garden centres).

Method

1 Place a leaf from the growing point of the pond weed in a drop of water on a microscope slide.
2 Focus the microscope on the slide on low power, and then on high power.

Patterns of growth

The two methods of cell division were described in the section on cell division in Chapter 1. From this you will know that mitosis is the type of cell division that leads to growth in both animals and plants. In principle the process is the same in animals and plants, but there are basic differences in the sites of mitosis in each.

In animals, all tissues undergo mitosis at some stage in their development. Some tissues are better at mitosis than others. For example, your skin has a special layer of cells that keeps replacing cells that are worn away throughout your life. The cells are replaced by mitosis. Nerve cells are not very good at replacing themselves. Even red blood cells are produced by mitosis in the bone marrow. This may seem strange, because red blood cells have no nuclei. In their early stages of development they have nuclei, but these are lost when the red blood cells become mature. Animals are able to repair most damaged tissues using mitosis.

Figure 3.5 A cross section of a tree clearly showing growth rings

In plants, the parts that undergo mitosis are limited to growing areas. Think of a tree. It starts life as a germinating seed a few millimetres in length and width. By the time it is fully grown, it may be 100 m tall with a diameter of 3 m. How does this happen? At the tips of the roots and stems there are areas of growth to which mitosis is confined. These regions are called **meristems**, and they are responsible for the growth of the plant downwards into the soil and upwards towards the sky. There are similar meristems in buds that give rise to side branches. Increases in diameter are due to a special meristem called the **cambium**, which adds layer upon layer of growth, year after year (see Figure 3.5). A similar method is used for the growth of bark on a tree.

Plants keep the ability to grow new parts, whereas animals generally cannot do so. Given the right conditions, cuttings from a parent plant will produce stems with buds and roots, and grow into mature plants. Animals are able to repair damaged tissues by growing new cells, but only invertebrates can regenerate major parts of the body. Some vertebrates such as lizards can grow new tails, but these are exceptions.

The size to which plants grow depends on:

- the ability to support themselves to have access to sunlight
- the maximum height to which water can be transported.

The size to which animals can grow depends on:

- the ability to support themselves so that they can move
- the efficiency of getting oxygen to the cells to release energy.

The largest animal ever to have lived is the blue whale. It relies on water to support itself and has the most extensive blood system in the animal kingdom for transporting oxygen-rich blood. The largest land-living plant in the world is the giant redwood tree. Raising water from its roots to its upper leaves is a limiting factor in its size and age.

Cell specialisation

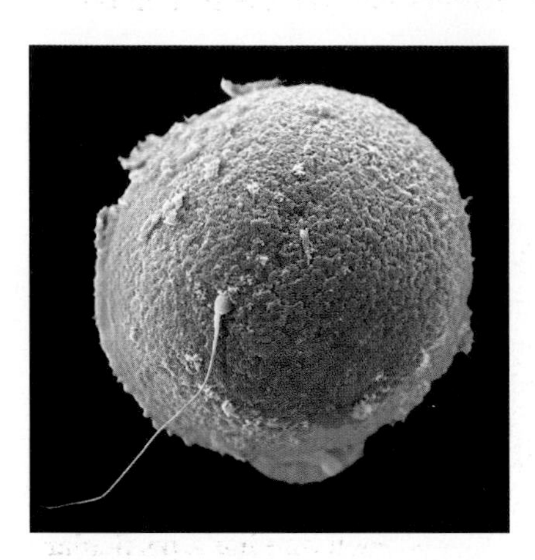

Figure 3.6 Egg from mother fuses with sperm cell from father.

The fate of most of our cells is decided from a very early stage in our development. At the time of fertilisation, we all begin life when an egg from our mother fuses with a sperm cell from our father (see Figure 3.6).

The entry of the sperm makes mitosis begin. The cell divides again and again until it forms hundreds of cells. Occasionally, one of the early divisions may divide the embryo into equal parts that separate. When this happens, identical twins, triplets, quadruplets or even more babies develop independently. This is a type of natural cloning.

After several divisions, a blackberry-shaped cluster of cells is formed. In current research programmes involving human embryos, the point at which the collection of cells should be called a human individual is a controversial issue. The fact is that there are no tissues or organs at this point.

When tissues and organs start to develop, they form from cells that become specialised into, for example, muscle, liver, kidney and skin cells. However, the fate of a group of cells called **stem cells** is uncertain. They remain unspecialised or **undifferentiated**. Scientists can take these cells and make them grow into different tissues that they need to replace damaged tissue in people suffering from certain disorders. Nervous tissue could be produced in this way to replace nerves that have been damaged. Insulin-producing cells could be produced in this way for people who suffer from diabetes.

The technology involved is controversial because it involves taking stem cells from the cluster of embryo cells that is produced after the first cell divisions after fertilisation. This could involve destroying the very beginnings of human life, and to some this is totally unacceptable. To others, there are no ethical problems with taking embryonic stem cells so that they can be used to cure human disorders.

Stem cells donated by adults are being investigated to see if they can help in treating certain disorders. In 2000, Italian scientists treated children with a form of leukaemia (blood cancer) with blood stem cells from bone marrow (not from embryos). The results were encouraging, with 56% of those treated still alive in 2005.

Enzymes: tools of a cellular factory

Cells are living chemical systems in which substances are constantly changing. Molecules react with each other; large molecules are built

Enzymes:
www.estrellamountain.edu

Biochemistry:
www.biology.arizona.edu

up and broken down. What starts these changes? What controls them? What keeps one change from interfering with another? The answer to all these questions is **enzymes**. They are proteins made by living organisms that speed up the rate of chemical reactions – they are catalysts. They can be extracted from organisms and still keep their properties. Enzymes are used in biological washing powders and also in certain food manufacturing processes.

It is important not to think of all enzymes as chemicals that break large molecules into small ones like digestive enzymes (see the section on digestion later in this chapter). There are over 100 000 chemical reactions going on in your body at any given moment, and almost all of them are controlled by enzymes, the vast majority of which have nothing to do with digestion. Most are involved in the release of energy in the process of **respiration** (see the section on respiration later in this chapter).

All enzymes have certain properties:

- They are proteins.
- They are **denatured** when they are heated. This means that their molecular structure is altered so that they can no longer work.
- They increase their rate of reaction when they are given energy in the form of heat, but only up to the temperature when they start to become denatured. At this point, their rate of reaction slows down.
- Their rate of reaction is altered when the pH (degree of acidity or alkalinity) is changed. Each enzyme has an optimum pH at which it works fastest.
- They are specific in their action, that is, each one has a particular job. For example, enzymes used for energy release cannot be used for anything else, such as digestion or building up proteins.

Did you know?

Catalase is used in industry to make foam rubber and polystyrene products. The reaction between the enzyme and a peroxide produces oxygen that makes the bubbles in the foam rubber or polystyrene. The same principle is used in the production of some first-aid dressings for military use. The oxygen produced kills certain bacteria that might otherwise cause gangrene in wounds.

The fastest working enzyme

Catalase is an enzyme that is found in all organisms. It works by breaking down a group of poisonous substances called peroxides that are waste products made in cells. The peroxides are broken down to the harmless substances oxygen and water. Catalase has to work quickly to prevent the peroxides from killing the cells that produce them. It has to be the fastest working enzyme.

Practical work

Demonstrating catalase activity

The activity of catalase can be demonstrated using hydrogen peroxide and either plant tissue such as a potato, or animal tissue such as liver. You will need a piece of fresh liver from a butcher, or a fresh potato, 3% (10 vol.) hydrogen peroxide, three test tubes, a splint to test for oxygen, water, a Bunsen burner, tongs and labels.

Method

1. Wear eye protection, and consider wearing a laboratory coat or overall to protect your clothes. You must have access to a damp cloth so that you can clear away any spillages quickly with water.
2. Take two pieces of tissue, each with a volume of about 1 cm^3. Place one in a test tube half-filled with water, and boil it, using tongs to hold the test tube.

3 Carefully pour 3% hydrogen peroxide into each of two test tubes until they are about half full. Label the tubes A and B.
4 Add a piece of unboiled tissue to tube A. Test any gas given off with a glowing splint.
5 Repeat stage 4 with test tube B and the boiled tissue.
6 Do not pour the contents of the test tubes down the sink. Place the test tubes in a plastic bowl.
7 Wash your hands carefully once you have finished your investigations.
8 Describe your observations.
9 Explain your observations.

Demonstrating catalase activity: an alternative procedure using a computer

1 Carry out stages 1–3 from the above method.
2 Connect two temperature sensors to a datalogging system, and place a sensor in each tube.
3 Start recording. You should have a graph trace on the computer screen. To one tube add a piece of unboiled tissue, and to the other add the boiled tissue.
4 Explain your observations.

You can plan and investigate the effect of temperature on enzyme action using the apparatus in Figure 3.7. Apply the procedure described above to plan an investigation of the effect of various temperatures on the rate of action of catalase.

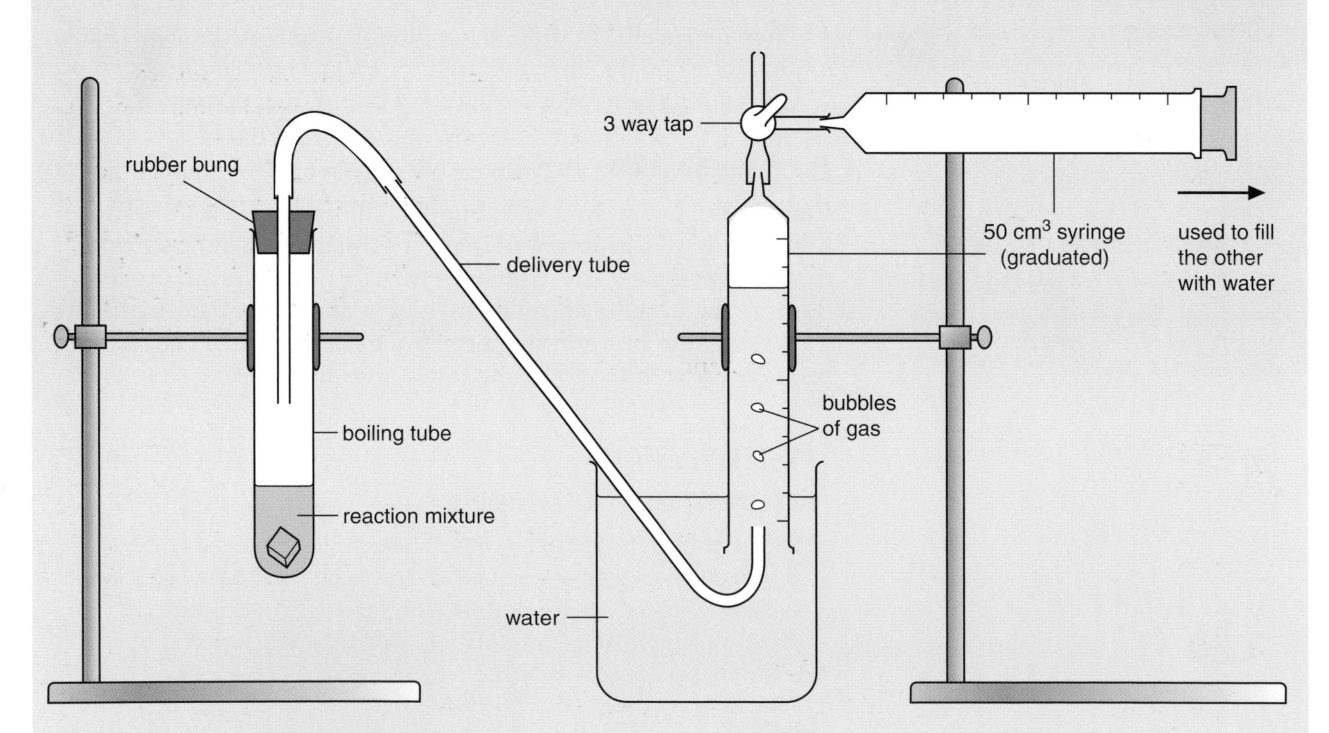

Figure 3.7 Apparatus used to test enzyme activity

Questions

1. What are the main features that distinguish animal cells from plant cells?
2. In what respect is the cell the basic unit of life?
3. Explain the importance of enzymes in the activities of cells.
4. What are the similarities and differences in the ways in which plants and animals grow?

Respiration

Energy release

Energy is needed for chemical reactions to take place in cells. Respiration is a process of releasing energy from glucose in cells. It can take place at normal body temperature because of enzymes, the biological catalysts that speed up reactions in the body. When oxygen is used in this process, it is called **aerobic respiration**, and this is the normal form of respiration that occurs in our cells.

H

However, it is possible for respiration to take place without oxygen. That process is called **anaerobic respiration**, and it releases only a fraction of the energy that is released when oxygen is present.

During anaerobic respiration, glucose is broken down with the aid of enzymes, but oxygen is not used. In some organisms such as yeasts, ethanol and carbon dioxide are formed. Much use of this is made in biotechnology, because it is the basis of baking and of alcoholic fermentation.

$$\underset{\text{glucose}}{C_6H_{12}O_6} \rightarrow \underset{\text{ethanol}}{2C_2H_5OH} + \underset{\text{carbon dioxide}}{2CO_2} + \text{energy}$$

In our muscle tissues, when no oxygen is available, glucose can be changed to lactic acid with the release of a little energy.

$$\underset{\text{glucose}}{C_6H_{12}O_6} \rightarrow \underset{\text{lactic acid}}{2C_3H_6O_3} + \text{energy}$$

This is useful during physical exercise, when muscle cells are sometimes starved of oxygen because the energy demand exceeds the supply of oxygen by the bloodstream. Under these circumstances, the body can still release energy for a short time, although there will be a build-up of lactic acid causing muscle fatigue. After such exertion, when at rest, we breathe more deeply and more rapidly. This is essential because a greater intake of oxygen is needed to pay off the oxygen debt caused by the extra demand. The fitter we are, the more efficient is our circulation, and the less lactic acid is produced.

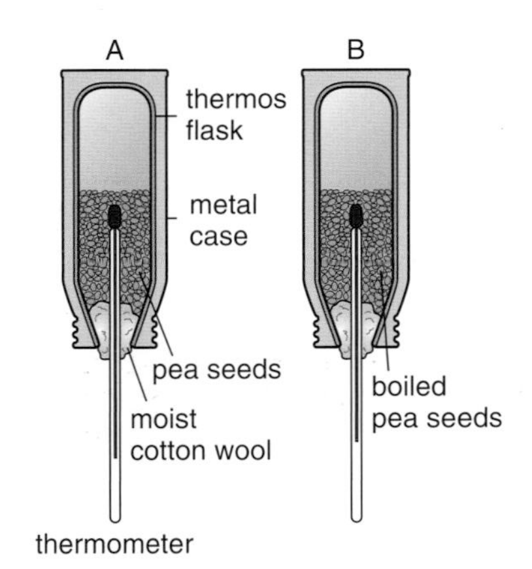

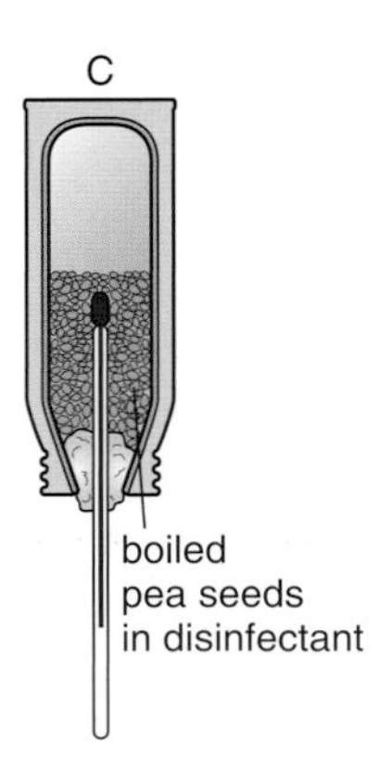

Figure 3.8 Apparatus used to show that, as germinating peas respire, they release energy as heat

Practical work

Energy release by respiring seeds

1. Set up three thermos flasks as shown in Figure 3.8.
2. Record the temperature of each flask at the beginning and end of a week.

Questions

- What is the purpose of the disinfectant?
- What is the advantage of using thermos flasks rather than beakers?
- Why are cotton wool plugs used rather than rubber bungs?
- Which one of the flasks is the control?
- Explain the results in each flask.

Energy release by respiring seeds: an alternative procedure using a computer

1 Set up the thermos flasks as shown in Figure 3.9.
2 Connect three temperature sensors to a datalogging system and place a sensor in each flask.
3 Start recording and continue to record for a day or more.
4 Sketch the graphs you would expect to obtain, and compare these with your actual results.
5 Add your comments to the graphs and print them out.

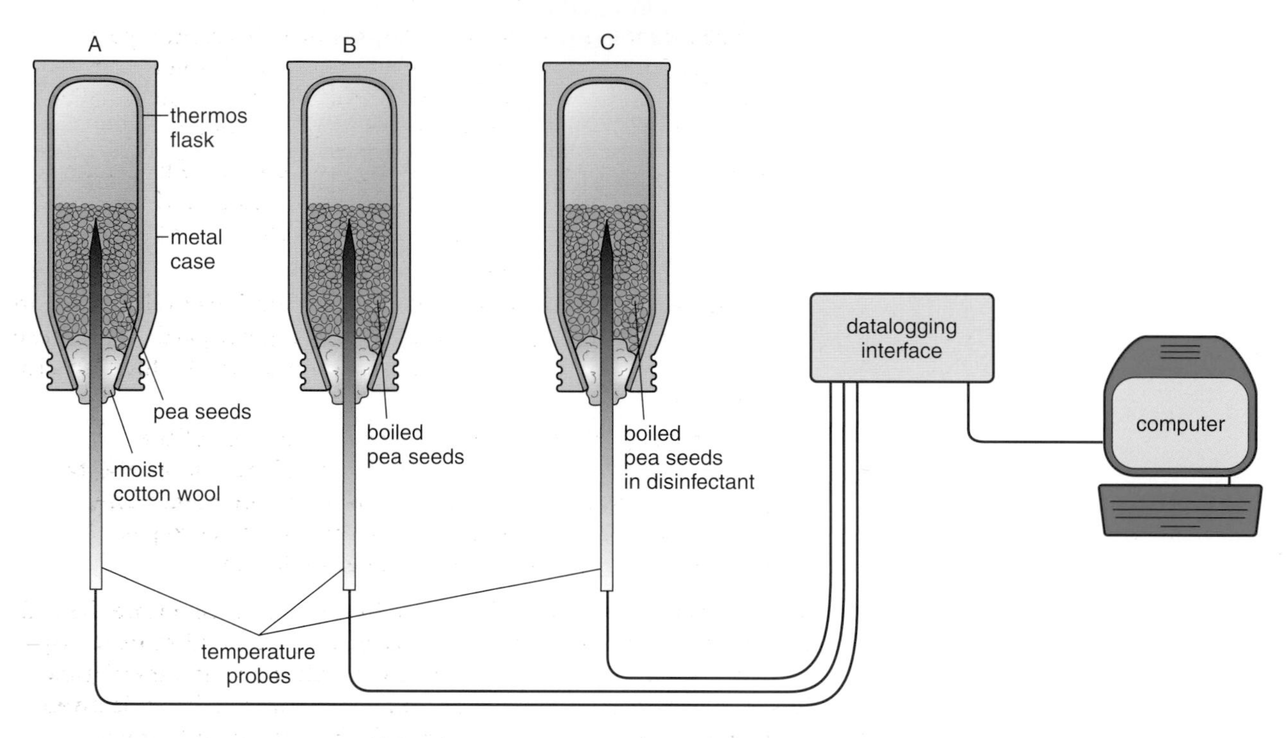

Figure 3.9 Using a computer to demonstrate the release of energy as heat in respiring seeds

Figure 3.10 This athlete will have built up an oxygen debt in his muscles.

Oxygen debt

During times of great muscular activity, the cells need more oxygen than the body can supply. The lungs cannot take in oxygen fast enough, and the blood cannot deliver it fast enough. When this happens, the cells switch to anaerobic respiration (see above). This means that oxygen is not used and lactic acid collects in the tissues, causing a feeling of fatigue and even pain. The build-up of lactic acid signals to the brain's breathing centre to increase the breathing rate to supply the tissues with more oxygen. If the heavy exercise continues, lactic acid keeps building up and causes an **oxygen debt**. This continues until the heavy exercise ends. Then, during a half-hour rest for example, some of the lactic acid is oxidised and some is converted to glycogen. The oxygen debt is paid and the body is ready for more exercise (see Figure 3.10).

H

Questions

5 What is the biological importance of respiration?

6 Describe the two types of respiration.

7 Describe how cells obtain the energy that they need for growth.

8 Give an account of the importance of making lactic acid in muscle cells during exercise.

How substances enter and leave cells

The importance of membranes

Life depends greatly on the movement of certain materials into and out of cells. Digested foods and oxygen are among the essential materials to be taken into cells. Waste products such as carbon dioxide must be removed. All materials going into or coming out of a cell must cross the cell membrane. Some molecules pass through freely, some less freely, and others not at all. In this way, the membrane selects materials for entry and exit. Several factors are important in determining this 'right of way':

- **The size of the particles**: Some large molecules cannot pass through the cell membrane. On the other hand, some large molecules pass through in greater numbers than much smaller ions (charged particles made of 'parts' of molecules).
- **Whether or not the particles will dissolve in water**: The liquids that bathe cells are usually solutions of molecules and ions in water. Substances that do not dissolve in water cannot pass through the cell membrane.
- **Conditions inside and outside the cell**: Conditions in the cell or in the cell's surroundings may affect the passage of particles. If there are more particles on the inside than on the outside, they tend to move outwards.
- **The structure of the cell membrane**: The surface of the membrane seems to have many tiny pores. There must also be spaces between the molecules that make up the membrane. These spaces may be too small for large molecules to pass through, but large enough for smaller molecules.

The rates and degrees at which certain substances penetrate the cell membrane vary. If a substance passes through a membrane, we say that the membrane is permeable to that substance. If a membrane lets some substances pass but not others, we say that it is **selectively permeable** or **differentially permeable**. The cell membrane is selectively permeable.

Diffusion

To understand better how substances pass through cell membranes, we need to understand more about molecules and their motion. In any substance, the molecules are constantly moving. This motion results from **kinetic energy** within the molecules themselves and not from outside forces. Molecular motion is slight in solids, but much greater in liquids and gases. The movement of the molecules is completely random. That is, the molecules move in straight lines until they collide with other molecules. Then they bounce off and move in a straight line again until they collide with still other molecules. It is easy to understand that this kind of movement results in a gradual spreading out of the molecules. In the end, they will all be spread evenly through any given space. This gradual, even spreading out of molecules is called **diffusion**.

As an example of diffusion, think of what happens if you open a bottle of perfume. As soon as you open the bottle, the molecules start diffusing into the air. Soon the people close to the perfume smell it, and, as diffusion continues, the smell becomes stronger. More and more molecules spread out at random among the molecules of the gases in the air. Finally, a state of equilibrium is reached. The molecules, although still in motion, are spread evenly among the gas molecules in the air. Actually, as the perfume diffused into the air, molecules of gases from the air also diffused into the perfume bottle. This brings us to a basic law of diffusion. According to this law, substances diffuse from areas of greater concentration to areas of lesser concentration. In this case, diffusion continues until the concentration of gas molecules from the air and the perfume are equal in all parts of the room.

Diffusion is affected by the following factors:

- **Concentration**: The greater the difference in concentration between two substances, the more rapidly diffusion takes place.
- **Temperature**: The higher the temperature, the greater is the speed of molecular motion.
- **Pressure**: The higher the pressure on particles, the greater is the speed of diffusion from a region of high pressure to a region of low pressure.

How does a membrane affect the movement of molecules in diffusion? The answer depends on the nature of the membrane and the substance being diffused. You can demonstrate this with **visking tubing** and glucose solution.

Practical work

Monitoring the diffusion of glucose through a membrane

Method

1 Make a sac out of visking tubing.
2 Fill the sac with glucose solution (see Figure 3.11).
3 Place the sac in water.

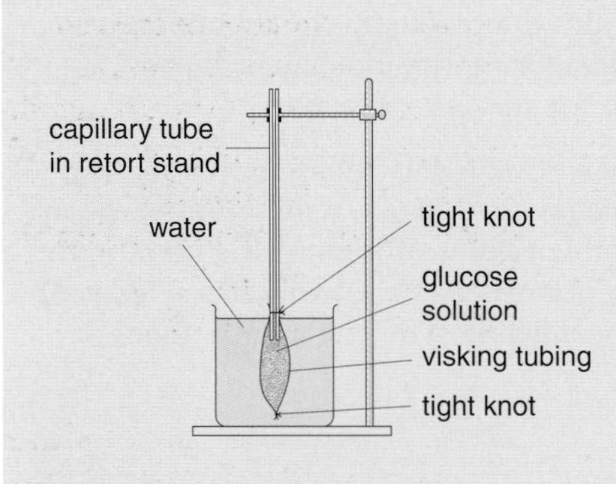

Figure 3.11 Apparatus required to monitor the diffusion of glucose through a membrane

Since the visking tubing is permeable to both water and glucose molecules, two things happen. Water molecules diffuse into the glucose solution and glucose molecules diffuse into the water. Eventually the concentration of glucose and water molecules is the same on both sides of the visking tubing. This is a state of **dynamic equilibrium**, because both types of molecule are passing in both directions at equal rates. The visking tubing has had little effect on diffusion in this case.

What if a cell membrane were completely permeable? The answer is simple: the cell would die. True, molecules of water and other substances could enter the cell more easily, but at the same time, the cell's own molecules would diffuse out into the surroundings. Clearly the cell membrane must be only selectively permeable if the cell is to survive.

Cells alive:
www.cellsalive.com

Cell biology:
www.biology.arizona.edu

Osmosis

You can demonstrate how a membrane can be selectively permeable by using larger molecules.

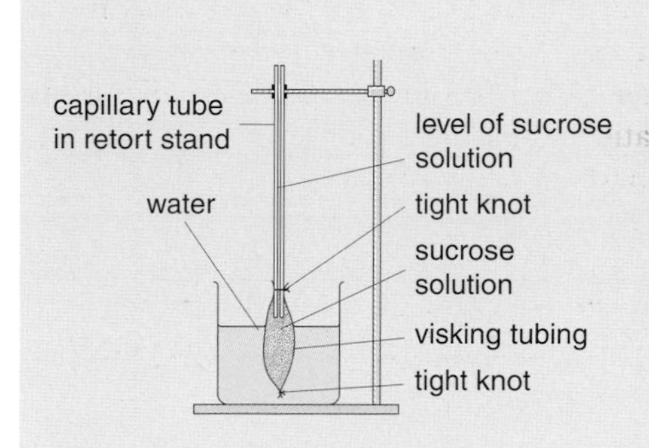

Figure 3.12 The glucose experiment can be repeated using sucrose solution.

Practical work

Monitoring the diffusion of sucrose through a membrane

Method

1 Make a sac out of visking tubing.
2 Fill the sac with sucrose solution (see Figure 3.12).
3 Place the sac in water.

Glucose molecules can pass through visking tubing but sucrose molecules cannot. This is because a sucrose molecule is almost twice as large as a glucose molecule. Water passes into the sucrose solution, but sucrose is not able to pass into the water. After about half an hour, the level of the solution in the tube has risen, while the level of water in the beaker has dropped.

This is an example of **osmosis**, where there is diffusion of water through a selectively permeable membrane from an area of greater concentration of water molecules to an area of lesser concentration of water molecules.

What does this mean for our cells? If the solution surrounding our red blood cells contained less water than the cytoplasm of the red blood cells, water would pass out of the cells faster than it would enter. The cells would then shrink and die. If the opposite occurred, and there was more water outside, then there would be a net gain of water and the cells would burst. Therefore, it is vital that the composition of our blood and other body fluids is regulated. This is an example of homeostasis, or regulation of our internal environment (see the section on homeostasis in Chapter 2).

Did you know?

It is possible to make ill-fitting leather shoes bigger by pushing potatoes to the ends of the toes. If they are left there for a few days, the shoes become a bigger size than before. It is possible to kill slugs in your garden by placing salt on them. They shrivel and die. Explain these two facts in terms of osmosis.

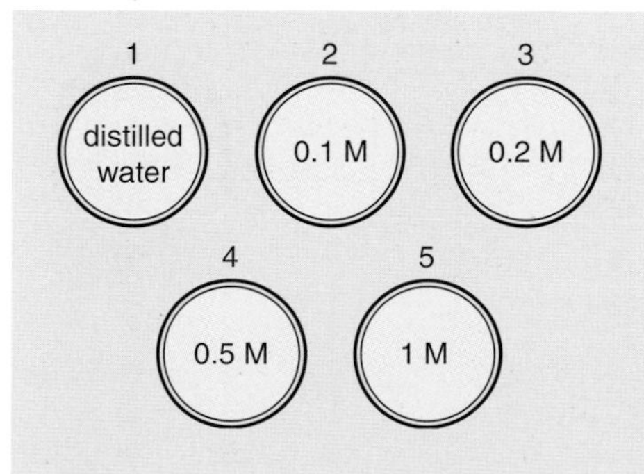

Figure 3.13 Labelling the Petri dishes

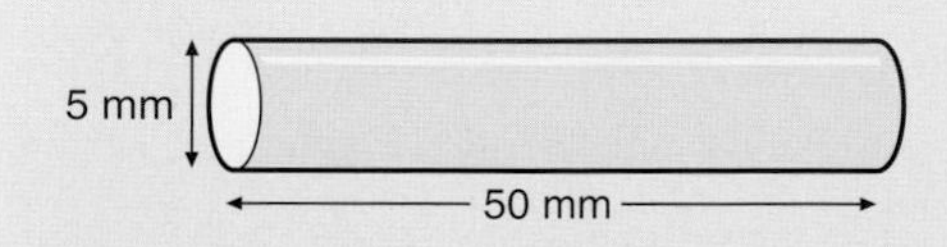

Figure 3.14 A potato cylinder

Practical work

Monitoring osmosis in a potato

Method

1 Label the lids of five Petri dishes as shown in Figure 3.13.
2 Into each Petri dish, put 30 cm^3 of sucrose solution in the concentration marked on the lid, with distilled water in the fifth dish (see Figure 3.13).
3 Put the lids on the dishes.
4 With a cork borer, carefully cut five cylinders from a potato so that they are each 50 mm long with 5 mm diameter (see Figure 3.14).
5 Weigh this batch of chips accurately on a top pan balance, and record the mass in a table of results.
6 Place these chips in the Petri dish marked distilled water, and put the lid back on the dish.
7 Repeat stages 4–6 for each of the other dishes.
8 *Leave the dishes for* 20 *minutes*.

H

9 Remove the chips from the distilled water. Blot them gently with a paper towel. Weigh them accurately, and record the mass in the table of results below.

Solution in Petri dish	Mass at start/g	Mass at end/g (after 20 min)	Change in mass/g (+/–)	Change/% (+/–)
0.1 M sucrose				
0.2 M sucrose				
0.5 M sucrose				
1 M sucrose				

Change in mass (%)

Concentration of sucrose (M)

$$\text{percentage change in mass} = \frac{\text{change in mass}}{\text{original mass}} \times 100$$

10 Repeat stage 9 for the chips in the other dishes.

11 Draw a graph of the percentage change in mass against concentration of sucrose. Draw the axes as shown in the diagram.

Questions

- Why is it important to calculate the percentage change in mass?
- In which solutions did water enter the cells of the potato?
- In which solutions did water leave the cells of the potato?

Monitoring osmosis: an alternative approach using a computer

Method

1 Fill a sac made of visking tubing with sucrose solution, and place this in a beaker of distilled water, as shown in Figure 3.15.

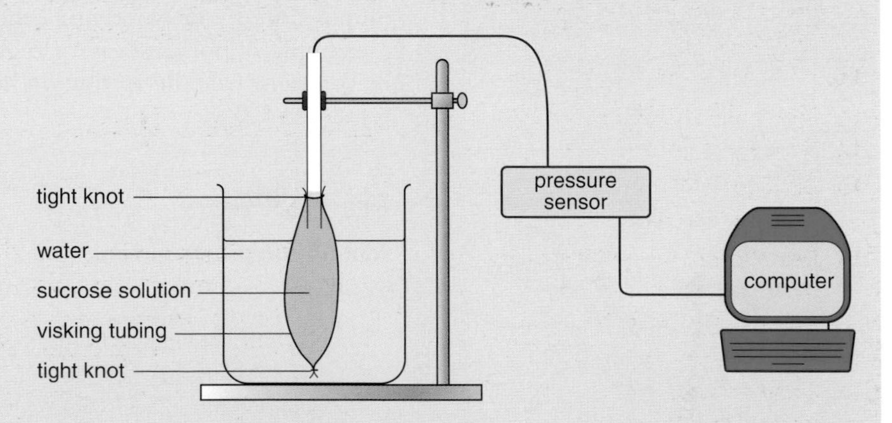

Figure 3.15 Using a computer to monitor osmosis

2 Fit a pressure sensor to the sac with a tube, and connect the sensor to a datalogging system.

3 Start recording to test the system. You should have a graph trace on the screen that deflects when you squeeze the sac gently.

4 Continue recording for an hour or more. While this is happening, write up the experiment.

5 Repeat the experiment with a different concentration of sucrose solution.

H

Questions

1 The graph trace should initially be a flat line, but it will change direction over time. What will cause the trace to change direction?

2 Draw a sketch graph to show how the graph will change during the experiment. As a clue, your sketch graph should have three sections that you can label: (a) an initial flat line, (b) a section where the pressure rises, (c) a section where the pressure no longer changes. Suggest how your graph explains what happens during osmosis.

3 You could use the above method to investigate osmosis using a more concentrated solution of sucrose. Sketch the graph you would expect to obtain then, and explain why it might be different from the original one.

4 Look at the results of your experiment. Change the scale of the graph to see the result more closely. Find and label parts of the graph that represent equilibrium and the flow of water into the sac.

More ideas for practical activities

(Apply the procedures used above.)

1 Plan an investigation to find out the concentration of sugar that is equivalent to the concentration of cell sap.

2 Beetroot tastes sweeter than potato. Compare the sugar concentrations of the cell sap of beetroot and the cell sap of potato.

Questions

9 Distinguish between permeable and selectively permeable membranes.

10 What is diffusion? What is the name for the balance that results from diffusion?

11 State the external factors that influence diffusion rates.

12 Define osmosis.

13 Distinguish between active transport and diffusion.

14 Describe what might happen to a cell if its membrane were permeable to all molecules.

15 What factors determine whether a particle will pass through a cell membrane?

Did you know?

The average person from a country in the developed world eats about 30 t (tonnes) of food in his or her lifetime, but the majority of us can maintain a body weight that is constant within 1%.

Active transport

H

In many cases, particles pass in or out of cells against a diffusion gradient, that is, against the direction of flow of particles. Sodium and potassium ions are examples of this. They are **actively transported** in or out of cells under certain circumstances. Diffusion does not use energy, whereas active transport requires energy released by the cell to 'pump' ions in one particular direction.

Animal nutrition

What is food?

Food is a collection of chemicals that are taken into an organism to provide growth and energy release, and to maintain all the life processes. By this definition, water, minerals and vitamins, as well as carbohydrates, fats and proteins, are food.

In the body, the process of preparing the foods to reach the body tissues occurs in a 9 m tube called the **alimentary canal**. Digestive enzymes break down bread, meat, milk, butter, etc. into smaller molecules. These can then pass into the bloodstream.

Foods needed in bulk

Carbohydrates and **fats** supply most of our energy.

Carbohydrates

In a **balanced** diet, more than half of your food should be carbohydrate. However, stored carbohydrate never makes up more

> **Did you know?**
>
> The average adult in Britain eats 43 kg of sugar every year. Sugar is a highly concentrated store of energy, and 43 kg of it could keep you walking briskly for over 2000 miles. Do you walk 2000 miles in a year? Even if you do, sugar is not the only energy-giving food that you eat. If eaten in excess, sugar leads to being overweight, because it can be changed into fat. It also causes tooth decay.

than 1% of your body weight because carbohydrates are mainly fuel from which your cells release energy. The energy is released during respiration (see the section on respiration earlier in this chapter).

You eat many different kinds of carbohydrates (see Figure 3.16). Some are digested easily and then travel to tissues with little chemical change. Others have to be broken down before your tissues can use them. Some carbohydrates, such as cellulose, are not digested at all. We need these types in our diets for bulk, or roughage. All carbohydrates enter body cells as glucose, ready to be used to release its energy in respiration.

Figure 3.16 Foods rich in carbohydrates

Starch makes up a large part of the carbohydrate in most diets. There is a lot of starch in potatoes and cereals. Starch is made up of huge chains of glucose units. During digestion, starch is broken down to glucose. Glucose is absorbed by the blood and carried to the body's cells after first going to the liver. Much of the glucose that goes through the blood to the liver is turned into glycogen. This can be turned back into glucose when the body needs it (see the sections on homeostasis and life-saving genetics in Chapter 2).

Cellulose is a complex carbohydrate that is found in the walls of all plant cells. We cannot digest cellulose, but it is important for the process of **digestion**. Cellulose helps the digestive system because it makes the muscles of the lining of the alimentary canal squeeze the food along.

Fats as an energy store

Fats and **oils** provide more than twice as much energy as carbohydrates. Common sources of fats and oils include butter, cream, cheese, margarine, vegetable oils and meats (see Figure 3.17).

Figure 3.17 Foods rich in fats and oils

During digestion, enzymes slowly break down fats. This happens in three stages. The result is one molecule of **glycerol** and three molecules of **fatty acids** from each fat molecule digested. Excess carbohydrates are converted into fats and stored under the skin and around the kidneys. Too much fat is not good for you: it leads to obesity, with all its problems of **cholesterol** build-up. So you should control how much carbohydrate and fat you eat. A balanced diet should roughly consist of 60% carbohydrate, 20% fat and 20% protein.

Proteins and their uses

Proteins are large molecules made of thousands of units called **amino acids**. They have to be broken down into these basic units during digestion. The amino acids are then carried to cells where they are built up again into human proteins. Proteins are used for growth and repair, but they are also essential because all enzymes and many hormones are proteins.

Figure 3.18 Foods rich in protein

When we take in more protein than we can use we cannot store it. After being digested to amino acids, any that are not wanted are changed to **urea** in the liver. This is transported by the blood to the kidneys for excretion in **urine**. Some of the best sources of protein include lean meat, eggs, milk, cheese, whole wheat, and beans (see Figure 3.18).

The stages of digestion

Why can your body not use most foods in the form in which you eat them? There are two reasons. First, many foods will not dissolve in water. This means that they cannot get into cells through cell membranes. Second, the foods you eat are made of molecules that are far too big to use. Cells cannot use them either to release energy

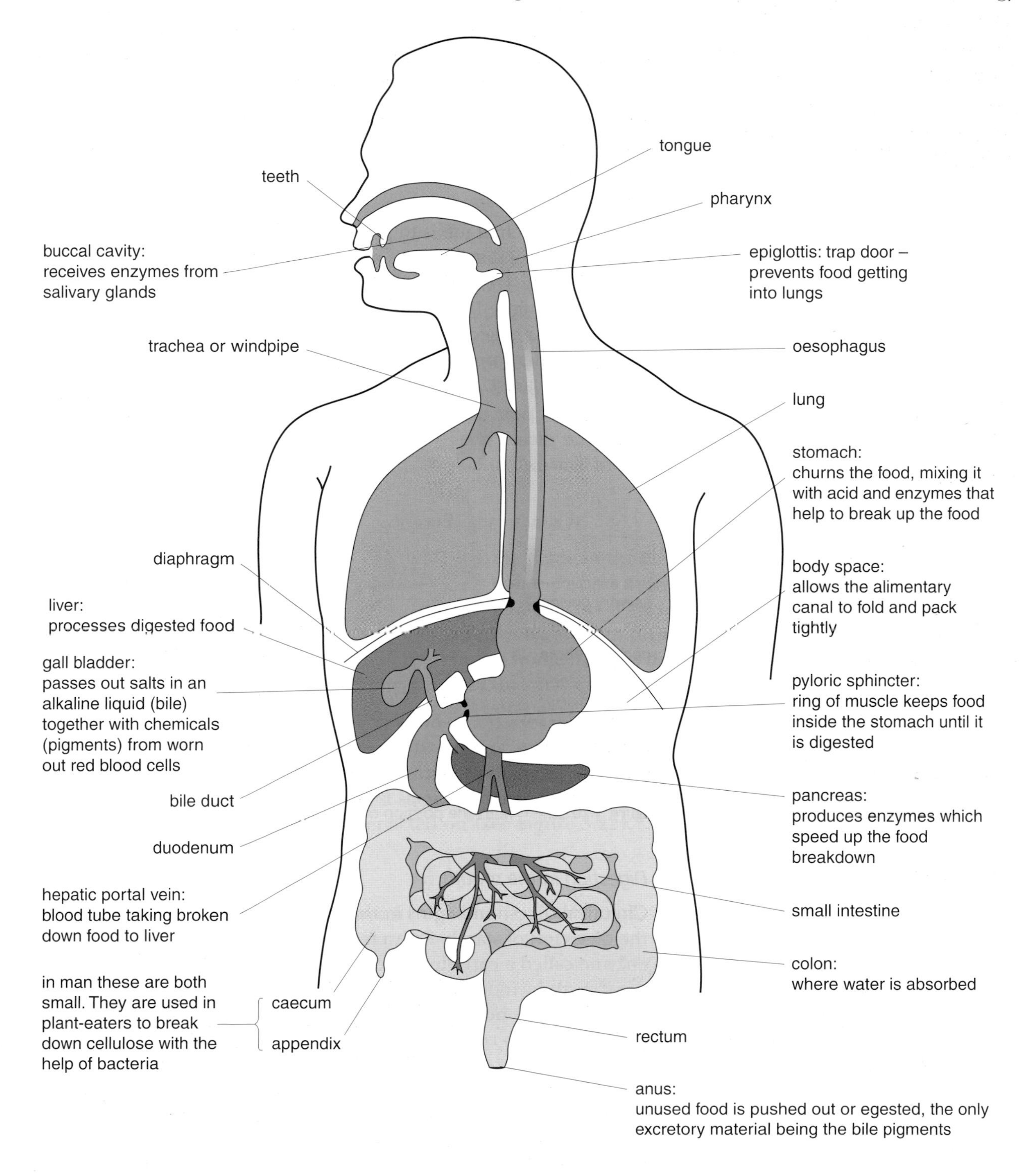

Figure 3.19 The human digestive system

Practical work

Monitoring photosynthesis

As oxygen is given off during photosynthesis, its rate of production can be used as a measure of photosynthetic activity. By counting bubbles produced in a given time or by measuring the volume of gas produced in a given time, the main limiting factors of photosynthesis (see the section on the raw materials of photosynthesis above in this chapter) can be investigated.

Method

1 By using either of the experimental set-ups in Figures 3.32 and 3.33, investigate the effects of light intensity, carbon dioxide concentration and temperature on the rate of photosynthesis of the Canadian pondweed *Elodea*.

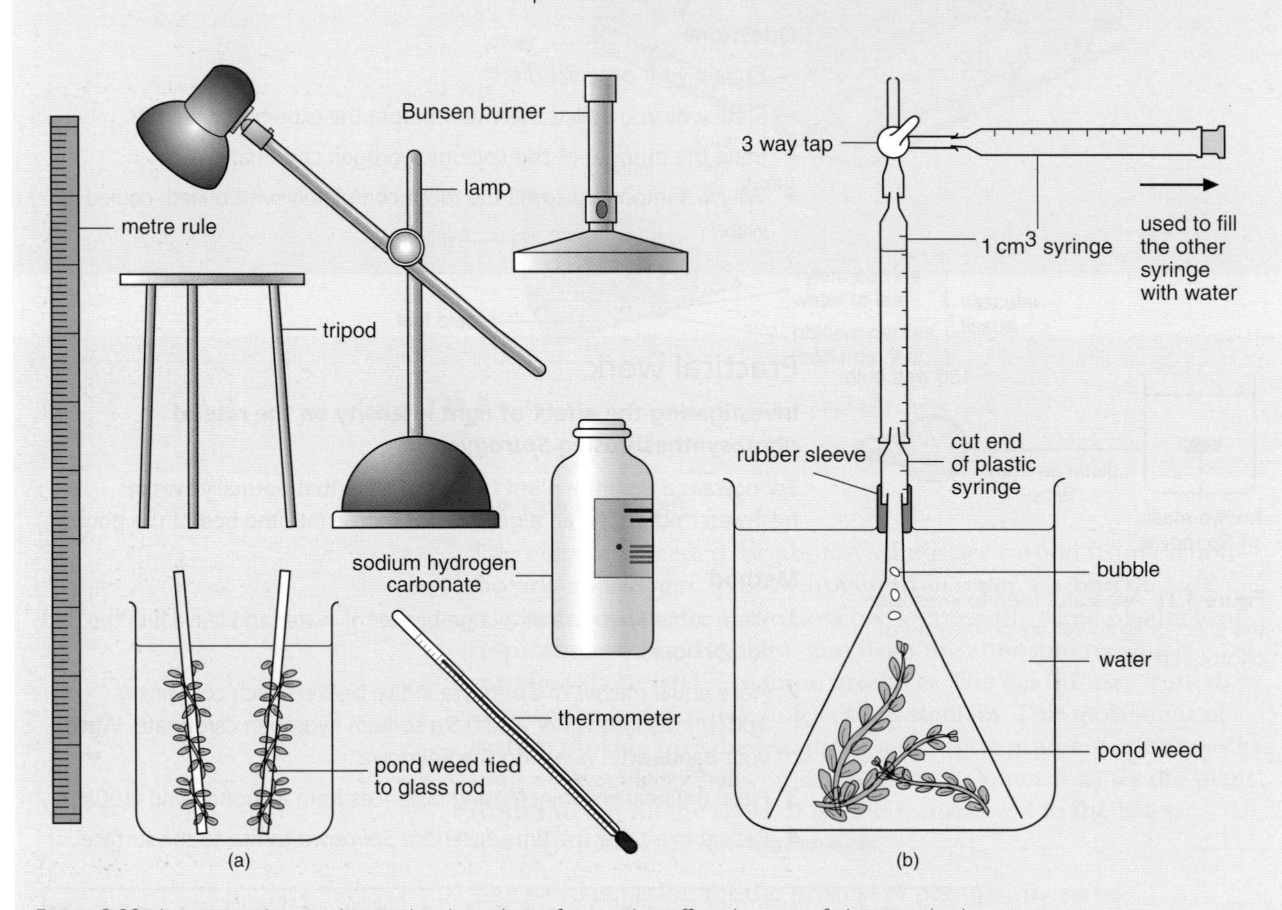

Figure 3.32 Apparatus that can be used to investigate factors that affect the rate of photosynthesis

2 Plot your results as graphs, and evaluate them in terms of valid conclusions, possible sources of error, and possible improvements.

By using coloured filters or coloured cellophane stuck on the beakers, you could investigate the effects of various wavelengths of light on the rate of photosynthesis.

Questions

- Do different colours of light affect what happens in photosynthesis?
- How does acid rain affect photosynthesis?

Monitoring photosynthesis: an alternative procedure using a computer

Oxygen production and carbon dioxide uptake can be monitored using sensors. Oxygen levels can be measured with an oxygen electrode, while changes in carbon dioxide level can be measured with a pH electrode. A datalogger can collect the readings automatically over a few days.

Method

1. To monitor oxygen production, place an oxygen electrode in a flask with sodium hydrogen carbonate solution and fresh *Elodea* (see Figure 3.33). Place the flask and a light sensor near a window.
2. Connect the sensors to a datalogging system and leave this to record for at least 24 hours.
3. Think about how the light intensity will change during your experiment, and then sketch a graph that shows this.
4. Next think about how the oxygen level will change, and add a sketch graph to show this.
5. Compare your predictions with your results.

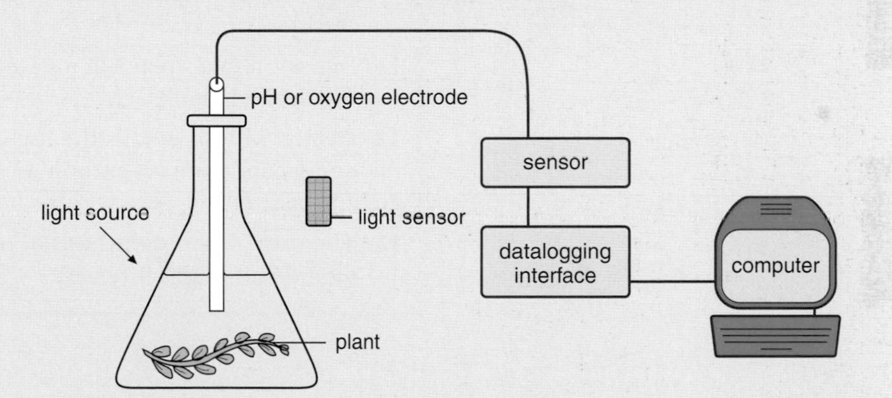

Figure 3.33 Using a computer to monitor photosynthesis

6. To monitor carbon dioxide uptake, place a pH electrode in a flask with sodium hydrogen carbonate solution and fresh *Elodea*.
7. Clamp the electrode close to the plant where it will monitor small changes in pH.
8. Place a light sensor nearby, and connect the sensor to a data-logging system.
9. Start recording, and continue for about 30 min.
10. For the first 10 min, cover the flask with foil. For the next 10 min, remove the foil, and for the final 10 min, switch on a bright light source.
11. Explain your results.

Questions

25 Explain why photosynthesis is essential to almost all forms of life.

26 Define photosynthesis using a word equation.

27 Give an account of the conditions that affect photosynthesis.

Summary

1. The cell is the basic unit of all living things.
2. There are similarities and differences in the structure and development of plant and animal cells
3. The stem cells of embryos keep their ability to develop into different tissues, so they can be used in the treatment of certain disorders.
4. Enzymes speed up the rate of chemical reactions taking place in cells.
5. Respiration is the release of energy from glucose in living cells. It can take place in the presence or absence of oxygen.
6. Diffusion is the movement of particles down a concentration gradient.
7. Osmosis is the diffusion of water molecules through a selectively permeable membrane.
8. Active transport is the movement of molecules against a concentration gradient, and it requires energy.
9. Animals require food for the release of energy and for growth.
10. Digestion is the breakdown of large insoluble molecules of food into small molecules of food that can be absorbed into the bloodstream.
11. Digestion takes place in a system composed of a variety of organs and glands.
12. Photosynthesis entails the use of chlorophyll to absorb light energy and convert carbon dioxide and water into glucose and oxygen.
13. The rate of photosynthesis depends on temperature, carbon dioxide and light intensity.

Chapter 4 Interdependence of organisms

By the end of this chapter you should:

- understand how food chains can be described quantitatively using pyramids of numbers and pyramids of biomass;
- understand how energy is transferred through an ecosystem;
- understand the role of microbes and other organisms in the decomposition of organic materials and in the cycling of carbon and nitrogen;
- know the positive and negative effects of human activity on the environment;
- know how food production can be managed to improve the efficiency of energy transfer.

Energy and nutrient transfer

Food chains

Plants are the first links in all **food chains** because they are **producers** – they change energy in sunlight into stored chemical energy. When plants are eaten by herbivores, some of the energy is passed to this next link, the **consumers** in the food chain. When the herbivore is eaten by a carnivore, the process of energy transfer is repeated. Energy passes in this way from carnivores to scavengers and decomposers, which feed on dead organisms. However, not all the energy stored by a herbivore is stored by a carnivore that feeds on it. Much is used in life processes such as movement, growth and reproduction. Some is also wasted as heat during respiration. Only leftover energy is stored by the carnivore.

Consider the food chain through which energy flows when we eat fish such as tuna. Energy from the Sun is first used by plant plankton (microscopic algae). It then passes to animal plankton, then to small fish, then to larger fish, then to tuna and then to us. There is usually no predator to eat us, so we are the top carnivores in this food chain.

plant plankton → animal plankton → small fish → large fish → tuna → humans

In fact, you could list hundreds of other species that could be food for the animals mentioned above, and draw arrows to show the energy flow. Your diagram would look more like a web than a chain (see Figure 4.1). For this reason, interlinked food chains are called **food webs**.

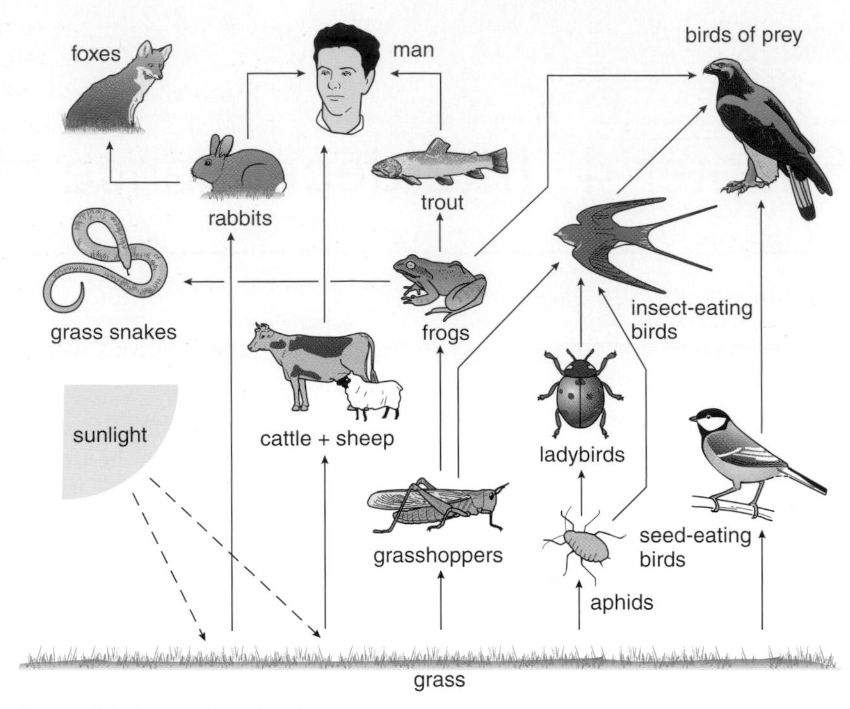

Figure 4.1 A typical food web

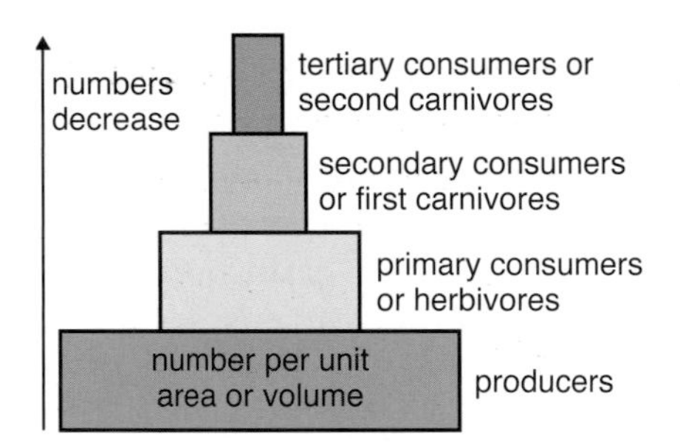

Figure 4.2 A general pyramid of numbers

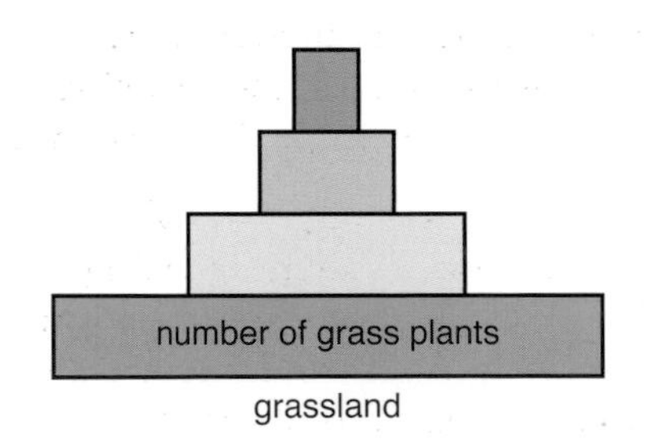

Figure 4.3 A pyramid of numbers for grassland

Food pyramids and energy transfer

Feeding relationships can be illustrated as pyramids (see Figure 4.2). These can tell us more about the energy that is available to organisms living in a measured area or volume. There are three ways to draw these pyramids:

- a pyramid of numbers that shows the number of organisms per unit area or volume in each feeding layer
- a pyramid of biomass that shows the dry mass of organic material per unit area or volume at each feeding level
- a pyramid of energy that shows the flow of energy through the feeding levels.

The first two methods can be misleading as true pictures of energy flow, for the following reasons.

Note that Figure 4.3 is a pyramid of numbers for a field of grass, and Figure 4.4 is for woodland. The shapes are totally different, even though they have been made by counting individual organisms. In the woodland, each tree can support thousands of animals, and so the base of the pyramid is smaller than the next level.

Note that Figure 4.6, the one with plant plankton as a base, has a larger first consumer level than base. This is because the biomass of plankton can change seasonally. No account has been taken of time. Figure 4.5, for woodland, is a more conventional shape.

The number of links in a food chain is limited by the chemical energy available. As the amount of chemical energy at each feeding level becomes smaller, so does the amount of living material that can be supported by that level. When the chemical energy decreases to nothing, the food chain (and the pyramid) ends.

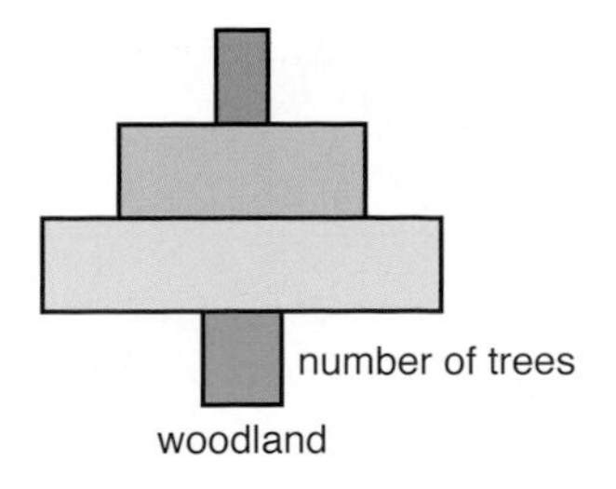

Figure 4.4 A pyramid of numbers for woodland

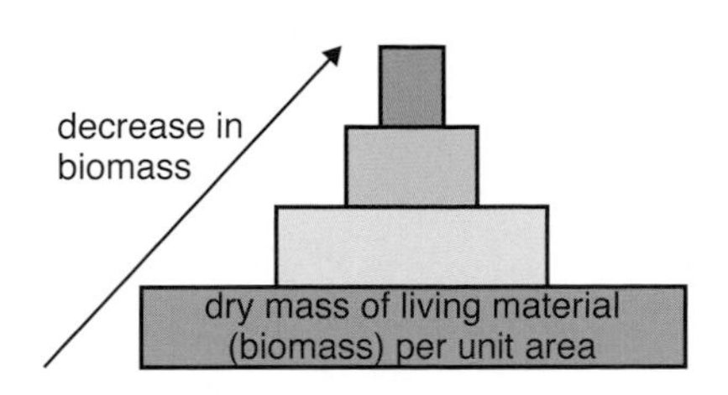

Figure 4.5 Pyramid of biomass for woodland

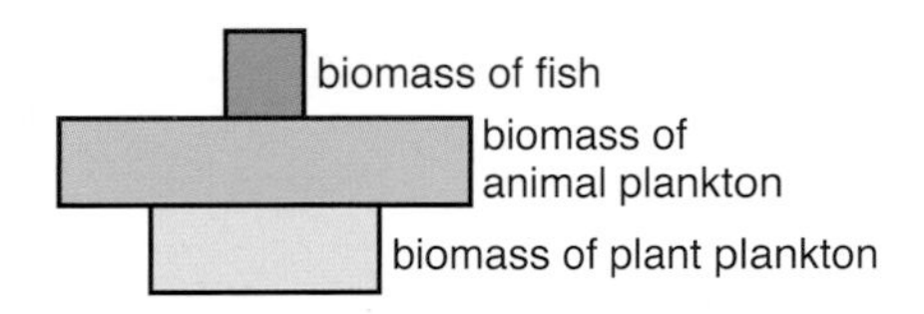

Figure 4.6 Pyramid of biomass for the sea

Customised modelling software shows these models more clearly. One example is *Creatures*, from Future Skill Software, Penrodyn, Pontrhydygroes, Ystrad Meurig, Dyfed SY25 6DP, Wales.

The plants at the beginning of the food chain are able to capture only 1% of the energy from sunlight. The transfer of this energy from one feeding level to the next is never 100% efficient. On average, about 10% of the energy from the previous level becomes available to the next as chemical energy stored in new body tissues. The remaining energy leaves the food chain in excreted waste products, as heat lost in respiration, or energy used in repair and maintenance of cells. An enormous biomass of producers is needed to support a small biomass of carnivores at the end of a food chain.

Studying population changes using software

Software allows biologists to model, or predict, how the numbers of animals in a population change over many generations. You may be able to use a spreadsheet to show the growth of a population of foxes and rabbits and to explore the effect of changing their birth rates and death rates. The spreadsheet will be able to show this in the form of a graph that shows numbers of animals against time.

The carbon cycle

It is unfortunate that when most people hear of microbes and bacteria, they immediately associate them with 'germs' and disease.

Questions

1 Starting with the food producers, name and define the various types of organism in a food chain with five links, ending with humans.
2 Explain what you would need to know to change a food chain into a pyramid of biomass.
3 Explain why photosynthesis is essential to almost all forms of life.
4 Draw a diagram of a food web that exists in your local area.

In fact, the vast majority of bacteria have nothing whatsoever to do with disease; most are completely harmless. Indeed many are so helpful to us that life as we know it would be impossible without them. It is just a relatively small number, called pathogens, that cause us harm. This is just as well, because we are constantly being colonised by millions of bacteria, no matter how hygienically we live.

Certain bacteria and fungi are essential for recycling materials in our environment. One group decomposes dead things, breaking them down to chemicals that can be used by another group to start the build-up process all over again. Without these **decomposers** we would simply be suffocated by millions and millions of years' accumulation of dead organisms. This can be illustrated by two cycles that are essential to our lives (see Figures 4.7 and 4.8).

The first is the **carbon cycle** (see Figure 4.7). Two basic life processes are involved in the carbon cycle. These are respiration and photosynthesis. Animals and plants take in oxygen for respiration (see the section on respiration in Chapter 3). During respiration glucose, containing carbon, is oxidised. As a result, carbon dioxide is released into the environment. If these processes continued, we would run out of oxygen, and carbon dioxide would continually build up in the atmosphere. Nature, however, has developed a neat

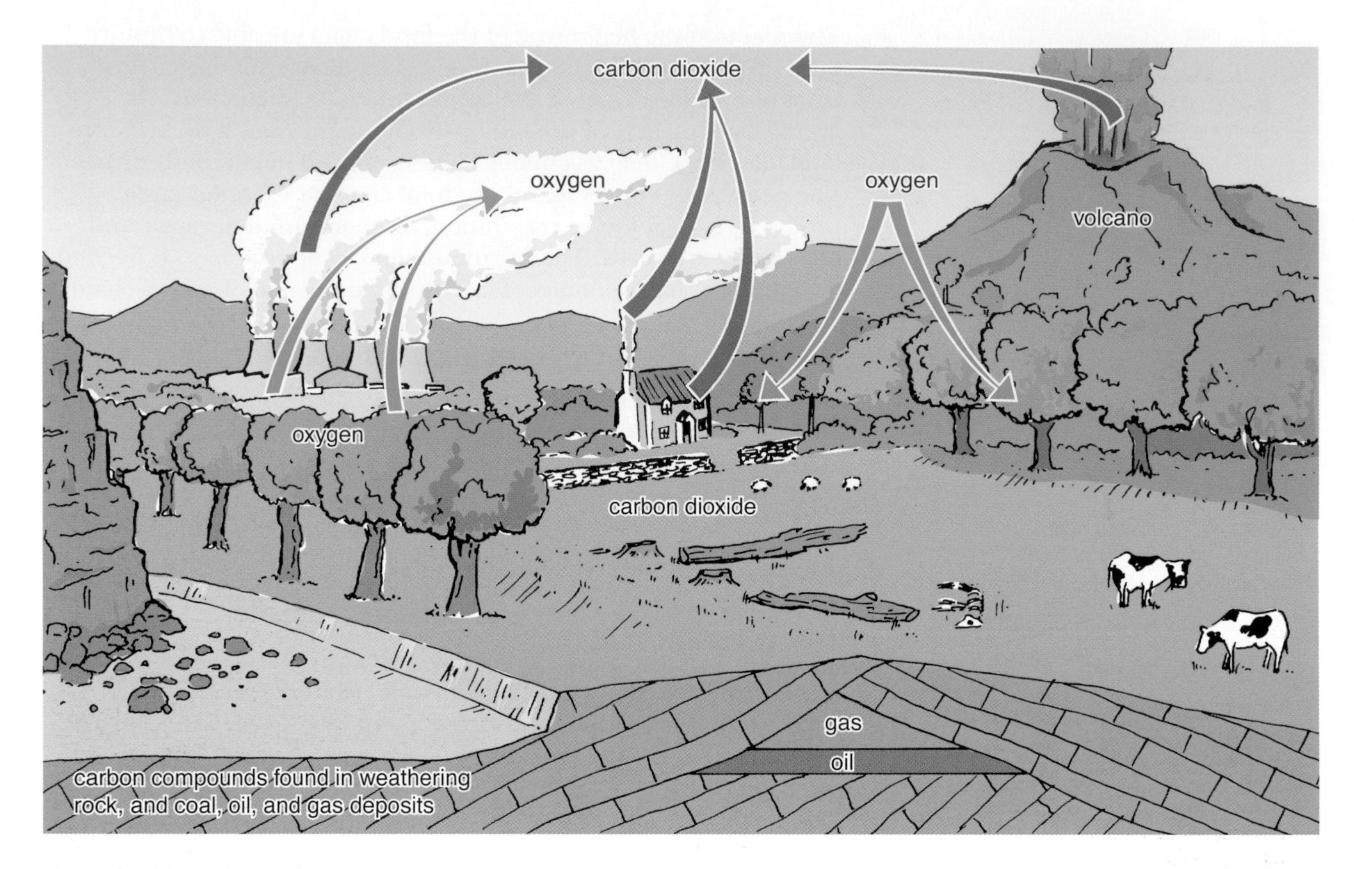

Figure 4.7 The carbon cycle

recycling trick to keep the oxygen and carbon dioxide levels in the atmosphere relatively constant.

When plants photosynthesise, they take in the carbon dioxide that would otherwise build up in the atmosphere, and they give out oxygen as a waste product. As plants photosynthesise, they make more than enough food for themselves. Plant-eating animals can take advantage of this and use the foods made in photosynthesis to make their own cells. These herbivores are, in turn, eaten by carnivores, and when any organism dies, the organic materials in it are broken down by essential fungi and **putrefying bacteria**. The carbon leaves the dead bodies during decay as carbon dioxide when the microorganisms respire. Burning fossil fuel also releases carbon dioxide into the atmosphere.

The nitrogen cycle

H

Another cycle, involving nitrogen (see Figure 4.8), also illustrates how bacteria are vital for our survival. The **nitrogen cycle** involves plants, animals and several kinds of bacteria. The plants' roots absorb nitrates from the soil to use in the manufacture of proteins. Why is all the nitrate in the world not used up by plants?

Herbivores obtain proteins by eating plants. Carnivores can obtain proteins by eating herbivores. Not all the protein ingested by these animals is used to build up their cells. All that is ingested is digested to amino acids (see the section on digestion in Chapter 3), but there

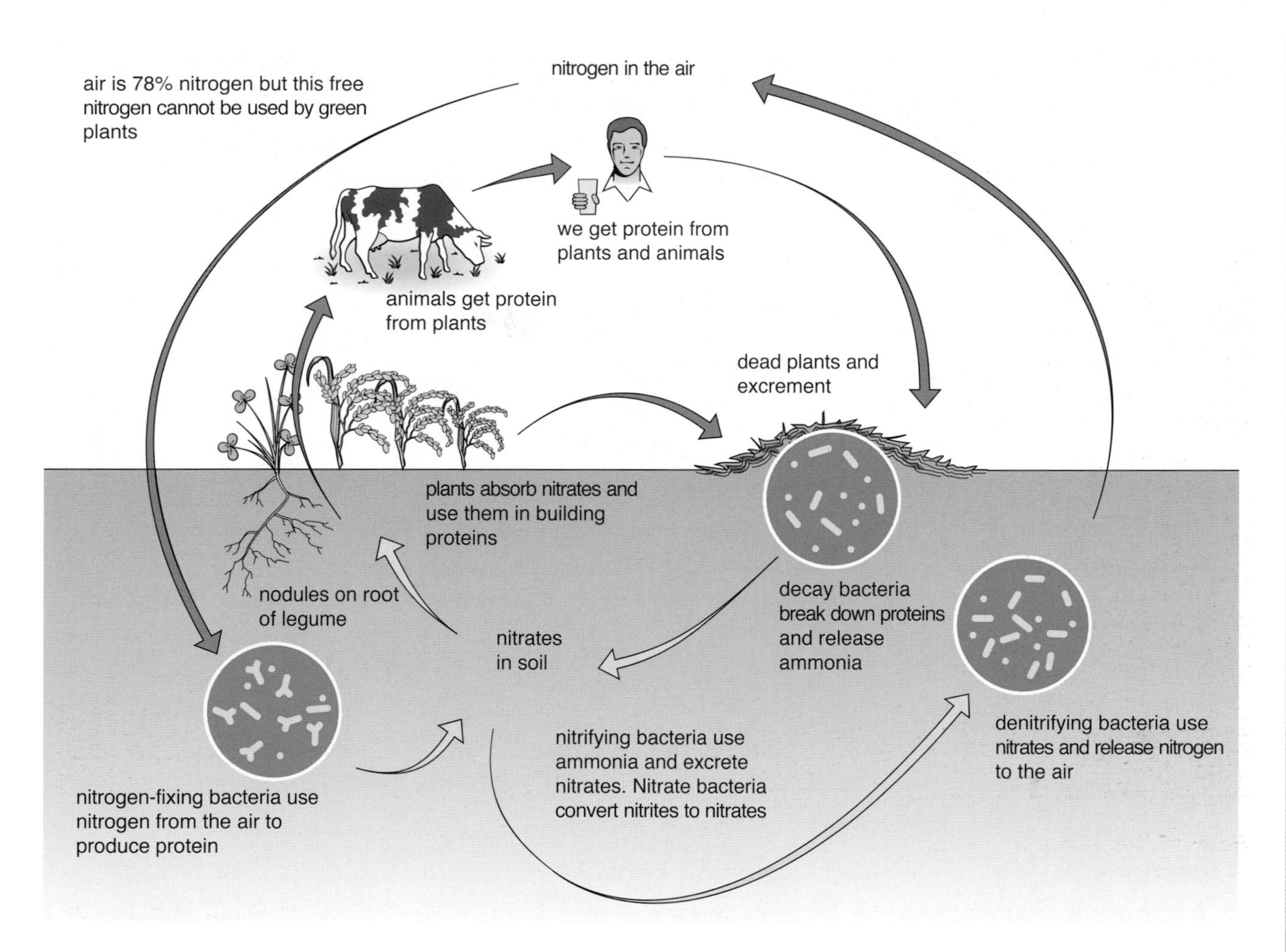

Figure 4.8 The nitrogen cycle

are more of these building blocks than the animal needs for building its own body. The excess amino acids change into urea, which is excreted in urine. Putrefying bacteria break down urea, together with the dead remains of animals and plants. **Ammonia** is formed, but this is so reactive that it combines with chemicals in the soil to become ammonium compounds. **Nitrifying bacteria** oxidise these compounds to form **nitrites** and then **nitrates**. The nitrates are absorbed by plant roots and the process begins all over again as plants make new protein.

About 78% of the atmosphere is nitrogen, but this cannot be used by plants. However, a type of bacterium that lives in the roots of **legumes** can use atmospheric nitrogen and convert it into compounds that can be used by plants. Legumes are pod-bearing plants such as peas, beans, clover and lupins. The bacteria get carbohydrates from the legumes, and supply them with a form of nitrogen that they can absorb in return. The process is called **nitrogen fixation**. A fourth group of bacteria called **denitrifying bacteria** release nitrogen from nitrates in the soil. These carry out the process in anaerobic conditions (see the section on energy release in Chapter 3), and are most abundant in tightly packed, waterlogged soil. If the soil is well drained, loss of nitrate in this way is minimal.

Activity

Look up the nitrogen cycle and the carbon cycle in a multimedia encyclopaedia or in the *Encyclopedia to the Environment: Biosphere* (available from AVP Software).

H

Did you know?

German scientists have perfected a technique to rid drinking water of nitrates with the use of bacteria. Agriculture has led to rising levels of nitrates in groundwater. Scientists believe that there may be a link between nitrates in drinking water and stomach cancer. High levels of nitrates can also cause **anaemia** (a shortage of red blood cells) in newborn infants. Scientists have successfully shown that a species of bacteria can eliminate nitrates from water. The water industry could find this process economical, because the bacteria convert the nitrates to harmless nitrogen.

Questions

5 Explain why agriculture has been responsible for an increase in nitrates in groundwater.

6 Which of the following types of bacterium is responsible for the above action?
- **a** putrefying
- **b** nitrifying
- **c** denitrifying
- **d** nitrogen-fixing.

Questions

7 Explain how the microbe spray cocktail acts as:
- **a** a decomposer
- **b** a fertiliser
- **c** a fungicide.

8 Suggest how the spray would protect the soil against erosion.

9 What is meant by 'the right conditions for the nitrogen-fixing bacteria'?

Did you know?

A cocktail of microbes may soon provide an economic alternative to burning straw, which often leads to air pollution. For hundreds of years, farmers have used fire to clear land for the next sowing of seeds. This kills off crop diseases. The only real alternatives to burning are either to leave the straw to rot where it is, or to plough it into the soil. Both methods can lead to an increase in diseases of crops. They also cause difficulties for seed-drilling machines.

Bacterial activity in the soil is normally limited by the lack of a ready source of energy. Straw contains large quantities of cellulose. Microbes can break down cellulose to simple sugars. These can then be used by nitrogen-fixing bacteria. Scientists have selected several types of microbe for this purpose. For example, there is a fungus that lives on other fungi that cause disease in root crops. This fungus can also break down cellulose. One nitrogen-fixing bacterium does the same. Yet another bacterium produces large amounts of gum which binds soil particles together. The oxygen consumption of this bacterium also helps to provide the right conditions for the nitrogen-fixing bacteria. These microbes are all mixed together in a cocktail that can be sprayed onto straw, and the straw is then ploughed into the soil.

Questions

10 What is the source of the carbon used by plants to make food by photosynthesis?

11 Explain how the constant level of carbon dioxide in the atmosphere is maintained.

12 Outline briefly the cycle of carbon between plants and animals.

13 By what means is atmospheric nitrogen made available to plants?

14 What is the alternative source of plant nitrogen, and how is it obtained?

15 Why do plants and animals need nitrogen?

16 Summarise the role of bacteria in the circulation of both carbon and nitrogen.

17 Justify the statement that 'we could not survive without microbes'.

H

Practical work

Observing and culturing nitrogen-fixing bacteria

1. Use a pea or bean plant. Carefully wash the roots with nodules on them.
2. Dip them in Domestos disinfectant for 5 min.
3. Your teacher will give you two sterilised slides.
4. Squash a nodule in a drop of water between the slides.
5. Make a very thin stained smear of a loopful of the squashed nodule and crystal violet (0.5 g in 100 cm^3 of distilled water).
6. Streak out a loopful of the remaining squashed nodule onto a glucose yeast-extract agar plate (see Table 4.1 for the recipe), and incubate at 25 °C.
7. A raised yellow sticky colony of nitrogen-fixing bacteria (*Rhizobium*) will grow in 1–2 days.

Table 4.1 Recipe for glucose yeast-extract agar

Ingredient	Quantity	Ingredient	Quantity
Peptone	10.0 g	Glucose	5.0 g
Marmite	10.0 g	Agar	15.0 g
K_2HPO_4 pH 7 buffer	5.0 g	Distilled water	1.0 dm^3

H

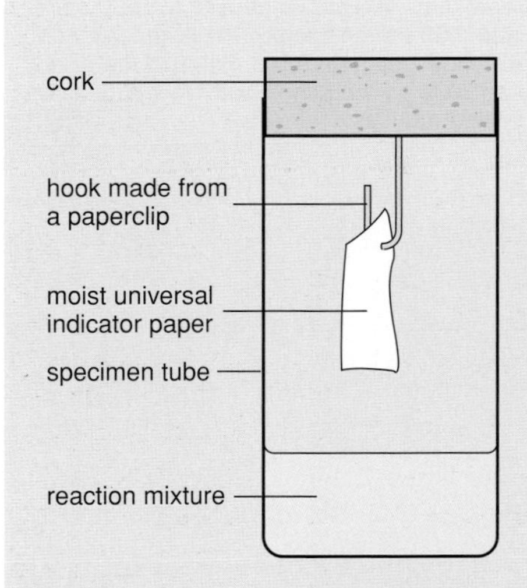

Figure 4.9 Apparatus to demonstrate how urease breaks down urea

Practical work

Demonstrating how the bacterial enzyme urease can cause decay

Some microbes produce an enzyme, **urease**, that breaks down urea in the soil to form ammonia and carbon dioxide. Ammonia is alkaline, and can be detected using universal indicator paper.

Method

1. Pour 20 cm^3 of urea (1 mol/dm^3) into a specimen tube.
2. Stick half a paperclip in a cork to form a hook with which to suspend a strip of moist universal indicator paper.
3. Add 5 cm^3 of 1% urease to the urea in the specimen tube, and place the cork with the suspended indicator paper in the tube as shown in Figure 4.9.
4. Set up a control tube with contents identical to those above but with 5 cm^3 boiled, cooled urease instead of the unboiled urease.
5. Leave both tubes for half an hour at room temperature.
6. Record any colour changes in the indicator paper, and compare these with the pH colour chart that accompanies the indicator paper.
7. Explain your observations.

The impact of human activity on the environment

Live now, pay later

The growth of human populations is undoubtedly the main environmental influence on problems related to overcrowding and pollution. Factors such as reproductive rates, death rates and disease all depend on the sizes of populations. We can determine the rate of human population increase by recording daily births and deaths. An estimated 270 000 babies are born in the world every day. There are an estimated 142 000 daily deaths. If you subtract numbers of deaths per day from the numbers of births per day, you arrive at the estimated daily rate of population increase. It is estimated that the human population increases by about 128 000 individuals each day. This rate will probably increase, because the more adults there are, the more babies are likely to be born. The world population is more than 6 000 000 000 people (see Figure 4.10). If our present rate of food production stays the same, there will not be enough food to feed everyone. Already, more than 20 000 people die from starvation or malnutrition each day.

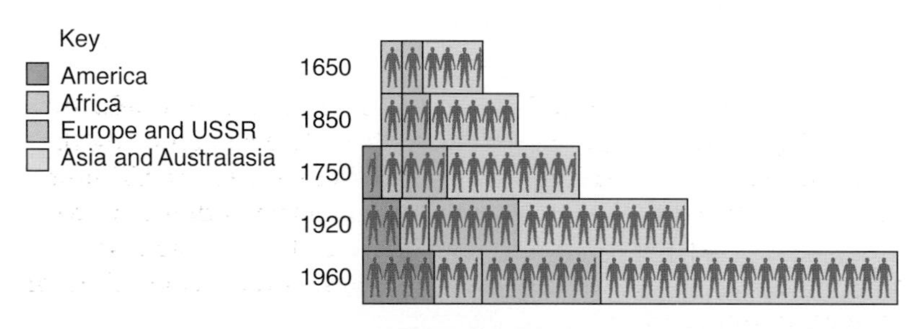

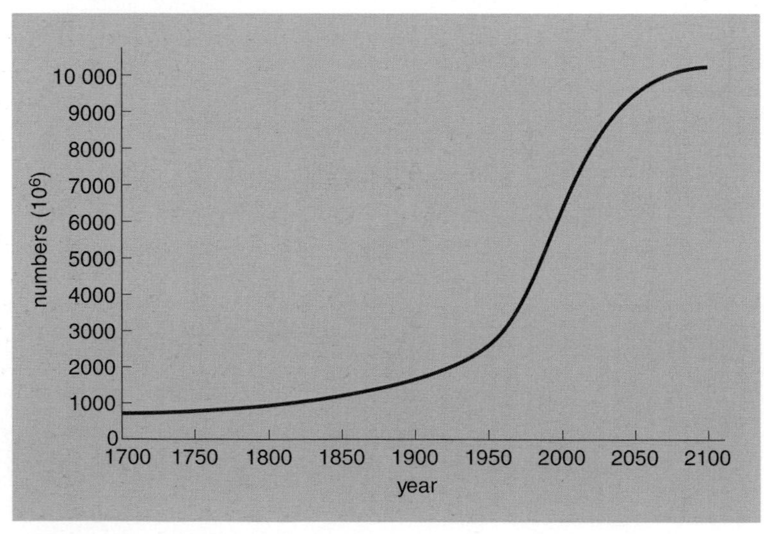

Figure 4.10 How the human population has changed over time

In addition to this problem of a population explosion, there are the many problems of pollution that come with the use of a limited space by an ever-increasing number of people. The biggest influence we have on the environment is its destruction through pollution.

A pollutant is a substance whose presence in the environment is harmful. Not only do some pollutants take the form of chemicals, but they also include heat and noise. Nature has always produced pollutants such as natural-oil seepages, volcanic gases, and products of combustion from forest fires. Nature has also dealt with these forms of pollution without leaving any permanent problems. However, since the evolution of humans, the level of pollution has increased to such an extent that it now threatens our health. We have perfected our ability to pollute our environment. Indeed, we are now so good at chemistry that we can make harmful materials that cannot be decomposed. They simply accumulate progressively, providing our present generation with problems and presenting future generations with even bigger ones.

Did you know?

Apart from the smallpox virus, mankind has not been able to make any harmful species extinct, even though many attempts have been made using pesticides, herbicides and antibiotics.

Air, water and land are polluted daily by thousands of tonnes of pollutants on a worldwide scale. It is only relatively recently that measures have been taken to tackle the problem. Unfortunately, this has been too late for some species that have become extinct as a result of the destruction of their habitats.

To build or not to build?

In the 21st century, the public's ideas of the interests of developers of buildings and roads are often out of date. Far from ignoring the effects of building on the environment, developers in the UK now have at least 23 separate legal regulations with which they have to comply when they begin work. The laws range from those that consider wildlife to those that cover pollution and consideration for ancient monuments.

UK legal regulations that cover living organisms in the environment include the:

- Wildlife and Countryside Act (1981)
- Environmental Protection Act (1990)
- Environmental Protection (Duty of Care) Regulations (1991)
- Protection of Badgers Act (1992)
- Environment Act (1995)
- Hedgerows Regulations (1997)
- Conservation (Natural Habitats) Regulations SI (2001).

During the planning process, Environmental Impact Assessments have to be carried out for all developments and shown to local authorities before building can go ahead. Failure to do this could result in very heavy fines for the companies concerned.

If rare species live in an area to be developed, then measures must be taken to protect the species and their habitat. All environmentally sensitive areas are treated in this way. If developments have involved quarrying, then when they have been completed, the area must be landscaped to simulate the original conditions as far as possible. Photographic evidence, before and after developments, is provided to allow people to assess any change in species types and numbers.

During road building, some protective measures used for wildlife include the provision of tunnels and water courses for species such as badgers, frogs and newts so that they can access their breeding sites. Rare plants are fenced off during developments to minimise disruption. Work is often suspended during the breeding seasons of sensitive species of butterflies and birds. These procedures are often expensive but are essential for the protection of habitats and to maintain biodiversity.

The environmental policies of the major building and road developers in the UK have many points in common. They are all aware of global concerns as well as the need to maintain local habitats. Industrial activity inevitably affects the environment, but developers try to keep the negative effects to a minimum. The aim is to create a balance between meeting the nation's needs and the need to protect the natural environment. All major building companies take environmental issues and factors into account when planning and developing new and existing activities. They broadly aim to supply society with the materials and services it needs while protecting the environment for everyone to enjoy in the present and the future.

Water pollution

Figure 4.11 An outlet pipe is discharging sewage, polluting the water.

Many of our lakes and rivers are polluted (see Figure 4.11). We have even polluted the oceans. When the forerunners of our cities first began to grow, they were small settlements, often built near large rivers. One reason for this was waste disposal. Inhabitants poured untreated sewage and other refuse into the water, which carried it to the sea. This was not a serious problem at first, when settlements were small and far apart. The pollutants were decomposed by natural processes.

Today the story is completely different. Natural decomposition cannot possibly deal with the volume of waste produced by the inhabitants of modern towns and cities. Strict laws on the use of waterways for sewage and other waste disposal are now enforced. Although untreated sewage is often poured into the sea, there are sewage works to deal with waste before it becomes effluent that is suitable for pouring into rivers. Sewage, when allowed into waterways, encourages bacteria to multiply. The bacteria use all the oxygen in the water. Sewage may also contaminate water with disease-causing microbes and so help to spread diseases such as typhoid and cholera. It is therefore dangerous to eat animals that come from polluted waters. Cases of typhoid, hepatitis and other infectious diseases have often been traced to mussels or other sea life that filter water and capture the harmful microbes.

Did you know?

In some waterways in Australia, blue-green algae multiply in the high summer temperatures and produce a poison that affects the livers of livestock that drink the water. Australian scientists have discovered a bacterium that feeds on the poisonous algae and could be used to control them without harmful chemicals being added to the environment.

The nitrates found in sewage act as fertiliser for microscopic plant life in the water into which sewage may be discharged. Algal blooms (a sudden increase in algae) then occur. The algae and other water plants reproduce to such an extent that they prevent light reaching the plants growing at the bottom of the lake or river. These die and decompose, allowing bacteria to multiply. The bacteria use up oxygen from the water. Fish and other animal life in the lake or river die because of the lack of oxygen (see Indicator species in Chapter 1).

The same problem of oxygen removal occurs when fertilisers are overused on crops. Fertiliser containing nitrates and phosphates encourages the growth of plants when it is washed into waterways by rain.

The millions of gallons of detergents reaching drains daily add to water pollution problems. Many contain phosphates, so they act as fertilisers, leading to problems similar to those posed by the over-fertilisation of crops.

Agriculture is also responsible for other forms of water pollution, including those caused by pesticides (see Figure 4.12). These are washed into water, and, although they are usually not in such a concentration that they are harmful to humans, they become highly toxic as they build up through food chains. Those animals at the top of the food chain can receive such a high concentration of pesticides that they are poisoned.

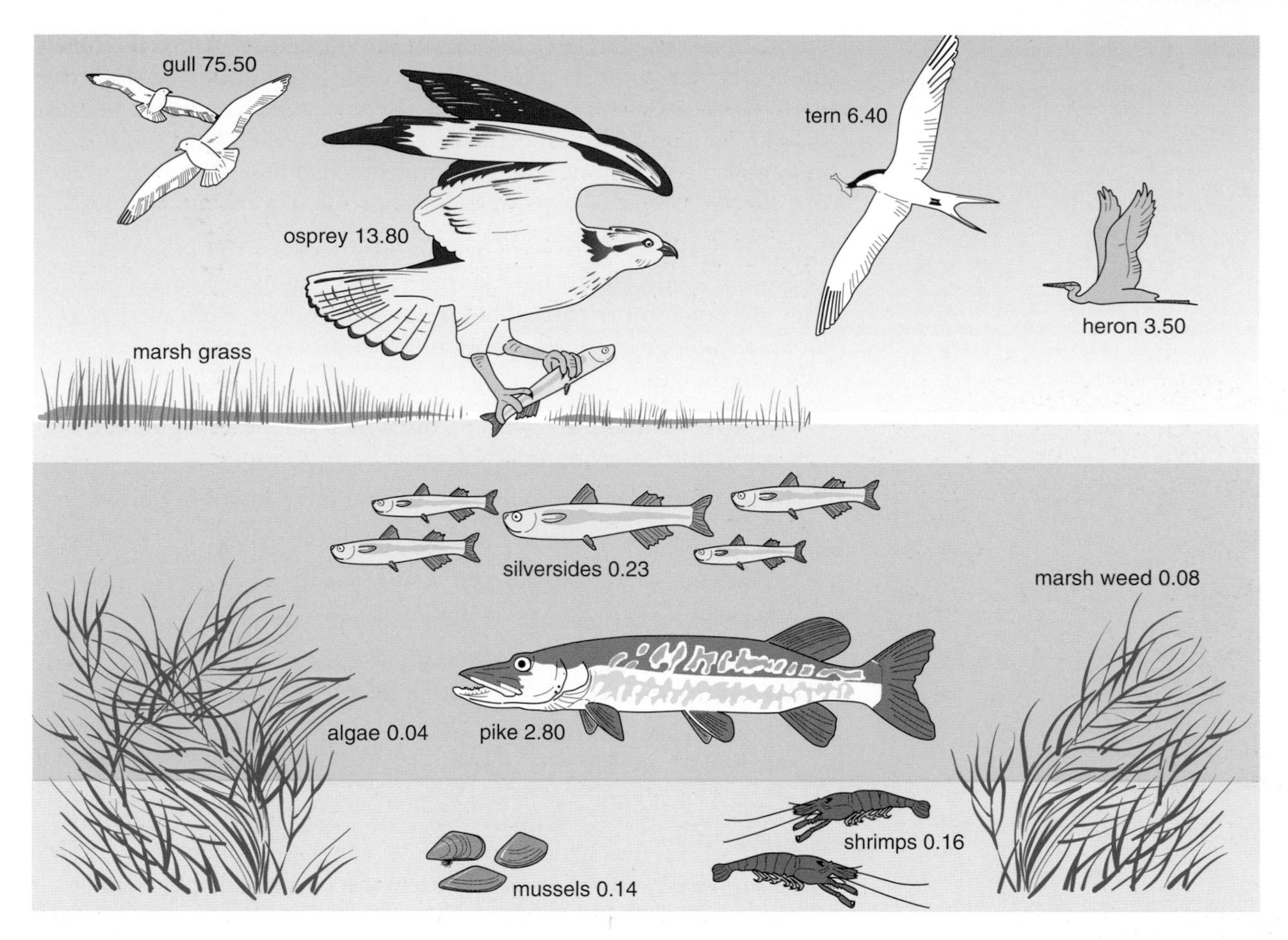

Figure 4.12 The concentration of DDT in a food chain; the DDT becomes stored in the tissues of the organisms at each feeding level of the food chain, and can build up to very high levels in the top carnivores (the figures indicate parts per million of DDT).

Helpful aliens

The problems of environmental pollution by pesticides are felt throughout the world. For many years, scientists have looked for alternative ways of controlling pests, and have used a method called **biological control**. This method does not use chemicals. It uses other organisms that are predators or parasites of the pest. Early attempts saw as many failures as successes. One of the first successes occurred in Australia. When people began to farm the land, they had to enclose vast areas so that the native Australian animals such as wallabies and kangaroos could not raid their crops. Fencing was expensive, so they made hedges out of the prickly pear cactus that they imported from South America.

The hedges kept out the native animals, but grew out of control and soon became a pest as a weed among the crops. Herbicides could not be used on the cactus, because they would damage the crops. A type of insect that normally fed on the cactus in South America was introduced to Australia and ate all the cactus. Fences proved to be the answer in the long run!

On the other hand, there have been many failures. One that is often quoted is the attempted control of an insect that was feeding on the coconut plantations of the West Indies. A type of toad was introduced to eat the insect pests. The toads were so successful that they almost totally wiped out the insect pest but then began to feed on other insects that were important in pollinating plants. By then the toads were so numerous that they had to be controlled. In order to do this, snakes were introduced. These began to be a pest because they killed native birds and chickens. In order to control them, a mammal called a mongoose was introduced. These killed the snakes, but also fed on birds' eggs, and so the problem went on, unsolved.

In modern times, there have been more successes than failures because scientists today have learned from past experiences, and they are much more careful before they introduce alien species to an area. Biological control programmes now take place in the following stages:

1 Biologists search the country from which the pest originally came for suitable natural predators or parasites to use.

2 Trials are carried out to make sure that the control organism:

a will attack only the pest and nothing else

b does not carry diseases that could spread

c can breed successfully in their new environment.

Care has to be taken to prevent any escape of the control organisms before trials are complete.

3 The control organisms are bred in large numbers. They are then released where the pest is present.

4 Progress is monitored to judge the success or failure of the programme.

Most modern methods of pest control use biological agents and some specific pesticides. This is called **integrated pest management**. A successful programme of biological control has a number of advantages over controlling pests with chemical pesticides:

- It is very specific and affects only the pest. No matter how carefully chemical pesticides are used, they affect other organisms.
- Once introduced, the control organism establishes itself and does not have to be reintroduced. Chemical pesticides must be used repeatedly, and in the long term this makes their use more expensive.
- Pests do not become resistant to predators. One of the reasons why we are constantly having to develop new pesticides is because pests rapidly evolve resistance to pesticides.

However, biological control has some disadvantages:

- There is often a time-lapse between the introduction of the control organism and a significant reduction in numbers of the pest. Chemical pesticides act much more quickly.
- An effective biological control programme keeps the numbers of pests at a low level. It rarely exterminates them completely.

Effective control programmes make use of a variety of methods. They often consist of the use of natural predators and parasites supported by the occasional use of specific pesticides.

Did you know?

Recycled sewage water will be the most important source of extra water in many developed countries in the next decade. Re-used water may provide one-fifth of total water supplies and one-third of the demand for irrigation.

Figure 4.13 The effects of oil pollution on a coastline

Heavy metals

Perhaps the most important long-term form of water pollution is due to industrial wastes such as heavy metals, for example mercury, lead, zinc and cadmium. These are often by-products of manufacturing processes in paper mills, steelworks, refineries and car factories. The concentration of heavy metals also increases through food chains until those feeders at the top are poisoned. The heavy metals interfere with the action of many vital enzymes in animals, causing death.

Non-biodegradable substances called polychlorinated biphenyls (PCBs) are used in the manufacture of paint and electrical equipment. There is no known method for removing them from water, and once they get into food chains, they accumulate from one feeding level to the next.

Pollution problems are also created by oil and other petroleum products. These come from refineries, drilling and pumping operations, shipyards and oil spills. The cumulative effects of car owners illegally dumping oil and the run-off of spilled oil on roads are responsible for the large volume of oil that finds its way into water courses. Many types of wildlife, particularly sea birds, are affected directly, but the most serious problems arise when links in food chains are killed off. In cases of major spills (see Figure 4.13), detergents used to clean up the problem are often toxic to the wildlife they are meant to save.

Did you know?

Lead shot used by hunters of water birds kills 25 000–30 000 birds each year in some European countries such as Spain. A 2-year survey carried out in the 1990s found that 3000 tonnes of lead shot reaches water in marshes and wetlands each year. Water birds consume the lead shot when they feed, and they are poisoned. Birds of prey have also died after eating the poisoned birds. Lead shot is now often replaced by steel shot.

Refuse and land pollution

Wherever humans go, they leave behind a trail of waste. In fact, archaeology is dependent on the discovery of materials left behind by previous civilisations. There is litter in the streets, on beaches, and along roadsides. Abandoned cars, refrigerators, and other useless rubbish are left as eyesores in areas used as scrapyards. Litter is a costly and serious aspect of a waste problem that we have made for ourselves. We live in a society with more and more disposable materials being made because we depend on a constant turnover of manufactured products to create jobs and wealth.

In the UK, we produce billions of tonnes of rubbish each year. It is estimated that over 2 kg of rubbish is produced per person per day in the UK. Many people take rubbish disposal for granted until something happens to prevent it being collected each week. Food scraps, newspapers, wood, lawn mowings, glass, cans, old appliances, tyres, furniture and many other items make up the contents of rubbish-collection trucks. These solid wastes are disposed of in several ways. Some are dumped at sea, several miles out. Some are burned in incinerators. Most are used as landfill (see Figure 4.14). Any method of waste disposal is certain to pollute the environment to some extent, but landfill is probably the safest. However problems have arisen when organic material is decomposed and there has been an accumulation of methane. Sometimes this has reached the surface and has ignited with disastrous consequences. Houses have exploded as a result. Good use is made of much of the methane produced in this way; it is collected and used as a fuel.

Figure 4.14 A landfill site

Did you know?

Bacteria that feed on sulphur are helping to reduce the mountains of used car tyres in the USA (see Figure 4.15). 200 000 000 used tyres must be disposed of each year. The bacteria alter the rubber so that it can be recycled.

Figure 4.15 Bacteria enable us to recycle these car tyres.

Figure 4.16 A recycling collection centre

Questions

18 List four types of water pollution, and suggest steps that would be necessary to solve the water pollution problem.

19 Explain how the decomposition of sewage in water kills aquatic life.

20 Explain how recycling can help solve waste disposal problems.

21 Explain how overproduction leads to the depletion of resources.

22 What are the possible consequences to biodiversity of increased industrialisation?

One solution to the problem of rubbish is to use it over again in the process of **recycling**. This reduces the production of waste materials that are a source of pollution. It also helps save our **non-replaceable resources** such as minerals and our **replaceable resources** such as trees.

It is possible to recycle paper, glass, and certain metals. In the past, cost and the abundance of natural resources limited the development of recycling technology. An efficient method of separating reusable wastes from the remaining wastes was also needed. Today, with the reduction in the amount of natural resources, we are forced to consider recycling. However, recycling is still far from being totally accepted. Lack of public interest, recycling centres, funding for research, recyclable goods and government incentives have slowed the process, although attitudes are now beginning to change.

Environmental groups have been active in setting up collection points (see Figure 4.16). The public is becoming increasingly aware of the need to take bottles, cans and paper to be recycled. It is also up to industry to start recycling on a large scale, and up to governments to propose legislation to help solve the problem.

Conservation – the public image

The word **conservation** presents varying images to various people. Conservationists usually aim to protect species or the places where species live. Conservation has in the past been placed rather low on governments' lists of official concerns. However, with the massive development of ecotourism throughout the world, it is now to the advantage of many countries to conserve their wildlife and habitats to generate considerable income.

With the launch of the World Conservation Strategy (WCS) in 1980, the relevant bodies hoped to improve the image of conservation. The strategy recognises the need for development, without which a large proportion of the world's population will remain poor. It argues that, for development to be long-lasting, it must rely on established conservation principles. Conservation and development are in fact dependent on each other. This is illustrated in developing countries where people are dependent on living organisms. For the 500 000 000 people who are underfed or the 1 500 000 000 people whose only fuel is animal dung or crop waste, conservation is essential.

Unfortunately, the poorest people in the world are compelled by their poverty to destroy the few resources available to them. They strip trees for fuel until the plants wither and die. The 400 000 000 tonnes of animal dung and crop wastes that are burned each year are badly needed as fertiliser and to prevent soil being eroded. The dung being burned would normally bind soil so that it would remain intact.

The WCS defines conservation as the management of the living systems of the planet so that it may provide lasting benefits to present and future generations. The aim of the WCS is to combine conservation and development to ensure the survival and well-being of all people.

Figure 4.17 Giant pandas live in bamboo forests that are under threat.

Figure 4.18 Grizzly bears still live in remote parts of North America.

Conserving species

The term 'biodiversity' is often used as a shortened form of 'biological diversity. It describes the richness of the natural world in terms of the variety of life on our planet.

We sometimes forget that, however varied nature is, it is not indestructible. Some species cannot cope with the changes we make to their environments. They include animals that specialise in eating one particular type of food. If the habitat containing their food is destroyed, then 'specialist' feeders die. Such specialists of the living world include the giant panda (see Figure 4.17) and the koala bear, and these are already threatened.

In the UK, conflict between conservationists and agriculture is seen in the persecution of the badger. Although there is little scientific evidence to support the point of view that badgers spread bovine tuberculosis, badgers are gassed or killed illegally with dogs. This is despite the fact that the badger is a protected species.

Preserving habitats

Nature conservation means preserving nature in its original form so that all species have a suitable habitat in which to live. Once a species has become extinct by destruction of its habitat, nothing can bring it back. Even gene technology has its limitations. DNA is biodegradable after a certain length of time, and the dream of *Jurassic Park* will have to remain a dream. However, some species on the brink of extinction have been saved by careful breeding programmes. This has happened with the wild horse, a few specimens of which can still be seen, but mainly in zoos. Since 1990, captive-breeding programmes have enabled Przewalski's wild horses to be reintroduced to Mongolia and to a national park in the Massif Central in France.

There are many mammals and birds that have only relatively recently come under pressure and have managed to survive to the present day in some countries. Lynx, bear and wolf disappeared from most of Europe within historical times. They were exterminated by our ancestors because they were regarded as competitors. However, these animals still live in other places such as North America (see Figure 4.18), and are being reintroduced to parts of Europe.

We have a duty to repair the damage already caused by our ancestors. There are species that we have tried repeatedly to eliminate, for example the malaria-carrying mosquito, the housefly and the locust. It is interesting that, with all of our advanced technology, we have not been able deliberately to eliminate any species, apart from the smallpox virus. However, as a result of overhunting and habitat destruction, we have increased the rate of extinction of many species of animals and plants. The dodo, the passenger pigeon, the Tasmanian wolf, and Steller's sea cow are some of the larger vertebrates that have been hunted to extinction. Many other less obvious animals and plants are probably becoming extinct before they become known to science. For some species on the brink of extinction, it may be possible to restore their natural habitats and reintroduce them after they have been bred in captivity. This has been attempted for some of the great apes, such as the gorilla and the orang-utan, and also for some birds, such as the Hawaiian goose.

Red kite feeding station:
www.gigrin.co.uk

Did you know?

In the eighteenth century, the red kite (see Figure 4.19) was one of the UK's most widespread and common birds. It was then thought to be a threat to livestock on farms and to game birds such as partridges. As a result, it was widely killed and almost became extinct. In 1905, only five wild red kites remained in Wales. The efforts of some very keen conservationists in the Welsh Kite Trust helped the red kite to recover its numbers. By 2002, the numbers had recovered to 300 breeding pairs.

Figure 4.19 A red kite in flight

Figure 4.20 A red squirrel

Figure 4.21 The American slipper limpet, *Crepidula fornicata*

Aliens in Wales

The grey squirrel

The red squirrel (see Figure 4.20) faces extinction in Wales. This is mainly because of the loss of its natural habitat and competition from an alien species. It has to compete with the larger and stronger grey squirrel. The grey squirrel is very widespread and is found throughout the UK. It was introduced to the UK from the USA in 1876. By 2004, the populations of red squirrels in Wales were confined to North Wales and Mid-Wales. Today, the largest colony can be found on Angelsey, an island separated from North Wales by two bridges. In 1997, it was feared that the population of red squirrels in Angelsey would become extinct, largely because of competition with the more numerous grey squirrel. However, through a programme of grey squirrel control by trapping, the population of red squirrels has been saved.

The threat from the grey squirrel began in the mid-1960s, when they reached the island by way of the road bridge across the Menai Straits that separates Angelsey from the mainland. Red squirrel numbers soon began to fall. During 1970–99, the pattern of extinction was similar to the pattern already seen elsewhere in the UK. By the winter of 1997–98, red squirrels could be found only in one forest, Mynydd Llwydiarth. Grey squirrel control took place, and because Anglesey is an island, there was much more chance of success than on the mainland. The rate at which grey squirrels were crossing to the island was much lower than the rate at which they were being controlled by trapping. European funding provided the resources necessary for an ongoing conservation project allowing a programme of grey squirrel trapping to take place. With the grey squirrel population greatly reduced, the red squirrel population increased rapidly, and by 2004, it had spread throughout the island.

The conservation of the native red squirrel has been highlighted by the European Squirrel Initiative (ESI), which represents the interests of landowners, conservationists and foresters seeking to protect the natural environment.

The American slipper limpet

Many other alien species besides the grey squirrel have accidentally reached Wales. Some have been very important because they have affected human interests in addition to altering the balance within the environment. One such example is a type of mollusc called the American slipper limpet *Crepidula fornicata* (see Figure 4.21).

These molluscs compete with commercially valuable oysters and mussels because they filter food from the sea. They are often so numerous that they smother large areas of oysters and mussels on natural and artificial beds. It then becomes impossible to use the oysters and mussels commercially. This alien pest is so firmly established on UK shores that it is now impossible to eliminate it. Any pesticide used would also affect mussels and oysters. It was accidentally introduced to the south and east coasts of the UK as long ago as the 1880s. By the 1950s, it had spread along the coast to Pembrokeshire in Wales, and in 1974 it had reached most of the rocky shores on the Bristol Channel as far as Fontygary near Barry in South Wales. There are very few rocky shores in Wales where the American slipper limpet cannot be found today.

Figure 4.22 Illegally imported animal products

Trading in endangered species

Each month, special agents of the United States Fish and Wildlife Service seize tens of thousands of furs that are illegally smuggled into the USA. In Europe, the situation is similar (see Figure 4.22). There are so many horror stories of wildlife such as chimpanzees, reptiles and parrots being found dead when smuggled into almost every European country.

Rhinoceros horn, tiger skins, snakeskins, ivory and other products from rare species are also smuggled across borders. Airports now have display cases that show the types of thing that it is illegal to smuggle in an attempt to reduce this shameful traffic. Governments are now taking action. Long prison sentences are given to those who are caught. The law enforcement relies on the **Convention on International Trade in Endangered Species** (**CITES**). The convention became law as long ago as 1975, and it now has about 100 member countries, including most of the important exporters and importers of wild animals and plants. The aim of the convention is to establish worldwide controls over trade in endangered species and their products. It recognises that this sort of trade is one of the major threats to the survival of species. Unfortunately, it has taken almost 35 years to see some signs that governments are treating the illegal trade in endangered species with the seriousness that it deserves.

Agents and organisations

Most countries in Europe now have nature conservation laws and protected areas. Protected areas differ greatly in their extent. The Greenland National Park is almost uninhabited, and its environment is totally controlled by climate, whereas the Camargue reserve in France is subject to many external influences. The International Union for Conservation of Nature (IUCN) has produced a report reviewing the various types of habitat protection that exist throughout the world.

- **Scientific reserves**: These are free from any human interference or internal artificial influences. They are areas set aside exclusively for scientific research, and they often protect species or habitats that are important for their biological value. Their size is determined by the area that can be preserved intact, for example the whole of Surtsey Island, or much smaller areas of wetlands in the UK.
- **National or state parks**: These parks serve some of the same purposes as reserves, that is, conservation of the environment and its species, for example the marine reserves around some islands in West Wales such as Skomer and Skokholm. Their regulations do not totally exclude the public or the provision of access routes such as the Pembrokeshire coastal path. The parks are generally zones to meet the aims of strict conservation, recreation and education.
- **Beauty spots or sites of national interest**: These sites, which are often spectacular, include gorges, canyons, waterfalls and caves. They are protected in the same way as historic monuments, but are accessible to the public. In Wales, the Brecon Beacons National Park is an example.

- **Managed nature reserves and sanctuaries**: These reserves are designed to protect a species or a community. They include forestry, game and fishing reserves. In such cases, the species must be managed to ensure that it is preserved. An example in North Wales is the Snowdonia National Park, with its mountains and moorland. It has particularly interesting plants that are maintained by grazing sheep, without which the environment would rapidly change.
- **Man-made landscapes and protected landscapes**: Landscapes that have been shaped by agricultural activities are destined to disappear once their economic value ends. Protecting such environments involves maintaining or reviving human activity. The regional national parks of Europe are good examples. In South Wales, the 202 ha (hectare) Pembrey Country Park is an excellent example of a man-made, protected landscape combining dunes and sandy beaches.

Seed banks and conservation

Seed banks are collections of stored seeds that aim to ensure that people in the future will be able to grow the varieties of plants that exist today. Seed banks are needed because of the possibility that useful plant types might become extinct. Some scientists suggest that as many as 34 000 plant species are threatened with extinction throughout the world. Given the correct conditions, it is possible to remove most of the water from many types of seed, and keep them at low temperatures so that they remain alive for many years. Disease-resistant types of cereal crop and those that can grow in extremes of climate can be preserved as seeds in this way.

Did you know?

In 1877, scientists in Austria sealed some wheat seeds in glass containers. In 1987, these seeds were exposed to air, water and a suitable temperature. As a result, they germinated. This proves that some seeds can remain alive for at least 110 years!

Seed banks were first set up in 1958 in the USA. The number increased each year, and they are now found throughout the world. In 1997, the Millennium Seed Bank Project was set up at the Royal Botanic Gardens in Kew, London. The main aim was to conserve, through international partnerships, over 24 000 species by 2010. Over 16 countries co-operate in this partnership. By 2002, the Food and Agriculture Organisation of the United Nations (FAO) had recorded over 19 000 species stored in over 400 seed banks.

Intensive food production and the environment

On a worldwide scale, agriculture is concentrating on a limited range of varieties of crops that are selectively bred for the needs of humans. The result is **monoculture**, uniform areas of the same variety grown on a large scale. Monocultures have both advantages and disadvantages.

The main advantage is that any genetically uniform crop has characteristics that the farmer can rely on, for example the degree of resistance to disease, the predicted yield, and the predicted need for fertiliser. However, a serious disadvantage is that a monoculture is about as unnatural an ecosystem as you can find, and it can be wiped out by a single pest or disease. Without the vast numbers of different species found in a natural environment, there are no useful animals to control the unwanted pests, and there is no recycling of essential nutrients (see the section on the carbon and nitrogen cycles earlier in Chapter 4). The danger of total

Figure 4.23 An engraving showing the Irish potato famine

devastation by a single pest was seen as long ago as 1844 in Ireland, when potato blight was responsible for the infamous famine that was the cause of one of the largest migrations of a population that Europe has ever seen (see Figure 4.23). At that time, the major part of the Irish population depended totally on one type of potato.

When monocultured crops are harvested, they take with them all the food they have produced and all the minerals they have absorbed. In order to maintain the fertility of the soil, farmers have to add artificial fertilisers containing nitrates and phosphates. These minerals can be washed or leached into waterways and give rise to pollution (see the section on water pollution earlier in Chapter 4). Many pesticides and herbicides can also cause environmental pollution by building up in food chains and causing havoc among animals and plants that normally live near farms.

As a long-term strategy, plant breeders must keep genetic variation so that new varieties are available if older ones die out owing to changes in the environment, such as climate or disease. Future effects of global warming may well demand new varieties, and so breeders are making great efforts to conserve older varieties so that their genes do not become extinct.

Varieties of the major crops of the world are kept in gene banks founded as long ago as 1971 by the Food and Agriculture Organisation of the United Nations. Today, increasing use is made of genetic engineering (see the section on genetic engineering in Chapter 1) to save rare genes. Theoretically, any gene can be introduced into crops so that future crops can have extra genes to control the characteristics that farmers require.

Did you know?

Australian genetic engineers have produced a variety of cotton that kills caterpillars that are threatening to wipe out the Australian cotton industry. They have introduced a gene for controlling the production of a natural insecticide, normally produced by a type of bacterium that lives in the soil. This could save many millions of dollars a year that would otherwise be spent on pesticides, and could also reduce the associated problems of pollution by pesticides being built up in food chains.

Increasing the yields of crops

Theoretically, it may seem easy to increase the yield of crops with modern techniques of selective breeding, genetic engineering and the use of pesticides and of fertiliser. The aim is to make the crop grow faster, taste better and be more resistant to climate changes and disease. Plant breeders select the best individual plants from many hundreds or even thousands growing under experimental conditions. Then they crossbreed individuals containing at least some of the desired characters. This is still a gamble, because they rely on chance to combine the best characteristics in the offspring. The chance of success increases with the number of crosses made.

Wheat has many wild grasses as relatives. Many of these have genes that control resistance to fungus disease. Therefore breeders cross the wild grass with the wheat to obtain the desired gene for fungus resistance. This is not as easy as it might seem, because the offspring may have many undesirable features of wild grass, for example low yield. The problem is solved by taking the first generation (**F1**) and crossbreeding it with the original wheat, keeping only those plants in the **F2** that have the resistance gene. By repeating such backcrossing, breeders gradually eliminate the contribution of the wild grass in all but those characters that are required, in this case, resistance to fungi.

Did you know?

In 1998, the UK's farmers grew more than 100 kg of potatoes for every person living here, over 6 700 000 t. Therefore they must be important! In fact, they provide a large part of our vitamin C and as much energy per hectare as any cereal crop.

Similar selective breeding has been carried out for many hundreds of years, and has resulted in the varieties of vegetables, fruits and flowers of today's markets. New techniques are now used to enable breeders to introduce genes from any organism into crop plants through genetic engineering (see the section on genetic engineering in Chapter 1). When this gene technology is combined with tissue culture on a grand scale, the result is the mass production of genetically identical plants.

This is an example of cloning. It is especially useful in commercially valuable crops that do not normally reproduce asexually. Coconuts and palm-oil plants form the basis of the economy of some countries, and are reproduced in this way. Pieces of plant tissue are kept alive in a culture solution that contains everything needed for natural plant growth. Some plants that produce natural drugs to cure some diseases can be cultured in this way, for example the cinchona tree, which makes quinine used to treat malaria.

Intensive farming of animals

Domesticated animals are improved in ways similar to the methods used for the increase of yields of crops. Farmers select individuals that have characteristics that they want, for example high meat and milk production in cattle, good egg production in chickens, and good wool and meat production in sheep. When they cross two animals that show desirable characters, they often see a combination of characters that they want in the offspring. The enormous varieties of domesticated animal bred for particular characteristics have been produced by this type of selective breeding over thousands of years. Today there are new technologies that speed up the process of selective breeding. They include the following:

- **Artificial insemination (AI)**: Sperms from a male animal that shows characteristics that are required by the farmer are artificially introduced to the female's reproductive system to fertilise an egg. Many females can be inseminated in this way using sperms of the same male.
- ***In vitro* fertilisation (IVF)**: Eggs are extracted from a female and fertilised in a Petri dish by sperms obtained from a male (see Figure 4.24). The fertilised eggs are then placed inside the uterus of the female.
- **Genetic engineering**: New DNA is introduced into young embryos produced by IVF. In this way, sheep have been given human DNA and produced a clotting factor in their milk that helps haemophiliacs. Also, animals have been given genes to produce more growth hormone so that milk production is increased.
- **Cloning**: This is an artificial way of producing twins. Young embryos have their cells separated and each cell can grow into an individual. Many genetically identical individual clones can then be produced once they have been transferred into females.

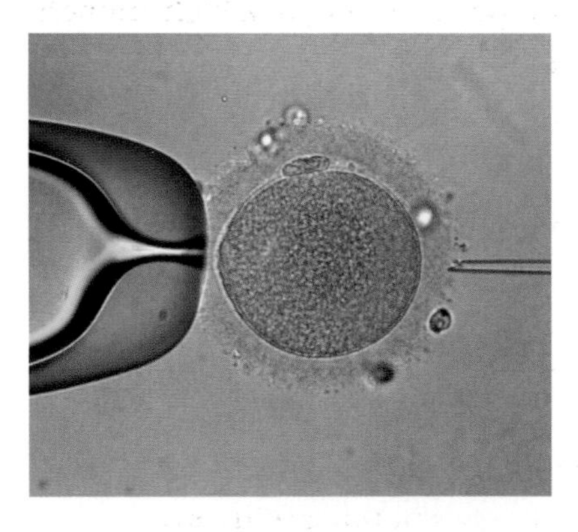

Figure 4.24 A needle about to inject DNA of a sperm into a human egg during the process of IVF

A duty of care for animal welfare

The figures for sales of animal products suggest that consumers want cheap but high-quality meat, eggs and milk. Most people want animals to be kept well, but many are unwilling to accept that good welfare costs money and that this extra cost will increase the price of the product. Farmers are under pressure to produce animal products cheaply, so many people question whether animal welfare is being put at risk.

Figure 4.25 Chickens are often reared intensively in battery farms.

Activity

Use the internet to look up further examples of the ideas in this section. For example, look up key words such as 'genetic manipulation', 'animal welfare', 'water pollution', 'oil pollution' and 'recycling'. Visit the websites of environmental pressure groups and UK Government environmental agencies.

Activity

Look at a multimedia encyclopaedia or in the *Encyclopedia to the Environment: Biosphere* about the environmental effects of human activities.

Hens in battery cages cannot dust-bath or scratch the ground as they do when they are in farmyards (see Figure 4.25). Are they suffering? Pigs confined to stalls with solid floors cannot root about in soil as they do naturally. It has been suggested that gnawing the confining bars is a sign of frustration or distress. However, proving in a court of law that animals suffer by being confined can be difficult.

Did you know?

Austrian scientists have developed an animal feed that could be used instead of antibiotics for factory-farmed livestock. Animals kept in confined conditions easily catch diseases caused by bacteria. Antibiotics are often given to control these diseases in pigs and chickens kept in batteries. This practice has been criticised because the antibiotics could cause bacteria to be resistant to them, and the resistant bacteria might be transferred to humans. In Austria, scientists have developed a milk-based feed that has been successful in keeping factory-farmed livestock free from bacterial disease. The feed works by controlling the growth of disease-causing organisms in the digestive tract. The feed contains a culture of bacteria that occur naturally in the digestive tract of calves. The bacteria help digestion and control the reproduction of harmful bacteria. The cultures are grown in ordinary cow's milk and dried at a temperature that does not destroy the bacteria. The powder can then be fed to the animals.

Battery-farmed animals in the UK are reared in a carefully controlled environment so that balanced diets can be given, temperature controlled, pests, predators and disease eliminated, and the minimum legally agreed space provided for individuals. In this way, the yields of hen's eggs and pork are kept at a high enough level to meet the demands of the public.

The UK Farm Animal Welfare Council (FAWC) states that good animal management systems should provide five freedoms:

- freedom from hunger and thirst: fresh water and a balanced diet
- freedom from discomfort: shelter and a resting area
- freedom from pain, injury and disease: prevention or rapid diagnosis of disease
- freedom to express normal behaviour: provision of sufficient space and the company of others of the same species
- freedom from fear and distress: treatment that avoids any suffering.

The UK Animal Health Act of 1981 safeguards the welfare of animals in markets and while they are being transported within the UK. The UK Slaughterhouse Act of 1974 specifies humane methods of slaughter.

Control of disease

In recent years there has been intense public interest in the spread of certain diseases between and within species of domestic animals. One such disease is **bovine tuberculosis (TB)**. The disease affects cattle and is caused by a bacterium. It can be easily spread by contact between infected cattle and so herds are regularly checked.

If infection is proved then a whole herd of cattle is usually slaughtered. There is concern that the bacterium may be spread to cattle by badgers if they share the same habitat. This has led to controversy over badger culls. Badgers are a protected species, but those who blame them for spreading the bovine TB bacterium to cattle seek to cull badgers in areas where they conflict with the interests of cattle farmers.

In regions where culling has occurred, some say that there has been a decrease in the disease but conservationists state that in these regions there has been no data collected on the methods and frequency of cattle transported between farms. The controversy will continue until evidence is obtained from valid and objective investigations. Such evidence is essential before a decision is made to cull badgers on a national scale.

Summary

1. Food chains begin with plants and represent the energy flow from one feeding level to the next.
2. Feeding relationships can be shown in the form of pyramids that demonstrate the rule that biomass decreases from the base of the pyramid to the apex because of the energy loss between feeding levels.
3. Most microbes are not harmful to mankind. Some are essential, for example those involved in natural cycles such as the carbon and nitrogen cycles.
4. A stable population density depends on a balance between the birth rate and the death rate. Present trends in human populations suggest that future generations will have difficulty in providing enough food for themselves.
5. The large numbers of people living in limited space, with consequent industrial demands, causes pollution problems.
6. Water is polluted by chemicals, including those produced as by-products of industry, fertilisers and detergents.
7. Some toxic chemicals are not broken down and accumulate in food chains.
8. Water is being used and polluted by the disposal of both biodegradable and non-biodegradable wastes.
9. We are now making a conscious effort to conserve wildlife and their habitats.
10. Meat production is kept at a high level by intensive farming and control of the environments of animals. Humans have a duty to manage food production in such a way that damage to the environment is minimal and animals do not suffer.

Chapter 5 Atomic structure and the Periodic Table, compounds and patterns in reactions

By the end of this chapter you should:

- know that elements are made up of atoms that have a small central nucleus made up of protons and neutrons;
- understand that most of the mass is concentrated in the positive nucleus, which is surrounded by light, negatively charged electrons;
- know that the electrons occupy energy levels (orbits or shells) around the nucleus;
- understand the terms 'mass number' and 'atomic number';
- know the maximum number of electrons that each orbit or energy level can hold;
- know the arrangement of the elements in the Periodic Table of elements;
- understand the formation of compounds;
- understand some chemistry of the elements in Groups 1 and 7 of the Periodic Table;
- know how to write simple chemical equations;
- know how to test for hydrogen and carbon dioxide;
- know how to test for chlorides and iodides;
- understand that elements are the basic building blocks of all substances and cannot be broken into anything simpler by chemical means;
- know that Mendeléev constructed the first periodic table by arranging the elements in order of relative atomic mass and leaving gaps for elements not discovered at that time;
- know that the modern Periodic Table arranges the elements in order of their atomic numbers;
- know that columns in the Periodic Table are called groups and rows are called periods;
- know the chemistry of the elements in two groups in the Periodic Table.

Atomic structure

The structure of atoms

Elements are the basic building blocks of matter. Elements cannot be broken into anything simpler by chemical means. Each element has its own symbol. We now know that elements are made up of **atoms**. The English chemist John Dalton (see Figure 5.1) put forward the first successful atomic theory at the beginning of the nineteenth century. He used it to explain observed facts about the behaviour of substances. This is how science works. Theories explain the observed facts.

Since then our knowledge has steadily grown. We now know that atoms are made up of even simpler particles:

- Each atom contains a small positively charged central region called the **nucleus**.

Figure 5.1 John Dalton

Did you know?

John Dalton suffered from colour blindness, which was sometimes called Daltonism.

Atomic structure:
www.chemguide.co.uk

General chemistry websites:
www.rsc.org
www.bbc.co.uk/schools
chemistry.about.com
www.sciencemuseum.org.uk
www.chemicalelements.com
www.schoolscience.co.uk

- The nucleus contains nearly all the mass of the atom.
- Light, negatively charged **electrons** surround the nucleus (see Figure 5.2) and are attracted to it. (Positive attracts negative.)

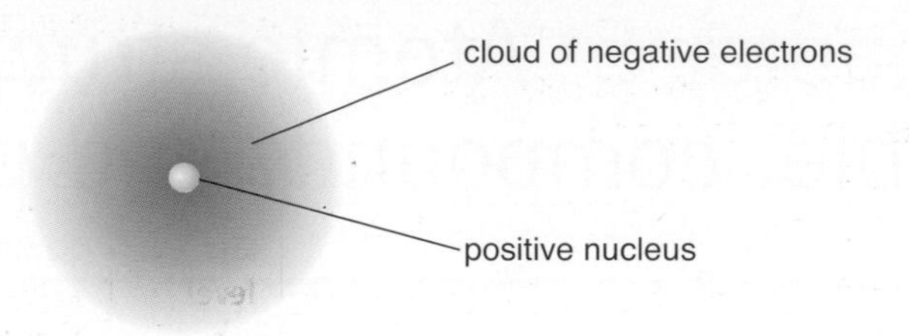

Figure 5.2 Model of the structure of the atom

The nucleus is made up of two types of particle: the **proton** and the **neutron**. The mass of the positive proton is 0.000 000 000 000 000 000 000 001 7 g, which is very small. The mass of a neutron is the same as that of the proton, but it has no electrical charge. For convenience, we can call the mass of a proton 1 unit and its charge +1, and describe the other particles relative to these values.

Table 5.1 Relative masses and charges of fundamental particles

Particle	Relative mass	Relative charge
Proton	1	+1
Neutron	1	0
Electron	Negligible	−1

Protons, neutrons and electrons (see Table 5.1) are called fundamental particles. The number of protons in the nucleus of an atom is called the **atomic number**. Each element has its own atomic number. For example, hydrogen has the atomic number 1. Lithium has the atomic number 3. Chlorine has the atomic number 17.

The number of protons plus the number of neutrons is called the **mass number**. The mass number and atomic number for an atom of sodium, Na, can be written as shown in Figure 5.3.

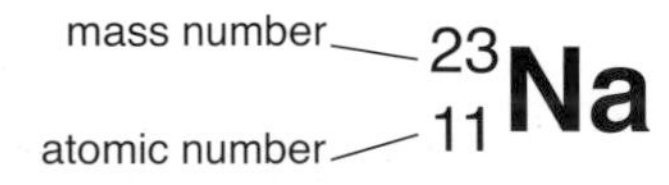

Figure 5.3 Mass number and atomic number

An atom is electrically neutral because **the number of electrons surrounding the nucleus equals the number of protons in the nucleus**. Where the number of electrons does not equal the number of protons, the particle is no longer an atom. It is called an **ion**, and it has an electrical charge.

Questions

1 How many neutrons are there in the nucleus of an atom of $^{23}_{11}Na$?

2 Write down the number of protons, neutrons and electrons in **each** of the following:
  a $^{4}_{2}He$
  b $^{15}_{7}N$
  c $^{27}_{13}Al$
  d $^{39}_{19}K$
  e $^{32}_{16}S$.

The electrons

The electrons around the nucleus of an atom occupy **energy levels** denoted by the symbol *n*, which has the value 1, 2, 3, 4, etc. These levels are sometimes called **orbits**, or **shells**. Each energy level can only hold a certain number of electrons (see Table 5.2).

Table 5.2 Electrons held by energy levels

Energy level *n*	Maximum number of electrons accommodated for elements hydrogen to calcium	The maximum number of electrons that can be accomodated in energy levels (*n* = 1–4)
1	2	2
2	8	8
3	8	18
4	2	32

Example
The element sodium, Na, has an atomic number of 11. This means that it has 11 protons in the nucleus, and so must have 11 electrons surrounding the nucleus. These 11 electrons are arranged as shown in Table 5.3.

Table 5.3 Arrangement of electrons in a sodium atom

Energy level (orbit or shell) *n*	Number of electrons
1	2
2	8
3	1
	11 electrons in all

The way in which electrons are arranged in an atom is known as the **electron configuration**. We sometimes picture this as shown in Figure 5.4, and we can write this electron configuration as 2.8.1.

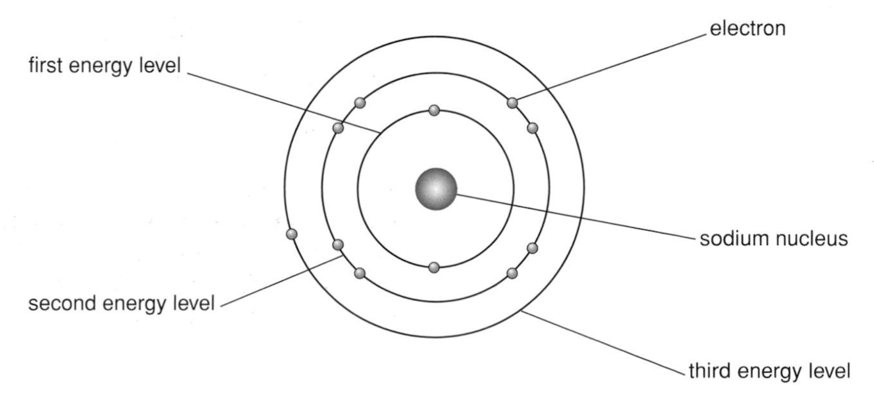

Figure 5.4 Electron configuration of a sodium atom

Table 5.4 Electron configurations for the first 20 elements

		Electron configuration			
Atomic number	**Element**	$n = 1$	$n = 2$	$n = 3$	$n = 4$
1	Hydrogen	1			
2	Helium	2			
3	Lithium	2	1		
4	Beryllium	2	2		
5	Boron	2	3		
6	Carbon	2	4		
7	Nitrogen	2	5		
8	Oxygen	2	6		
9	Fluorine	2	7		
10	Neon	2	8		
11	Sodium	2	8	1	
12	Magnesium	2	8	2	
13	Aluminium	2	8	3	
14	Silicon	2	8	4	
15	Phosphorus	2	8	5	
16	Sulphur	2	8	6	
17	Chlorine	2	8	7	
18	Argon	2	8	8	
19	Potassium	2	8	8	1
20	Calcium	2	8	8	2

The electron configurations for the first 20 elements are given in Table 5.4.

Periodic Table:
www.webelements.com
www.chemicalelements.com
www.chemsoc.org

The Periodic Table of the elements

The first successful Periodic Table of the elements was devised by Dmitri Mendeléev in 1869. Dmitri Ivanovich Mendeléev (1834–1907) was a Russian chemist. Mendeléev arranged the elements in order of increasing atomic weights (relative atomic masses) and discovered that the elements showed similar properties at regular intervals. To make this *periodicity* fit the observed facts he had to leave gaps in his table for elements that had not been discovered at the time, such as germanium, gallium and scandium (see Figure 5.5). Mendeléev predicted the properties of these elements. For example, for the element we now know as germanium he predicted:

- it should have an relative atomic mass of 72 (actually 72.6)
- its density would be 5.5 (actually 5.47)
- that it would form a liquid chloride, XCl_4, which would boil below 100 °C (actually germanium forms $GeCl_4$, which boils at 84 °C).

Series	Group I	Group II	Group III	Group IV	Group V	Group VI	Group VII	Group VIII
1	1 H							
2	Li 2	Be	B	C	N	O	F	
3	3 Na	Mg	Al	Si	P	S	Cl	Fe Co Ni
4	K 4	Ca	?	Ti	V	Cr	Mn	
5	5 Cu	Zn	?	?	As	Se	Br	Ru Rh Pd
6	Rb 6	Sr	?	Zr	Nb	Mo	?	

Figure 5.5 Part of an early form of Mendeléev 's Periodic Table showing how he left gaps for undiscovered elements

The first periodic tables arranged the elements in order of increasing atomic weight. The term atomic weight is no longer in use. Chemists now use the term *relative atomic mass.* When the elements were arranged in order of their atomic numbers, the inconsistencies found in the earlier periodic tables based on atomic weights or relative atomic masses were corrected. Nowadays the elements are arranged in order of their atomic numbers.

A modern Periodic Table is shown in Figure 5.6. (The Periodic Table provided by WJEC to students in its examinations is on p. 309.) Each column is called a **group**. Each horizontal row of elements is called a **period**. The first energy level can hold only up to two electrons, and so hydrogen and helium are usually placed as a separate period of two elements.

Activity

Enter 'Periodic Table' in your search engine and find a variety of interactive versions of the Periodic Table.

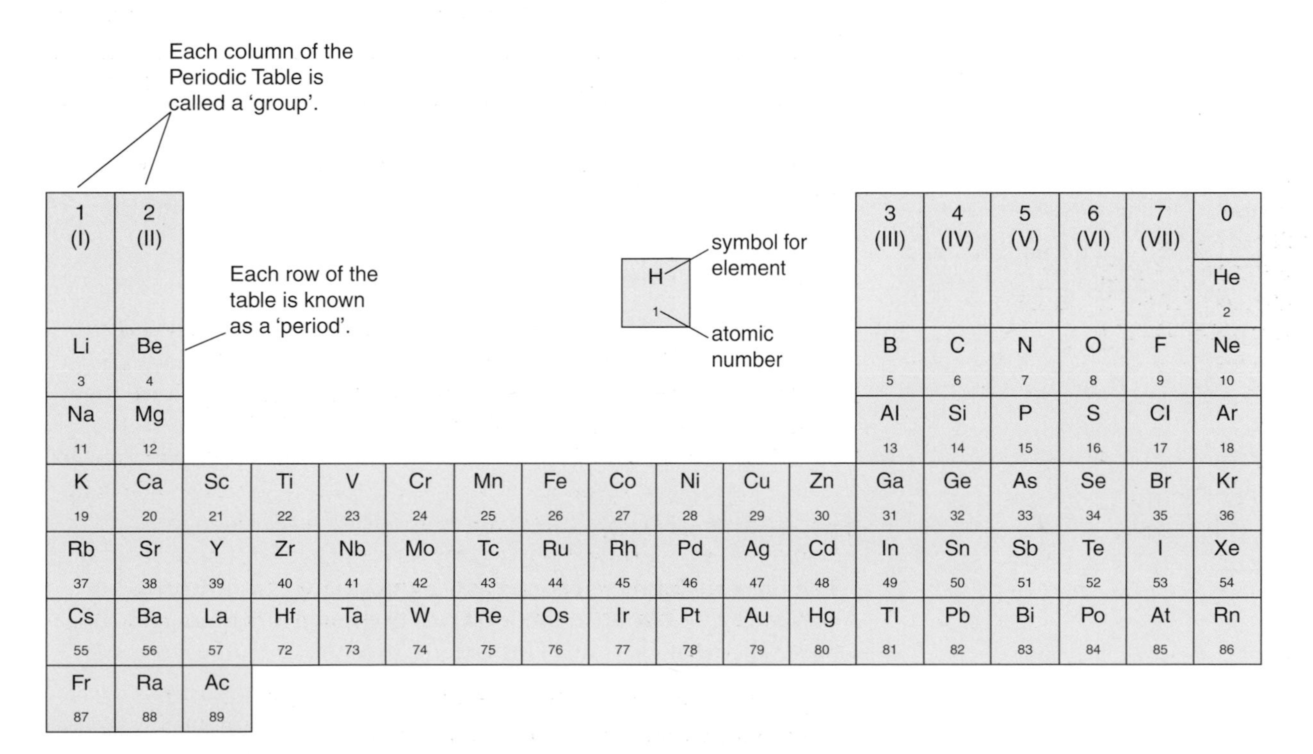

1 (I)	2 (II)											3 (III)	4 (IV)	5 (V)	6 (VI)	7 (VII)	0
																	He 2
Li 3	Be 4											B 5	C 6	N 7	O 8	F 9	Ne 10
Na 11	Mg 12											Al 13	Si 14	P 15	S 16	Cl 17	Ar 18
K 19	Ca 20	Sc 21	Ti 22	V 23	Cr 24	Mn 25	Fe 26	Co 27	Ni 28	Cu 29	Zn 30	Ga 31	Ge 32	As 33	Se 34	Br 35	Kr 36
Rb 37	Sr 38	Y 39	Zr 40	Nb 41	Mo 42	Tc 43	Ru 44	Rh 45	Pd 46	Ag 47	Cd 48	In 49	Sn 50	Sb 51	Te 52	I 53	Xe 54
Cs 55	Ba 56	La 57	Hf 72	Ta 73	W 74	Re 75	Os 76	Ir 77	Pt 78	Au 79	Hg 80	Tl 81	Pb 82	Bi 83	Po 84	At 85	Rn 86
Fr 87	Ra 88	Ac 89															

Figure 5.6 The modern Periodic Table, showing symbols and atomic numbers (elements 58–71, the lanthanides or rare earth elements, and elements with atomic number greater than 89 have been omitted for simplicity).

Some periodic tables number the groups 1, 2, 3, 4, etc. In others, the groups are labelled with Roman numerals: I, II, III, IV, etc.

Apart from Group 0, the noble gases, the number of the group shows the number of electrons in the outer energy level. In Group 0, helium has two outer electrons and all the other noble gases have eight outer electrons (an octet).

The Periodic Table shows that the properties of the chemical elements depend on their atomic numbers. The **elements in a group** have similar chemical properties and their physical properties gradually change down the group. For instance, **metallic character**:

- increases down a group
- decreases across a period.

A good example is Group 4, which moves from the non-metal carbon at the top to the metal lead at the bottom. In the period Na–Ar, sodium is a metal but chlorine is a non-metal. Note that most of the elements are metals; all non-metals occur on the right of the table.

Metals have the following properties:

- good conductors of electricity (**high electrical conductivity**)
- good conductors of heat (**high thermal conductivity**)
- **malleable** (can be beaten into sheets)
- **ductile** (can be drawn in wires)
- **lustrous** (freshly exposed surfaces are shiny)
- usually high densities
- high melting points
- high boiling points.

Non-metals are brittle, dull and of low density. Some elements lie between the two, for example silicon and germanium are **semi-conductors** and are very important in electronics. The **graphite** form of carbon has all the characteristics of a non-metal except that it is a good conductor of electricity.

Did you know?

Germanium in Group 4 was used in the first successful transistors in electronic circuits. Silicon is the basis of silicon chips, which are the heart of modern computers.

Some groups in the Periodic Table have special names:

- The Group 1 elements are called the **alkali metals**.
- The Group 7 elements are called the **halogens** (which means 'salt-formers').
- The Group 0 elements are called the **noble gases**.

All the elements in a group have similar chemical and physical properties, and there is a gradual change in properties as we go down the group. The noble gases are very unreactive because they have a complete first energy level. They all have eight outer electrons, except for helium which has two outer electrons.

Reactivity within a group

The reactivity of the elements in a group changes as the group is descended. In Group 1, the least reactive metal is lithium and the most reactive is caesium. This is demonstrated by the reactions of the elements with air and with water.

Rubidium and caesium, which are not available in school laboratories, react extremely violently with both air and water. Samples of lithium, sodium and potassium react slowly with air and can be demonstrated to react with water with increasing vigour from lithium to potassium. In Group 7, the most reactive element is fluorine and the least reactive is iodine (astatine is radioactive and available in only small quantities).

These trends can be explained in terms of the size of the atoms. In Group 1, the outer electron gets further away from the positive nucleus as the group is descended, and as a result the outer electron is lost more easily. This means that it is easier for caesium to form a positive ion than for the elements above it. In the same way, in Group 7, the outer electrons become further away from the positive nucleus as the group is descended, so that iodine forms the iodide ion less easily than fluorine forms the fluoride ion by gaining an electron.

Did you know?

If the distance between a negative and a positive charge is doubled then the force of attraction is only one-quarter of its original value.

Group 1: The alkali metals

The alkali metals of Group 1 are lithium, sodium, potassium, rubidium and caesium. They are all very reactive soft metals that react with air. This means that they must be stored carefully in a non-reactive oil to prevent this from happening (see Figure 5.7). The Group 1 metals also react with halogens such as chlorine gas (see Figure 5.13). Sodium forms sodium chloride. The test for a chloride ion is given on page 118.

Figure 5.7 Sodium metal stored in oil

Sodium can easily be cut with a knife to reveal a shiny surface that soon tarnishes as it begins to react with the air.

sodium + oxygen → sodium oxide

$4Na(s) + O_2(g) \rightarrow 2Na_2O(s)$

H

The reaction does not stop there. The oxide reacts to form sodium hydroxide that absorbs water and carbon dioxide from the air, and after some time a white solid of sodium carbonate forms.

Practical work

Reactions of Group 1 elements with water

The reactivity of the alkali metals increases in vigour from lithium to potassium. The elements rubidium and caesium are not available in school laboratories and their reaction with water is explosive.

These experiments are not to be undertaken by students. In demonstrations of the reactions of metals and water, there must be a safety screen between the demonstration and the students. The students must wear goggles.

Demonstrations

1 **Lithium** reacts relatively quietly with water, giving off bubbles of gas. The gas is hydrogen.
2 When a small piece of **sodium** is added to water, it floats and starts to melt, moving across the water rapidly and giving off hydrogen gas that ignites with a yellow flame (see Figure 5.8). The sodium finally disappears. The resulting solution contains sodium hydroxide.

   sodium + water → sodium hydroxide + hydrogen

   $2Na(s) + 2H_2O(l) \rightarrow 2NaOH(aq) + H_2(g)$
3 The reaction of **potassium** with water is more violent than that of sodium. The hydrogen liberated by the potassium immediately ignites

Figure 5.8 Sodium reacting with water

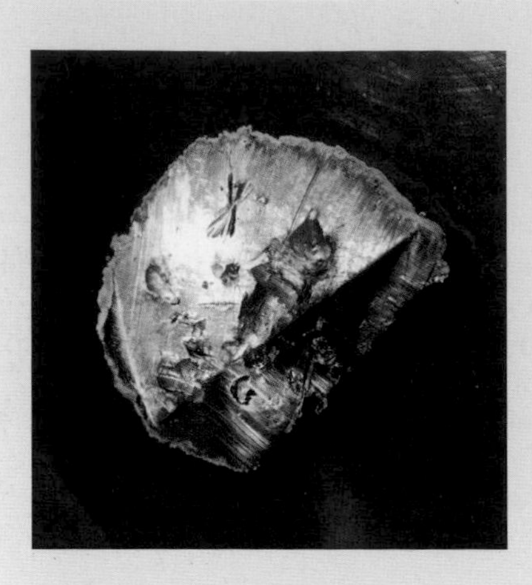

Figure 5.9 Metallic potassium just removed from its oil and cut with a knife

with a lilac flame (see Figure 5.9). Molten potassium may splutter around the reaction.

Table 5.5 shows data for the Group 1 elements.

Table 5.5 The melting points and boiling points of the Group I elements

	Melting point (°C)	Boiling point (°C)	Density (g/cm^3)
Li	181	1326	0.53
Na	98	883	0.97
K	63	760	0.86
Rb	39	686	1.53
Cs	29	669	1.88

Group 1 – The Alkali Metals, Visual Elements:
www.chemsoc.org

Alkali metal:
en.wikipedia.org

Question

3 Write word and symbol equations for the reactions of lithium and potassium with water.

Did you know?

Platinum wire is very expensive, but wire made from the alloy nichrome is much cheaper and is disposable. Nichrome wire is used in electrical elements.

Practical work

Carry out the flame test on compounds of:

1 lithium
2 copper
3 strontium
4 calcium.

Practical work

The flame test

A number of elements emit coloured light when heated in a Bunsen burner flame. There are two methods of carrying out flame tests.

Method 1

1 Wear goggles.
2 Prepare a small amount of the chloride of the metal you are testing, moistened with a drop or two of concentrated hydrochloric acid.
3 Take a piece of platinum or nichrome wire, and dip the end into the chloride of the metal.
4 Hold the wire in a roaring Bunsen burner flame (see Figure 5.10).
5 Look for a distinct colour in the flame. Record what it is.

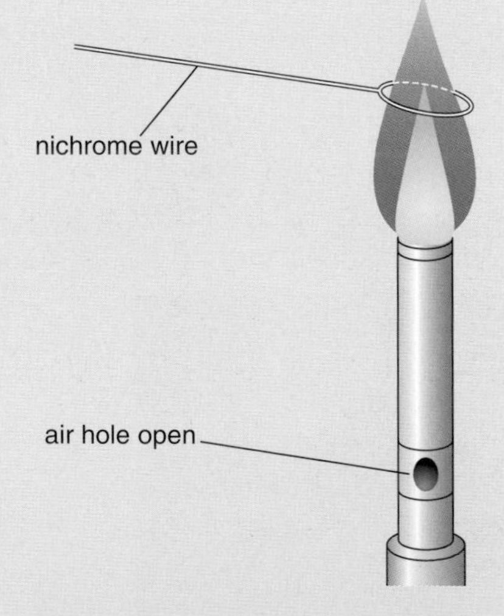

Figure 5.10 The flame test

Method 2

1 Wear goggles.
2 Spray fine droplets of an aqueous solution of the chloride of the metal you are testing into the roaring Bunsen burner flame using an atomiser. Spray the solution away from yourself.
3 Look for a distinct colour in the flame. Record what it is.

Try the above experiments with sodium chloride and potassium chloride. You should observe the following.

- Sodium compounds have a **golden-yellow** (almost orange) colour.
- Potassium compounds have a distinct **lilac** colour.

Figure 5.11 Samples of (a) chlorine, (b) bromine and (c) iodine. **Volatility** is how readily elements turn to vapour. Note the volatility of bromine and iodine. Although bromine is a liquid, we can see a lot of bromine vapour above the liquid.

Group 7: The halogens

The word halogen means *salt-former*. The elements of Group 7 are non-metals (see Figure 5.11).

Bromine and iodine vaporise very easily. They are said to be **volatile**. The vapours are poisonous. Notice how the physical properties change as the group is descended (see Table 5.6).

Activity

Predict the physical properties of astatine.

Table 5.6 Halogens

	Physical state at room temperature	Colour of vapour
Chlorine	Yellow-green gas	Yellow-green
Bromine	Red-brown liquid	Red-brown
Iodine	Grey shiny solid	Purple

Did you know?

Bromine used to be manufactured from seawater at Amlwch in Anglesey at the Octel works. Production stopped in the early 2000s when the US owners, The Great Lakes Chemical Corporation, closed the plant, relying on the Dead Sea as its source of bromine. Dow Corning is another example of a chemical works and is situated in the Vale of Glamorgan.

Figure 5.12 Dow Corning chemical works, Vale of Glamorgan

The Group 7 elements are very reactive, with the chemical reactivity decreasing down the group. Chlorine reacts readily with many metals.

A reaction with sodium can be carried out in a gas jar of chlorine in a fume cupboard (see Figure 5.13 left on page 110).

sodium + chlorine → sodium chloride

$$2Na(s) + Cl_2(g) \rightarrow 2NaCl(s)$$

With iron, the product is iron(III) chloride, which forms in the tube as a blackish solid which turns brown if left in moist air (see Figure 5.13 right).

H

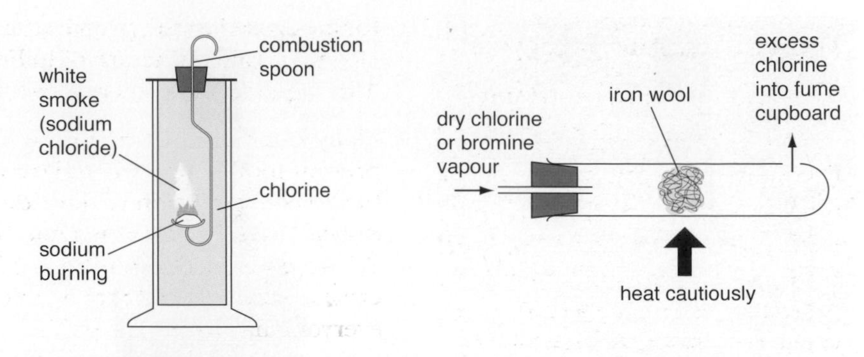

Figure 5.13 Reacting metals with chlorine

Displacement reactions

A more reactive halogen can displace a less reactive element from a solution of its salts. Chlorine, the most reactive halogen, will displace both bromine and iodine from solutions of bromides and iodides.

chlorine + sodium bromide solution → bromine + sodium chloride solution

$$Cl_2(g) + 2NaBr(aq) \rightarrow Br_2(l) + 2NaCl(aq)$$

chlorine + sodium iodide solution → iodine + sodium chloride solution

$$Cl_2(g) + 2NaI(aq) \rightarrow I_2(s) + 2NaCl(aq)$$

H

In the same way, bromine, which is more reactive than iodine, displaces iodine from iodide solutions.

bromine + sodium iodide solution → iodine + sodium bromide solution

$$Br_2(l) + 2NaI(aq) \rightarrow I_2(s) + 2NaBr(aq)$$

H

Halogens:
en.wikipedia.org

Practical work

Halogen displacement reactions

1. Wear safety spectacles. Do not attempt to smell the contents of the test tubes. Wash the products down the sink after you have made the observations. Pupils with asthma should use a fume cupboard if the contents are heated.
2. Place some dilute sodium or potassium bromide solution in a test tube, and add chlorine water drop by drop, shaking the tube gently.
3. Notice the development of the brownish colour of bromine.
4. Repeat the experiment using dilute sodium or potassium iodide solution.
5. Notice the development of a yellow-brown colour as iodine is formed.
6. If you add too much chlorine water, you may see a grey-black precipitate of iodine form.
7. If you warm the contents of the tube gently, you will see the purple colour of iodine vapour.

Uses of halides

Chlorine is a major industrial chemical. It is used in the manufacture of hydrochloric acid, bleach, pesticides, PVC (poly(chloroethene)), and sterilisation of the water supply. These uses depend on chlorine being reactive and being a good oxidising agent.

Did you know?

Some water authorities are replacing chlorine as the disinfecting agent in public water with a chemical compound called chloramine, $ClNH_2$. Welsh Water has started adding chloramine in the Anglesey area. In an upgrade of the Alaw and Cefni Water Treatment Works, new equipment has been installed at the plant to add chloramine to the treated water.

Iodine is used in the preparation of antiseptics. Iodine dissolved in alcohol is called tincture of iodine and it is used as an antiseptic. Although iodine is less reactive than chlorine it can still kill bacteria.

Many water authorities add fluoride ions to drinking water to prevent tooth decay (dental caries). However, it is known that too high a concentration of fluoride in drinking water can cause tooth decay. Therefore there is some controversy about the fluoridation of drinking water. The addition of fluorine to the water supply is an ethical matter since there is no choice available to the individual – everyone has to use the public water supply.

In some areas surveys have been carried out on the frequency of dental caries in a specific age group, such as five-year-olds. These data have been used as a baseline to measure the effectiveness of the addition of fluorine to the water supply. In 2002, there was a call from doctors and dentists for the National Assembly for Wales to force Welsh Water to add fluoride to its water. The *Western Mail* reported in April 2005 that the Assembly had no plans for adding fluoride to drinking water. Research (National Statistics, 2003) has shown that Welsh children have a higher incidence of tooth decay than children in the English Midlands where fluoride is added. The introduction of fluoridation in Anglesey was claimed to produce a dramatic reduction in dental decay. However the addition of fluorine remains controversial. Too high a concentration of fluoride (even that sometimes found in natural waters) can harm teeth. Fluoride has been linked to bone cancer and other illnesses. In a paper published in June 2004, the British Medical Association supported fluoridation of mains water supplies in England and Wales. It could find no convincing evidence of any adverse risk to human health through water fluoridation.

Fluoridation:
www.bfsweb.org

Group 0: The noble gases

The noble gases are unreactive elements. Molecules of the noble gases are single atoms, unlike the molecules of, say, oxygen, which have two atoms (O_2). They occur in the atmosphere in very small amounts, argon being the most abundant.

Noble gases:
en.wikipedia.org

Helium is used in weather balloons. It is not quite as efficient in balloons as hydrogen, but is safer since hydrogen can form explosive mixtures with air and is flammable. It is also sometimes used, mixed with oxygen, in underwater breathing apparatus.

Neon is used in neon lights in advertising and art displays.

Argon is used to fill light bulbs. Since it is unreactive, it does not react with the metal filament in the bulb. In some welding processes, it is necessary to have an inert atmosphere to protect the metal from oxidation, and argon is suitable for this.

Chemical compounds and formulae

Compounds are formed when two or more elements combine together. A chemical compound can be represented by a formula that shows the ratio of the atoms of the elements it contains. Well known compounds and their formulae are:

- water: H_2O
- carbon dioxide: CO_2
- sodium chloride: NaCl.

Sodium chloride is different from the other two in that it contains ions. Ions are electrically charged particles.

Table 5.7 Formulae of some positive ions

Ion	Formula
Group 1	
Lithium	Li^+
Sodium	Na^+
Potassium	K^+
Group 2	
Magnesium	Mg^{2+}
Calcium	Ca^{2+}
Strontium	Sr^{2+}

Table 5.8 Formulae of some negative ions

Ion	Formula
Group 6	
Oxide	O^{2-}
Group 7	
Fluoride	F^-
Chloride	Cl^-
Bromide	Br^-
Iodide	I^-

Writing the formulae of simple ionic compounds

The formulae of simple ionic compounds can be written using the formulae of ions given in Tables 5.7 and 5.8. Compounds are neutral, so the number of positive charges must balance the number of negative charges.

Sodium chloride is easy. The sodium ion is Na^+ and the chloride ion is Cl^-. The formula is $(Na^+)(Cl^-)$, which we usually write as **NaCl**.

The calcium ion is Ca^{2+} and the chloride ion is Cl^-. The formula for **calcium chloride** is $(Ca^{2+})(2Cl^-)$, which we usually write as **$CaCl_2$**.

Question

4 Complete this table.

Compound	Formula of positive ion	Formula of negative ion	Formula of compound
Calcium bromide	Ca^{2+}	Br^-	$CaBr_2$
Sodium oxide			
Magnesium bromide			
Potassium chloride			
Calcium oxide			
Sodium iodide			
Potassium iodide			

Question

5 Write down the formulae of the following:
- sodium hydroxide
- calcium hydroxide
- iron(III) oxide
- barium chloride
- copper(II) sulphate
- ammonium sulphate
- magnesium sulphate
- sodium carbonate
- aluminium oxide
- sodium phosphate.

Hint: Don't forget how to use brackets.

Writing the formulae of other ionic compounds

H

The formulae of other compounds can be written from the formulae of their ions. The formulae of some common ions will be given in the examinations, as shown in Tables 5.9 and 5.10.

Table 5.9 Formulae for positive ions

Charge: +1		Charge: +2		Charge: +3	
Sodium	Na^+	Magnesium	Mg^{2+}	Aluminium	Al^{3+}
Potassium	K^+	Calcium	Ca^{2+}	Iron(III)	Fe^{3+}
Lithium	Li^+	Barium	Ba^{2+}	Chromium(III)	Cr^{3+}
Ammonium	NH_4^+	Copper(II)	Cu^{2+}		
Silver	Ag^+	Lead(II)	Pb^{2+}		
		Iron(II)	Fe^{2+}		

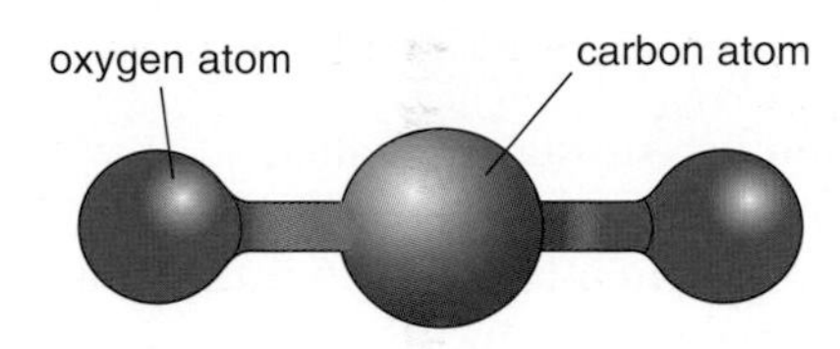

Figure 5.14 A model of the carbon dioxide molecule

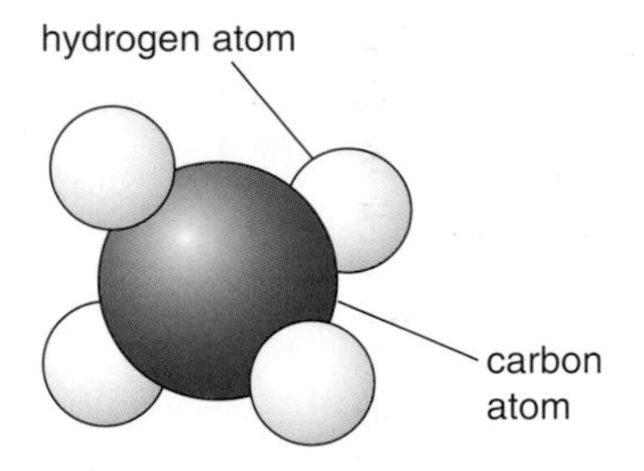

Figure 5.15 Model of a methane molecule

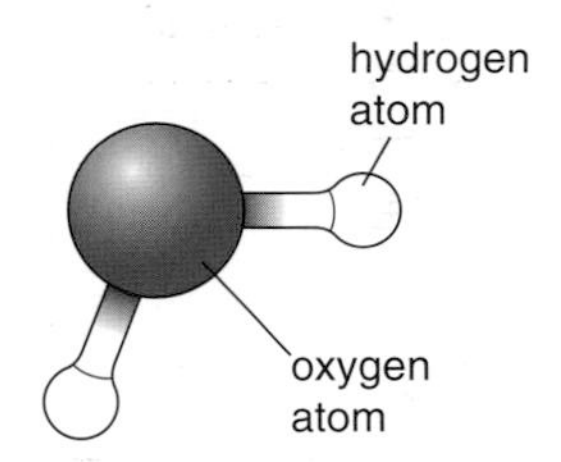

Figure 5.16 Model of a water molecule

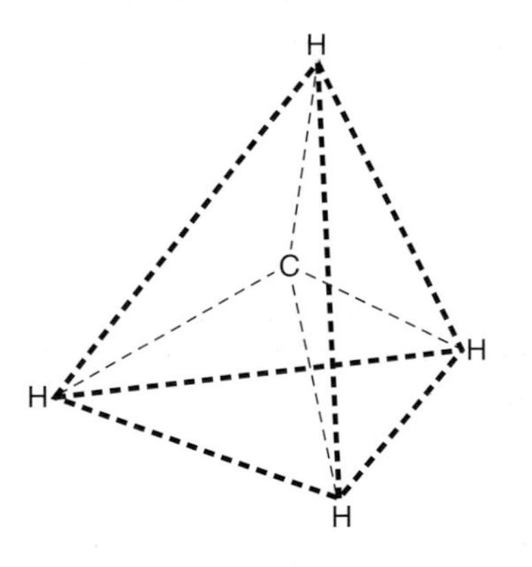

Figure 5.18 The tetrahedral methane molecule

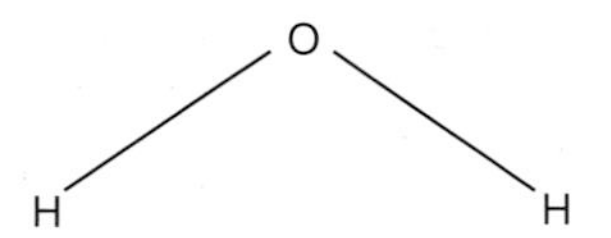

Figure 5.19 Water

Table 5.10 Formulae for negative ions

Charge: –1		Charge: –2		Charge: –3	
Chloride	Cl^-	Oxide	O^{2-}	Phosphate	PO_4^{3-}
Bromide	Br^-	Sulphate	SO_4^{2-}		
Iodide	I^-	Carbonate	CO_3^{2-}		
Hydroxide	OH^-				
Nitrate	NO_3^-				

Example

Find the formula of calcium nitrate. The calcium ion is Ca^{2+} and the nitrate ion is NO_3^-. The formula is $(Ca^{2+})(2NO_3^-)$, which we usually write as $\mathbf{Ca(NO_3)_2}$. **Notice how the brackets are used: $\mathbf{Ca(NO_3)_2}$, *not* $\mathbf{CaNO_{32}}$.**

Molecules

Not all compounds are ionic. Many compounds exist as atoms joined together to form a neutral particle called a **molecule**. Common molecules are carbon dioxide (CO_2) (see Figure 5.14), ammonia (NH_3), methane (CH_4) (see Figure 5.15) and water (H_2O) (see Figure 5.16).

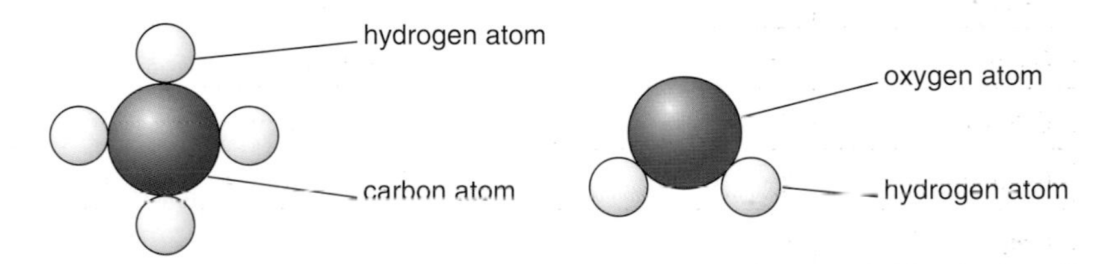

Figure 5.17 Space filler diagrams of methane and water

These molecules can be represented very simply as space filler diagrams (see Figure 5.17). In fact, the molecules exist as three-dimensional structures. You can see this best by making the molecules using model kits (see Figures 5.14 and 5.16).

Methane is a central carbon atom surrounded by four hydrogen atoms at the corners of a tetrahedron (see Figure 5.18).

Water is planar and can be drawn as shown in Figure 5.19.

Carbon dioxide is linear (all three atoms are in a straight line). It can be written as O = C = O.

Did you know?

The reaction between concentrated sulphuric acid and water is very exothermic. The water can react violently and dangerously with concentrated sulphuric acid. When you are diluting concentrated sulphuric acid, **always add the acid *slowly* to the water, stirring carefully.**

A **mnemonic** is a remembering aid. Here is one: 'Always do as you oughta.

Add the acid to the water.'

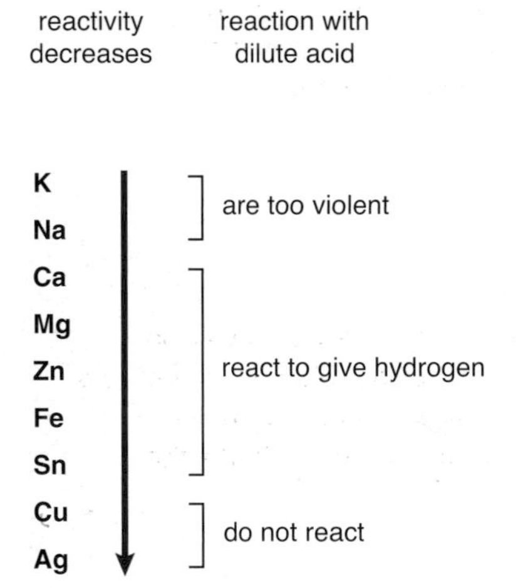

Figure 5.20 The reactivity series

Did you know?

Dilute acids and alkalis are often called 'bench reagents'. The word *dilute* means that the reagent has been prepared with the correct concentration for normal laboratory practical work.

Always read the label on the bottle carefully so you know you have the correct chemical. Eye protection must be worn. Never use a bottle labelled concentrated when you should use dilute.

Chemical reactions

A chemical reaction takes place when two substances called **reactants** undergo change to form new substances called **products**. Sometimes a single reactant will change when heated. An indication that a chemical reaction has taken place may include:

- a colour change
- a gas given off
- a precipitate formed
- a change in temperature.

The formation of products in a chemical reaction involves a rearrangement of the atoms but no atoms are created or destroyed.

If the reaction involves a rise in temperature, it is said to be an **exothermic reaction**. The most common exothermic reaction is burning a fuel, for example burning coal on a fire or glucose in the body. If the reaction involves a fall in temperature, it is said to be an **endothermic reaction**.

There are often similarities in a series of reactions. Dilute acids react with carbonates to give off carbon dioxide. This is useful in identifying some rocks. If a sample of a rock fizzes when dilute hydrochloric acid is added, the rock is probably a carbonate. Calcium carbonate occurs naturally as the rocks limestone, chalk and coral.

When dilute acids are added to alkalis, there is a rise in temperature as the acid neutralises the alkali. A common alkali is sodium hydroxide solution. Reactive metals always give hydrogen gas with dilute hydrochloric acid and dilute sulphuric acid (see Figure 5.20).

Practical work

Chemical reactions

1. Wear goggles.
2. Place a small amount of calcium carbonate in a test tube. Add dilute hydrochloric acid.
3. Repeat the experiment with dilute sulphuric acid and dilute nitric acid.
4. Repeat the experiments with a small amount of magnesium carbonate.

1. Place 25 cm^3 of dilute sodium hydroxide in a polystyrene cup.
2. Take its temperature with a thermometer. Now add 25 cm^3 of dilute hydrochloric acid. Stir, and then retake the temperature.
3. Repeat with dilute sulphuric acid and sodium hydroxide.

Equations

A chemical equation is a summary of a chemical reaction. **A chemical equation can be written only if the reactants and products are known.** The simplest type of equation is a word equation.

Example

When dilute hydrochloric acid is added to a piece of magnesium ribbon, hydrogen gas is formed with magnesium chloride solution.

Questions

6 Write equations for the following reactions:

a Copper(II) carbonate reacts with hydrochloric acid to form copper(II) chloride, water and carbon dioxide.

b Calcium hydroxide reacts with hydrochloric acid to form calcium chloride and water.

c Sodium hydroxide reacts with sulphuric acid to form sodium sulphate and water.

d When it is heated, copper(II) carbonate forms copper(II) oxide and carbon dioxide.

H

magnesium + hydrochloric acid → magnesium chloride + hydrogen

It is now possible to re-write the equation using formulae.

$$Mg + 2HCl \rightarrow MgCl_2 + H_2$$

By placing the '2' in front of the HCl, we have balanced the equation.

A balanced equation has the same number of atoms of each element on both sides of the arrow. The only way to balance equations is to put the appropriate numbers **in front** of a formula in the equation. **Never try to alter a correct formula to balance an equation.**

More examples

zinc + hydrochloric acid → zinc chloride + hydrogen

$$Zn + 2HCl \rightarrow ZnCl_2 + H_2$$

copper oxide + hydrochloric acid → copper chloride + water

$$CuO + 2HCl \rightarrow CuCl_2 + H_2O$$

sodium hydroxide + hydrochloric acid → sodium chloride + water

$$NaOH + HCl \rightarrow NaCl + H_2O$$

sodium carbonate + hydrochloric acid → sodium chloride + water + carbon dioxide

$$Na_2CO_3 + 2HCl \rightarrow 2NaCl + H_2O + CO_2$$

Practical work

Tests for carbon dioxide

When you bubble carbon dioxide into limewater (calcium hydroxide solution), a white precipitate of calcium hydroxide is formed. If you pass the carbon dioxide through for a long time, the white precipitate disappears to leave a colourless solution. **Never contaminate the limewater with acid**.

Method 1

1. Wear eye protection.
2. Add dilute hydrochloric acid to the suspected carbonate. Place the stopper in the tube (see Figure 5.21).
3. Pass the gas through a small volume of limewater.
4. A white precipitate, indicated by cloudiness, should confirm the presence of a carbonate. Some people say 'the limewater goes milky'.

Method 2

1. Suck up the suspected carbon dioxide in a clean teat pipette, and then force the gas through no more than 1 cm^3 of limewater in a clean test tube. Cleanliness is essential.

Test for hydrogen

1. Wear eye protection.
2. Apply a burning splint to a test tube containing hydrogen.
3. You should hear a squeaky pop. This is a small explosion as the hydrogen combines with the oxygen in the air to form water.

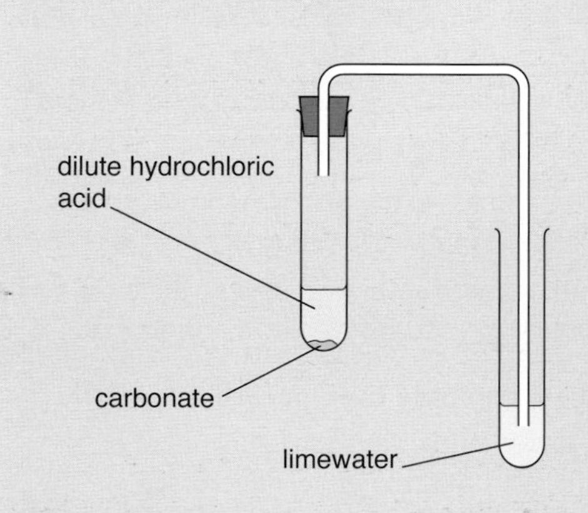

Figure 5.21 Testing for carbon dioxide

Did you know?

- All sodium, potassium and ammonium salts are *soluble*.
- All nitrates are *soluble*.
- All sulphates are soluble, *except* for barium sulphate, lead sulphate and calcium sulphate, which is slightly soluble
- All chlorides are soluble *except* for lead chloride and silver chloride.
- All carbonates are *insoluble*, except for those of sodium, potassium and ammonium

Salts

Bases

Bases are oxides or hydroxides of metals that react with an acid to give a salt plus water.

base + acid → salt + water

A solution formed when a basic oxide dissolves in water is called an alkali.

Practical work

Making soluble salts from an acid and a base

Method 1: From an acid and an alkali

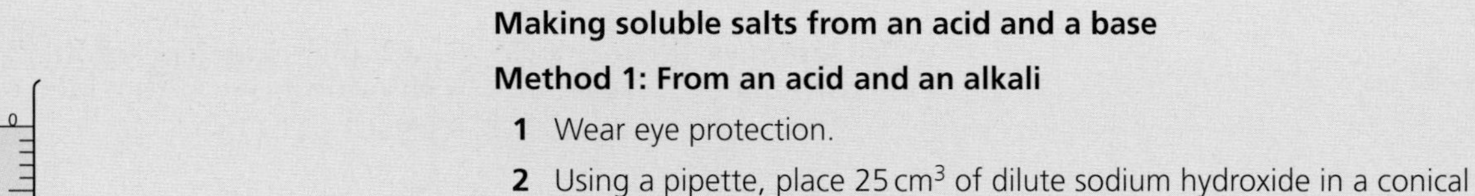

1. Wear eye protection.
2. Using a pipette, place 25 cm^3 of dilute sodium hydroxide in a conical flask (see Figure 5.22).
3. Fill the burette with dilute hydrochloric acid.
4. You need to find out how much acid is needed to neutralise the sodium hydroxide solution exactly.
5. For this you need an **indicator**. Although universal indicator could be used, it is better to use a solution of a coloured compound, methyl orange. Add about four drops of methyl orange solution to the sodium hydroxide. The mixture should turn **yellow**.
6. Now add the acid slowly from the burette until one drop turns the contents of the conical flask from **yellow** to **orange**.
7. Note the burette reading.
8. If you add more acid, the mixture will turn **red**. You may need to repeat this a few times until you get it spot on.
9. You can now repeat the process without methyl orange, but adding exactly the same volume of acid. You now have a solution of sodium chloride.

   sodium hydroxide + hydrochloric acid → sodium chloride + water

   $NaOH + HCl \rightarrow NaCl + H_2O$

10. Evaporate some of the water and allow the rest to stand. After a time, white cubic crystals of sodium chloride will be formed.

    Sodium chloride is **common** or **table salt**. In chemistry, the word *salt* is used to describe a compound formed from an acid and a base.

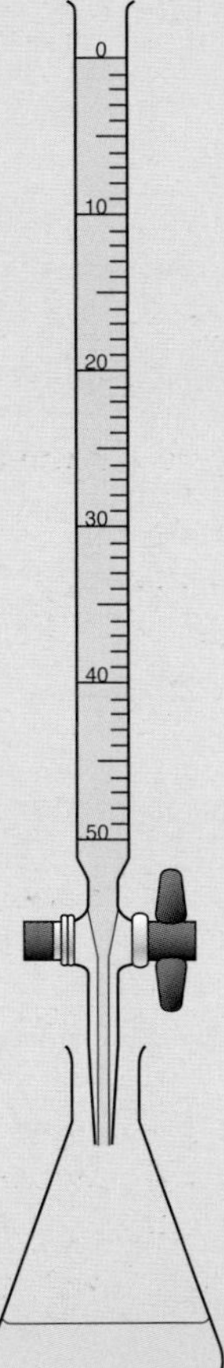

Figure 5.22 Titration using a burette

Method 2: From an acid and an insoluble base or carbonate

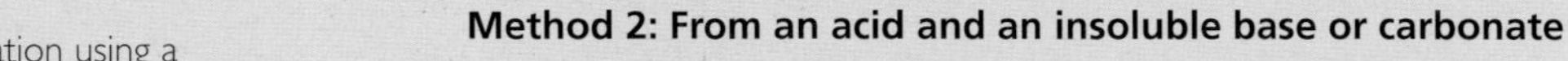

1. Place some acid in a small beaker and add the insoluble base or carbonate until no more will react. Warm gently if necessary.
2. The acid should be completely neutralised. Filter off the excess base or carbonate.
3. Now evaporate off some of the water from the salt solution and allow to stand (see Figure 5.23). Crystals of the salt will appear. Remove the crystals and dry them between filter papers.

H

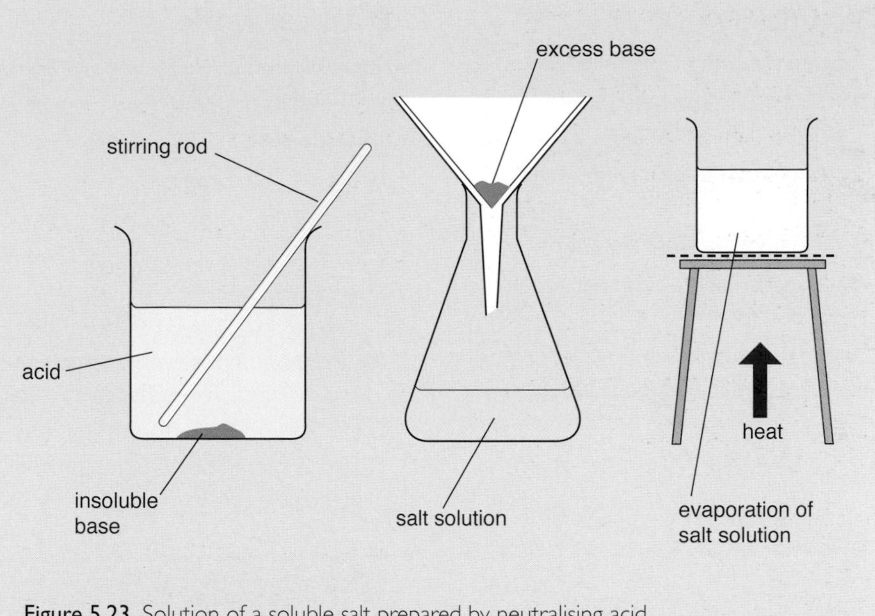

Figure 5.23 Solution of a soluble salt prepared by neutralising acid

Did you know?

Many crystals contain water of crystallisation. For example, blue copper sulphate crystals have the formula $CuSO_4.5H_2O$. To get good crystals, only some of the water is evaporated and the crystals are allowed to form from a concentrated solution.

Figure 5.24 Small copper sulphate crystals

Crystals

Examples

Copper chloride can be made from copper oxide and dilute hydrochloric acid.

copper oxide + hydrochloric acid → copper chloride + water

$$CuO + 2HCl \rightarrow CuCl_2 + H_2O$$

To make copper sulphate crystals, use copper oxide or copper carbonate and dilute sulphuric acid (see Figure 5.24).

copper oxide + sulphuric acid → copper sulphate + water

$$CuO + H_2SO_4 \rightarrow CuSO_4 + H_2O$$

copper carbonate + sulphuric acid → copper sulphate + water + carbon dioxide

$$CuCO_3 + H_2SO_4 \rightarrow CuSO_4 + H_2O + CO_2$$

How to make insoluble salts

Insoluble salts are formed by **precipitation**, in which two solutions are mixed and the insoluble salt forms a solid that is called a **precipitate**. The solid can be filtered off and then washed with water, dried and stored.

Examples of precipitation reactions

sodium carbonate + calcium chloride → calcium carbonate + sodium chloride

barium chloride + sodium sulphate → barium sulphate + sodium chloride

potassium iodide + lead nitrate → lead iodide + potassium nitrate

Practical work

Prepare samples of the following:

1 magnesium sulphate from magnesium carbonate
2 lead sulphate from lead nitrate solution
3 zinc sulphate from zinc oxide
4 copper sulphate from copper carbonate
5 potassium chloride from potassium hydroxide solution.

Practical work

H

Applications of precipitation reactions are the tests for chlorides, iodides and sulphates in solution.

Chloride test

1 Wear eye protection.
2 Place $2\,cm^3$ of dilute sodium chloride solution in a test tube.
3 Add $2\,cm^3$ of dilute nitric acid followed by $2\,cm^3$ of dilute silver nitrate solution.
4 A white precipitate of silver chloride should form that turns darker in sunlight and dissolves when dilute ammonia solution is added in excess. This confirms the presence of a chloride.
5 The nitric acid can be left out, but when the solution is an unknown, then it should be added to prevent other ions such as carbonate and hydroxide interfering with the silver nitrate reaction.

Did you know?

Silver chloride and silver bromide can be used in photography, since light causes them to form silver.

Iodide test

1 Carry this out in exactly the same way as the chloride test above.
2 A yellow precipitate of silver iodide should confirm the presence of iodide. This yellow precipitate is insoluble in ammonia solution.

Questions

7 The elements of Groups 1 and 7 are shown below. Complete the tables by writing the symbols of the elements and by ticking for both groups
 a the most reactive element
 b the element with the smallest atom.

Element	Symbol	Most reactive	Smallest atom
Lithium			
Sodium			
Potassium			
Rubidium			
Caesium			

Element	Symbol	Most reactive	Smallest atom
Fluorine			
Chlorine			
Bromine			
Iodine			
Astatine			

Potassium	K
Sodium	Na
Calcium	Ca
Magnesium	Mg
Zinc	Zn
Iron	Fe
Tin	Sn
Copper	Cu
Silver	Ag

8 **a** Sodium reacts with chlorine. State the name of the product and give its formula.
b Give **two** uses for chlorine.
c Give **one** advantage and **one** disadvantage of fluorine being added to the public water supply.

9 Copper(II) oxide is an insoluble black solid, but it reacts with warm dilute sulphuric acid to form copper sulphate solution. Describe, **in detail**, how you would make some pure copper(II) sulphate crystals in the laboratory from copper(II) oxide and dilute sulphuric acid.

10 The reactivity series of the elements is shown in the table opposite.
a Write the symbol of the least reactive element.
b Write the symbol for **one** element that will not react with dilute sulphuric acid.
c Write symbols for **two** elements in the same group in the Periodic Table.
d Give the symbol of one element which is stored in oil, and give a reason for this method of storage.

Summary

1. Atoms are made up of fundamental particles: protons, neutrons and electrons.
2. The mass of an atom is concentrated in the positive nucleus made up of protons and neutrons.
3. The number of protons in the nucleus is called the atomic number.
4. The number of protons plus the number of neutrons in the nucleus is called the mass number.
5. The light, negative electrons surround the nucleus of an atom in energy levels or shells.
6. The arrangement of the electrons is called the electron configuration.
7. The chemical properties of an element depend on its electron configuration.
8. Dmitri Mendeléev constructed the first Periodic Table of the elements, arranging elements in order of relative atomic mass.
9. The modern Periodic Table arranges the elements in order of their atomic numbers.
10. A column of the Periodic Table is called a group and a row is called a period.
11. The properties of the elements in a group are similar, but change gradually as the group is descended, as shown by the displacement reactions of Group 7 elements.
12. Group 1 is made up of reactive metals called alkali metals.
13. Group 7 is made up of reactive non-metals called halogens.
14. Group 0 is made up of very unreactive elements called noble gases.

15 Flame tests can identify metal ions and precipitation reactions can identify chlorides and sulphates in solution.

16 The common uses of elements such as Group 7 and Group 0 elements depends on their properties.

17 The decision to use fluoride in drinking water is based on evidence that it prevents tooth decay, but the decision can be controversial.

18 A chemical equation is a shorthand summary of a chemical reaction.

19 A balanced chemical equation uses the formulae of the reactants and the products.

20 Acids react with bases to form a salt and water.

21 Bases are usually the oxides, hydroxides or carbonates of metals.

22 A solution of a soluble base in water is called an alkali.

23 Soluble salts can be made by titration or reaction of an insoluble base or carbonate with an acid.

Chapter 6 Rates of reaction, the production and use of fuels and Earth science

By the end of this chapter you should:

- understand how chemists can control the rate of chemical change when making new materials;
- understand the nature of nanoparticles;
- know about the production and combustion of fuels;
- know how useful materials are obtained from the Earth;
- understand the beneficial and possible harmful effects of the use of materials;
- understand the response of the scientific community to understanding and addressing problems that arise;
- know about changes in the Earth's surface and atmosphere over geological time.

Making new materials and the rates of chemical reactions

Raw materials for the chemical industry are obtained from the atmosphere, the Earth's crust and the sea.

Examples

The gas nitrogen is obtained from the atmosphere and used for the manufacture of ammonia, which, in turn, is the starting material for fertilisers, nylon, nitric acid and many other materials. Petroleum or crude oil is obtained from below the Earth's surface and processed to give fuels, plastics, polymer fibres, etc.

Metals are obtained from rocks called **ores** that occur in the Earth's surface.

The elements bromine and magnesium can be extracted from sea water.

All the useful materials are produced by treating the raw materials with chemical and physical processes.

The chemical industry tries to maximise the efficiency of all the processes it uses to make products. This involves studies of:

- the **kinetics** of chemical reactions (how fast they go); *time equals money!*
- how far chemical reactions proceed; the more product made, the greater the profit
- the energy needed or given off in reactions
- waste products (their safe disposal so that the threat to the environment is minimal)
- by-products (these are useful substances formed alongside the main product that can be sold to increase profit).

Mineral resources issues:
www.natural-resources.org

Did you know?

Did you know that alcohol produced by the fermentation of plants may be added to petrol to produce 'gasohol', a fuel for cars. As the plants use the energy of the Sun in growing, this alcohol is a renewable source of energy. Since 1998, some US cars have been designed to run on an 85% alcohol (ethanol) 15% petrol mix.

Ethanol fuel:
en.wikipedia.org

Did you know?

Barium chloride solution can be used to test for sulphate ions in solution.

Barium chloride solution is added to a solution suspected of containing sulphate ions, which has been acidified with dilute hydrochloric acid. If a while precipitate is formed, then sulphate ions are present.

Rates of chemical change

Chemical reactions take place over a period of time. Some reactions are very fast. These include precipitation reactions. When barium chloride solution is added to sodium sulphate solution, a white precipitate appears rapidly. This is because the oppositely charged barium ions and sulphate ions attract each other to form insoluble barium sulphate.

barium + sodium sulphate → barium sulphate + sodium chloride chloride

$$BaCl_2(aq) + Na_2SO_4(aq) \rightarrow BaSO_4(s) + 2NaCl(aq)$$

Other reactions occur much more slowly. The formation of rust on a car takes place over a considerable time. The fermentation of carbohydrates to form the alcohol in beers and wines takes place over a lengthy period. When wine is left exposed to the atmosphere, the ethanol (alcohol) in it is slowly changed to vinegar. *Vinegar is a solution of ethanoic (acetic) acid.*

Following the rate of a reaction

You can follow reactions by measuring the **rate** at which the reacting chemicals disappear or by measuring the rate at which products are formed using an observable property, for example change in volume, change in pressure or change in colour.

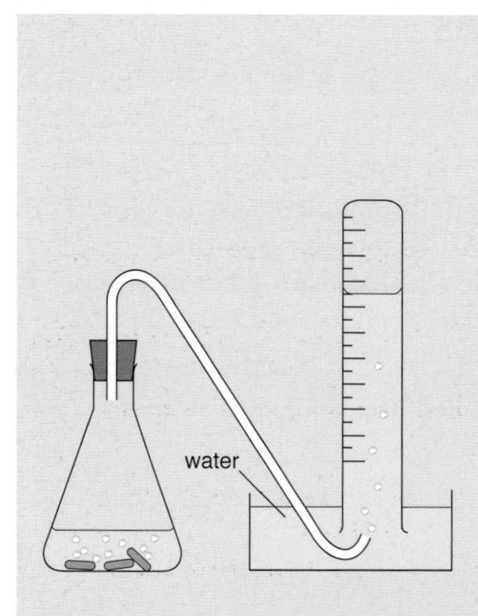

Figure 6.2 Reaction: graduated tube

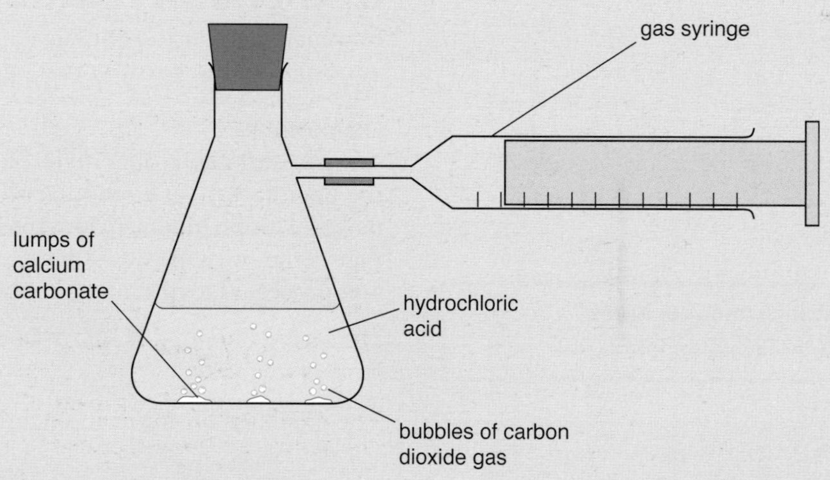

Figure 6.1 The reaction between calcium carbonate and hydrochloric acid in which carbon dioxide is formed: gas syringe

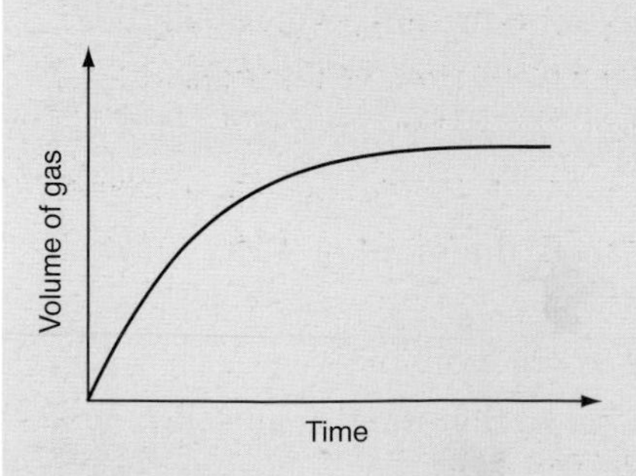

Figure 6.3 The rate at which a gas is given off

Practical work

Measuring the rate at which a gas is evolved from a chemical reaction

1 Wear eye protection.
2 At a specific time, pour the hydrochloric acid quickly into the flask onto the calcium carbonate and then replace the stopper. (See Figures 6.1 and 6.2, and use one of the two types of apparatus. One uses a gas syringe, and the other a graduated tube that is initially full of water.)
3 Measure the volume of gas in the syringe (or graduated tube) every 30 seconds.

Rates of reaction:
www.gcsechemistry.com

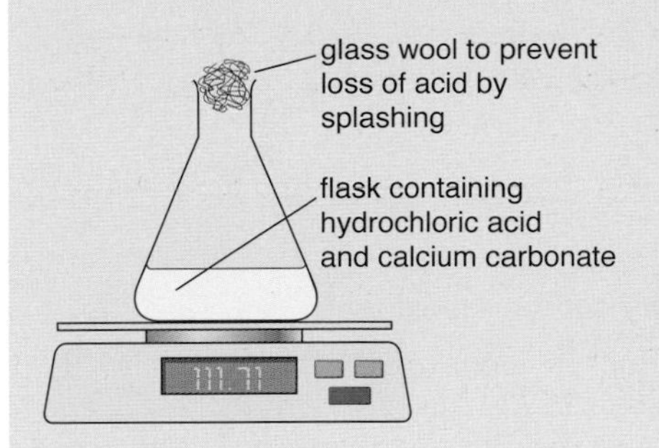

Figure 6.4 Flask being weighed

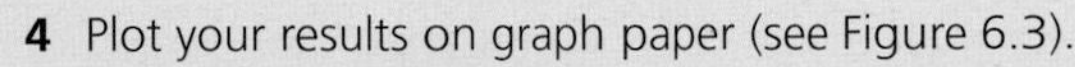

4 Plot your results on graph paper (see Figure 6.3).
5 Mark on the graph:
   a the point where the reaction is fastest
   b the point where the reaction is complete.

Recording loss of mass

The reaction above can be followed by recording loss of mass.

Method

1 Record at regular intervals the total mass of a flask, dilute hydrochloric acid, lumps of calcium carbonate, and cotton wool to prevent loss of mass by splashing, on an accurate top-pan balance (see Figure 6.4).
2 You can connect modern balances to a computer equipped with the correct interface and software, and the results can be displayed on the VDU (see Figure 6.5).

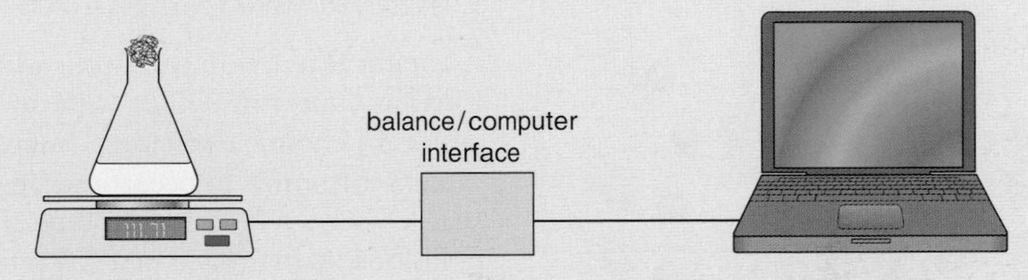

Figure 6.5 Datalogging the loss of mass

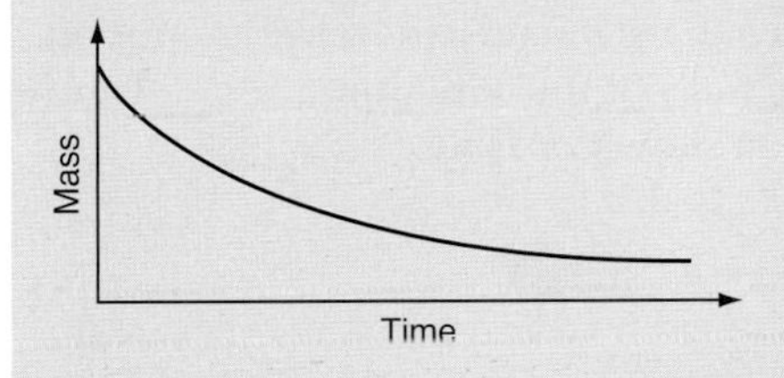

Figure 6.6 Graph of total mass against time

3 Using your results, plot a graph of total mass against time (see Figure 6.6).
4 Mark on the graph the points where:
   a the reaction is fastest
   b the reaction is complete.

The disappearing cross

1 Wear eye protection.
2 Place a measured volume of sodium thiosulphate solution (a colourless liquid) in a flask.
3 Draw a cross on a piece of white paper, and place the glass flask on top (see Figure 6.7).
4 At a given time, add hydrochloric acid to the sodium thiosulphate solution. Swirl the flask and place it back over the cross
5 A precipitate of sulphur is slowly formed. This causes the reaction mixture to become cloudy and eventually opaque.
6 Sulphur dioxide (this is toxic) is produced as well. You must not inhale the vapours from this reaction.
7 View the cross from above, and record the time when the cross can no longer be seen. The shorter this time is, the faster the reaction is occurring. It is a good approximation to say that the rate of reaction is inversely proportional to the time.

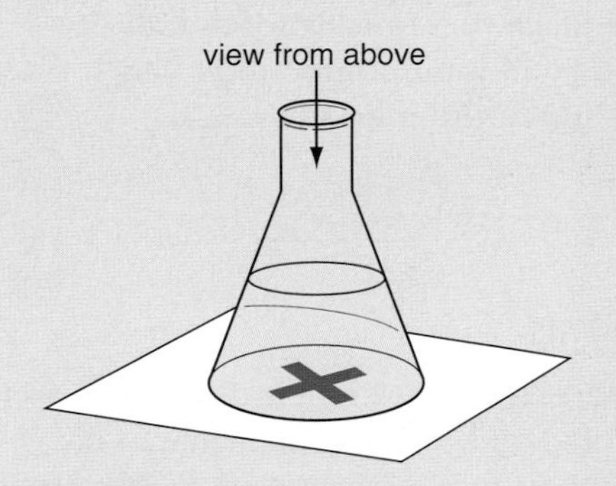

Figure 6.7 Flask on cross

The results of class experiments in which various students carry out the same experiment with different values for one of the variables can be collected together and entered into a spreadsheet. They can then be plotted graphically using a computer.

You are likely to encounter this experiment in your coursework. All these experiments can be used to investigate the factors that alter the rate at which a chemical reaction proceeds:

- changes in concentration
- changes in temperature
- changes in the surface area of solid reactants
- adding a catalyst.

Activity

Plan experiments to show how the rate of a chemical reaction is affected by:

- changes in concentration
- changes in temperature
- changes in the surface area of solid reactants.

Your plan should include:

- the apparatus you would use
- the reactants
- the precise way you would carry out the experiments
- how you would present the results of your experiments.

Figure 6.9 Moving gas particles collide frequently.

How reactions take place

A chemical reaction can occur when reacting molecules, atoms or ions encounter or collide with one another. Not every collision results in chemical reaction, but when the collision has enough energy for bonds to break and be reformed, then a reaction takes place. Such successful collisions are a small fraction of the total collisions taking place in a given time.

The easiest reactions to visualise are those that take place in the gaseous state. The molecules of a gas are in constant random motion, colliding with themselves and the walls of the containing vessel. Figure 6.9 represents the gaseous reaction shown in Figure 6.8.

Figure 6.8 Molecules react when they collide.

Ionic reactions in solution take place very quickly when ions of opposite charge come together. Solid ionic compounds, which react in solution, often show no reaction in the solid state.

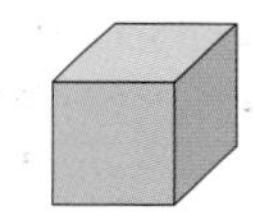

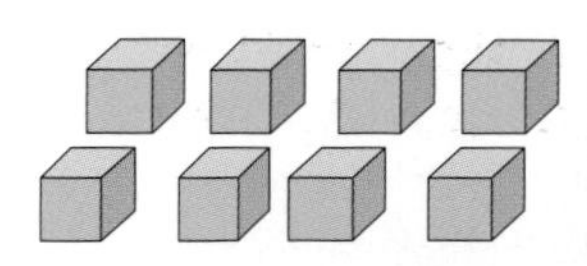

dividing further into 64 even smaller cubes gives a total surface area of 24 cm^2

Figure 6.10 Division of a cube increases its surface area.

Factors affecting the rate of a chemical reaction

The physical state of the reactants

The greater the surface area, the faster the reaction! Finely divided solids react more quickly than lumps of a solid. The reason is that in the finely divided or powder form, the surface area exposed to reaction is greater. This means that more collisions between the reactants can take place in a given time, and so there are more successful collisions and the reaction takes place more quickly.

The increase in surface area when a material is finely divided is shown in Figure 6.10. Work out what the total surface area would be if the original cube were divided in this way five times.

In flour mills there is a risk of explosions. Flour is made of starch, which can be burnt. When fine flour particles are mixed with air, a very large surface area is exposed to the oxygen in the air. There have been many cases of sparks setting off a combustion reaction, which then proceeds explosively.

Many solids show no reaction in the solid state but react in solution. This applies to ionic compounds in particular. In solution, the ions are free to move and interact.

The concentration of the reactants

The greater the concentration, the greater the rate of reaction! When the concentration of a reactant is increased, there are more particles in a given volume. Therefore there are more collisions per unit time, and more successful collisions, so the rate of the reaction increases.

When magnesium ribbon reacts with hydrochloric acid, the following reaction takes place.

magnesium metal + hydrochloric acid → magnesium chloride + hydrogen

$$Mg(s) + 2HCl(aq) \rightarrow MgCl_2(aq) + H_2(g)$$

H

Did you know?

Cooking involves chemical reactions. Vegetables cook faster in a pressure cooker because water boils at above 100 °C under pressure.

Practical work

The rates of chemical reactions

You are given some 3 cm lengths of magnesium ribbon and some dilute hydrochloric acid. Devise an experiment to investigate how the rate of this reaction varies with the concentration of the hydrochloric acid. Your plan should include the equipment you would use, and how you would record your results.

The effect of an increase in temperature

The higher the temperature, the greater the rate of a chemical reaction! The particles in gases and liquids are in constant motion. The average energy of the particles depends on the temperature; the greater the temperature, the greater the average energy of the particles.

Not all particles have the same energy. Only a fraction of the particles have sufficient energy to react. As the temperature increases, the number of particles with sufficient energy to react increases, and there are more successful collisions and the reaction goes more quickly. For many reactions, a 10°C rise in temperature doubles the rate of reaction.

Practical work

1 It is easy to study rates of reaction at home. Measure the time taken for a soluble indigestion tablet to dissolve in:
   a a glass of cold water
   b a glass of hot water.
2 Compare these times with those taken for a powdered tablet to dissolve. (No one in the household should drink any of these solutions.)

Questions

1 The graph shows the rate at which the volume of carbon dioxide is evolved when a lump of calcium carbonate is added to 100 cm³ of

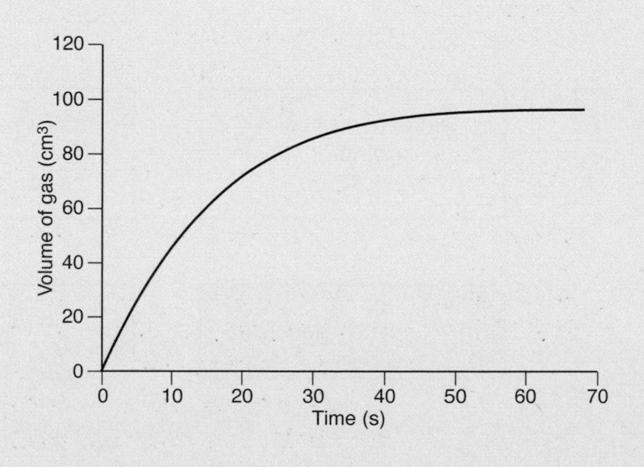

hydrochloric acid. When no more gas is given off some of the calcium carbonate remains in the flask.

a Sketch the graph that would be expected if:
  i the same mass of powdered calcium carbonate had been used with 100 cm^3 of the acid at the same temperature
  ii the acid had been at a higher temperature when the lump of calcium carbonate was added
  iii 50 cm^3 of acid were used with the same mass of powdered calcium carbonate.

b Mark on the graph:
  i the point when the reaction is fastest
  ii the point when the reaction is complete.

Did you know?

Catalysts are used extensively in industry. Catalysts are used in the manufacture of ammonia, sulphuric acid, margarine, polythene, and a host of other important chemicals.

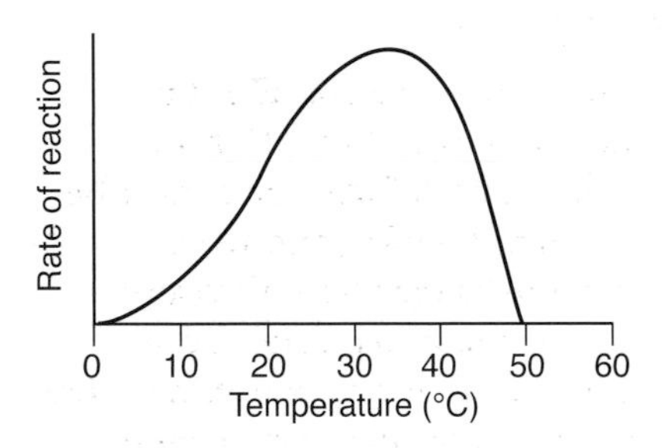

Figure 6.11 Enzyme-catalysed reaction

Catalysts

Catalysts are substances that speed up a chemical reaction but remain chemically unchanged at the end of the reaction. Many catalysts work only for one particular reaction.

A solution of hydrogen peroxide decomposes rapidly into water and oxygen when in contact with manganese(IV) oxide.

Note that the manganese(IV) oxide catalyst does not appear in the equation.

hydrogen peroxide solution → water + oxygen

$$2H_2O_2(aq) \rightarrow 2H_2O(l) + O_2(g)$$

In living systems, biological catalysts are produced by cells and are called **enzymes**. The enzymes produced by yeast are used in the brewing and baking industries. Enzymes from yeast change carbohydrates into alcohol (ethanol) and carbon dioxide. Bubbles of carbon dioxide make the dough rise. In the dairy industry, the enzyme called rennin (or chymosin), which is found in rennet, is used to produce curds from the milk. This is the first stage in cheese-making.

When enzyme-catalysed reactions are heated, the reaction rate increases at first, but then decreases because enzymes are destroyed at higher temperatures (see Figure 6.11).

Catalysts are very important in industry, and there is a constant search for new and improved catalysts. Catalysts allow reactions to

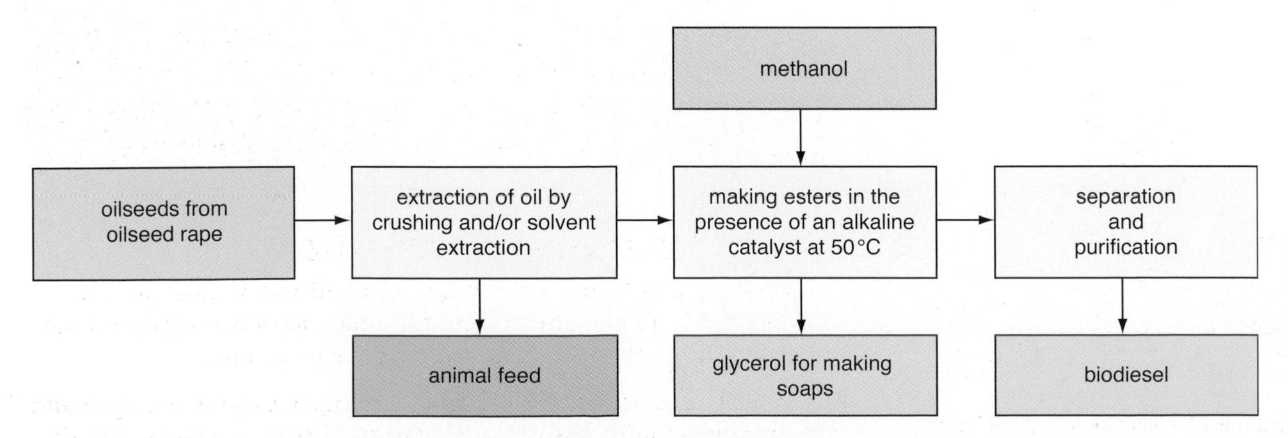

Figure 6.12 The reaction that breaks down vegetable oil into biodiesel and glycerol needs a catalyst.

proceed at a good rate at a lower temperature and hence they save energy.

Some catalysts allow products to be created from renewable resources, for example fuel produced from renewable vegetable matter. Catalysts often enable a product to be formed in fewer stages. The important production of **biodiesel** from renewable resources (see Figure 6.12) uses an alkaline catalyst in one of its stages.

Did you know?

Biological washing powders and liquids contain enzymes. Why are these detergents used in low-temperature washes?

Nanoparticles

A nanometre (nm) is 1.0×10^{-9} m, or 0.000 000 001 m.

Scientists have recently developed small particles of nanometre size. When reduced to this size, particles often display new and different properties to the same substance in bulk form.

One new advance in this area has been the production of silver particles on a nanoscale. These particles have been found to have anti-bacterial, anti-fungal and anti-viral properties. It is thought that they work through the production of silver ions. It is hoped that they may be effective against MRSA (methicillin-resistant *Staphylococcus aureus*). This infection is antibiotic-resistant and is commonly acquired in hospital. It can be fatal.

Nano-sized silver particles are currently being used in the linings of refrigerators to make them self-sterilising.

One firm is reported to have produced a device for treating urinary infections by inserting a biodegradable plastic device covered with nano-sized silver particles into the urinary tract.

Nanoscience is a new science, and there are concerns about its applications. Since a substance in the nano form has different properties from the same substance in the bulk form, care must be exercised. Nanoparticles may pass through the skin and have adverse biological effects. Carbon nanotubes may have the same adverse properties as asbestos fibres. Since nanoparticles are so small, they may be easily dispersed into the environment. Much that is written is speculation, and research is continuing to determine what dangers really exist.

Nanoscience:
www.wellcome.ac.uk/bigpicture

Did you know?

Nanoparticles have different properties from larger masses of the same material, possibly because each particle is only a few hundred atoms. Gold particles are totally inactive until they are less than about 8 nm in diameter but nano-sized gold particles are used as catalysts to oxidise pollutants. Their high surface area means you need only a tiny amount of gold – which is cheaper!

Fossil fuels and combustion

There are three forms of fossil fuel: coal, oil and natural gas. The three were formed many hundreds of millions of years ago, in what is called the Carboniferous Period of geological time.

- **Coal**: Coal is made up of carbon, hydrogen, oxygen, nitrogen and varying amounts of sulphur. There are three main types of coal: anthracite, bituminous coal and lignite. Coal was formed from the decay of prehistoric plant material, under the influence of

Figure 6.13 Cefn Coed Colliery, South Wales, 1930, then the deepest anthracite mine in the world

Figure 6.14 Texaco Oil Refinery, Pembroke, Wales

heat and pressure. Evidence of the plant origin can often be found as fossils in coal strata (see Figure 6.13).

- **Oil**: Petroleum formation occurred in several steps. There was a lot of organic matter from organisms such as plankton, bacteria, small animals and algae. This organic matter was buried quickly before oxidation took place. Chemical reactions under heat and pressure slowly transformed the organic material into the hydrocarbons found in petroleum (see Figure 6.14).
- **Natural gas**: This was formed via the same processes as petroleum, but it could also seep through porous rock and collect in pockets in the Earth's crust. It mainly consists of methane (CH_4).

As all three fossil fuels were formed hundreds of millions of years ago, they are **finite resources**, and already supplies are dwindling in various parts of the world.

Fractional distillation of crude oil

Rates of reaction: www.schoolscience.co.uk

Petroleum or crude oil is a complicated mixture of hydrocarbons. Hydrocarbons are compounds containing carbon and hydrogen only. For useful materials to be obtained from crude oil, the oil must be subjected to several processes. The first of these is **fractional distillation**. This involves separating the complex mixture of hydrocarbons in crude oil into simpler mixtures of hydrocarbons (fractions), depending upon their boiling points (see Figure 6.15).

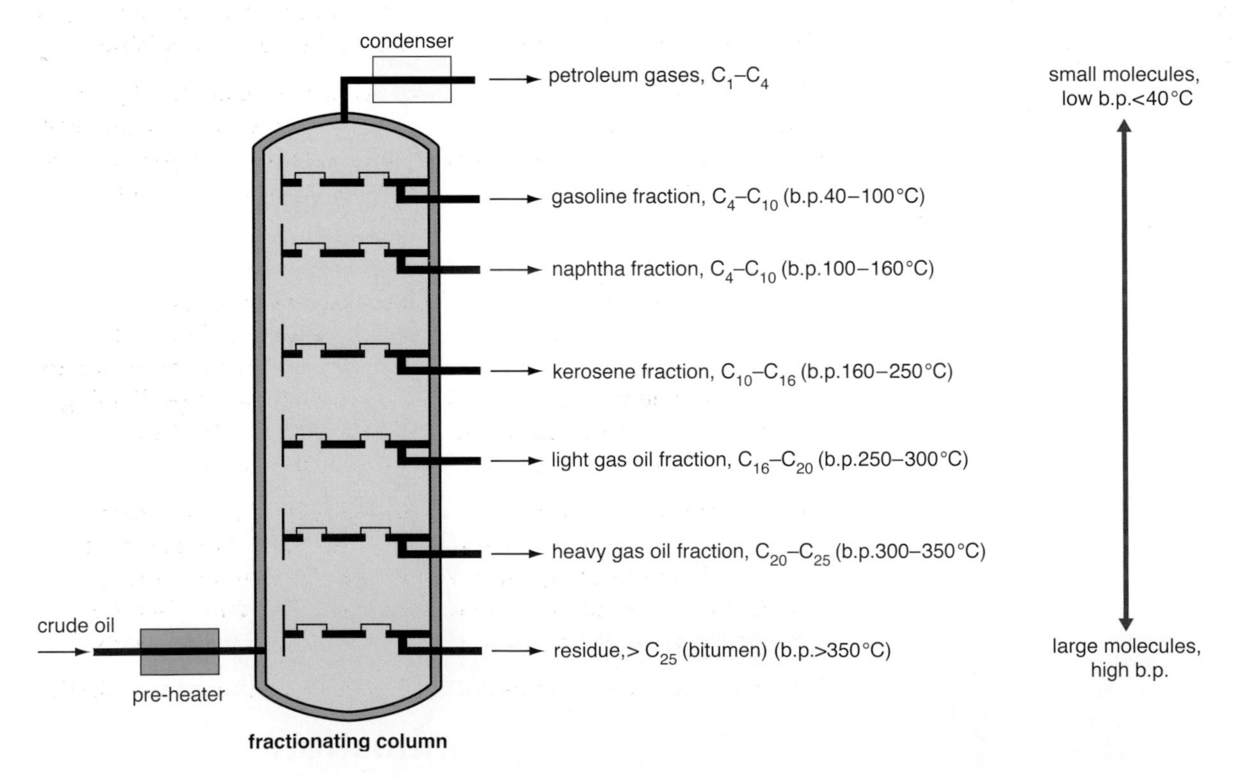

Figure 6.15 The fractional distillation of crude oil

Crude oil boils when it enters the base of the fractionating column. The temperature of the column decreases with height. As the vaporised crude oil rises up the fractionating column, it passes through bubble cap plates that collect condensed liquid at that temperature and allow vapours of liquids with lower boiling points to move higher up the

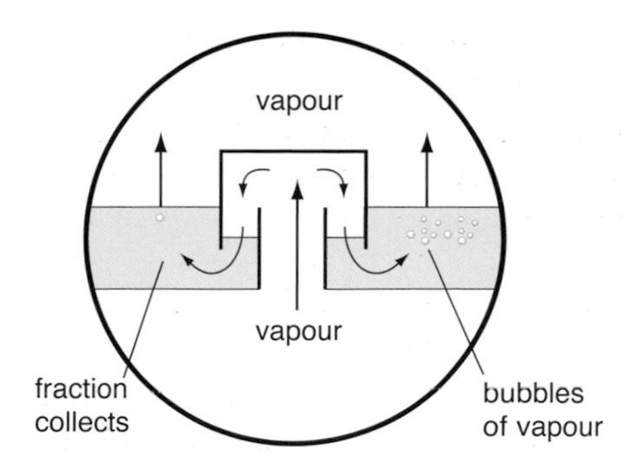

Figure 6.16 Bubble cap plates collect liquid and let vapour pass.

column. Each plate contains many bubble caps like the one shown in Figure 6.16.

Some of the products from the fractional distillation of oil are fuels. These include the lowest boiling point fractions, gases and gasoline.

How the use of fossil fuels has changed over time

Coal

The demand for coal in the UK increased greatly at the time of the Industrial Revolution. Oil was extracted mainly from animal and vegetable sources. Mining on a significant scale started at the end of the eighteenth century. Coal was needed for the developing iron and steel industries, for steam power, for the railways, and as cheap domestic fuel.

It was found that when coal is heated in the absence of air, it forms a gaseous fuel called **coal gas**. The residue is **coke**, which was used in the iron and steel industry. Another by-product was **coal-tar**, which was a very important source of chemicals until it was replaced by oil. Throughout the nineteenth century and early twentieth century, coal gas was used as a fuel for lighting, and to some extent for domestic cooking. This continued until the switch to natural gas in the 1960s.

Several regions of the UK were rich in underground coal deposits, none more so than Wales. In 1889 there were about a dozen working mines in the Monmouth and Edge of Glamorganshire area alone (see Table 6.1), and there were small working collieries all over Wales.

Table 6.1 Collieries in the Llanelly area, 1889

Bigyn	Great Mointain
Brynwyn	Machynis
Cae	Mardaffer
Caebad	Oldcastle and Bres
Creswyddy	Pencoed
Cwmawr	Telsamey
Gelli	

The closure of deep mines in the UK was virtually complete by the year 2000; even those remaining have an uncertain future. Open-cast mining takes place in some areas. In this method, outcrops of coal are removed, and the topsoil is replaced as the landscape is restored.

Oil

Although oil had been known about since ancient times, it was not until the 1850s that the shortage of whale oil in the USA made people look for other sources of oil. In places, petroleum seeped to the surface, and was used to make kerosene for oil lamps. Drilling for oil began in Pennsylvania in the USA in the 1850s. Other products, such as gasoline (petrol), were thrown away.

The invention of the internal combustion engine and the petrol-driven motor car transformed the oil industry. We now rely on oil for the bulk of our energy needs, but we know that it is a finite resource that will run out. There are people who think that the hydrocarbons in oil are too rare and useful to be burnt as fuels.

Products from oil refineries are used to make plastics and synthetic clothes.

Questions

2 Plastics made from the products from crude oil are extensively used for domestic products. They include plastic bottles, wrapping materials and ballpoint pens. Write a short account of the **advantages** of the plastics used for these products, and also point out any **disadvantages** of using such materials.

Natural gas

The discovery and exploitation of natural gas reserves began on a large scale in the UK in the 1960s. Natural gas power stations have replaced coal-fired power stations, and the demand for gas is very high. Natural gas is a finite resource. The price of gas increased greatly in 2005 and early 2006.

Another look at combustion reactions

A chemical reaction involves the formation of new substances (the **products**) from the reacting substances (the **reactants**). **Combustion** is a reaction of a fuel with oxygen.

In the case of the complete combustion of hydrocarbons, the reactants are the hydrocarbon and oxygen, and the products are carbon dioxide and water. The combustion reaction involves the breaking of bonds in the reactants and the reforming of bonds to form the products. A hydrocarbon molecule contains only carbon and hydrogen atoms bonded together. During the reaction, those atoms form bonds with oxygen atoms.

Consider the combustion of methane (see Figure 6.17):

methane + oxygen → carbon dioxide + water

$$CH_4 + 2O_2 \rightarrow CO_2 + 2H_2O$$

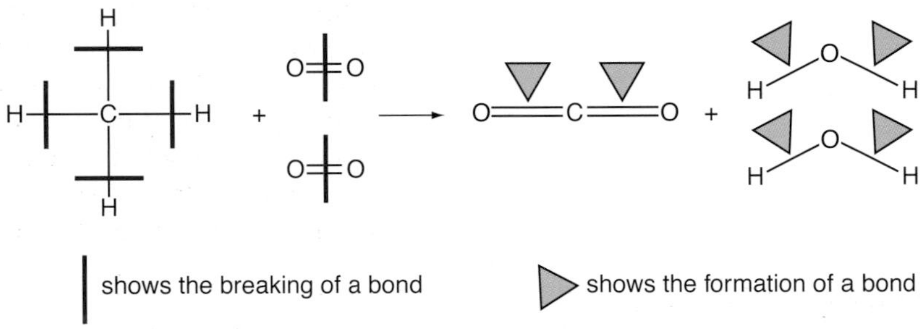

H

shows the breaking of a bond — shows the formation of a bond

Figure 6.17 Combustion of methane involves the breaking and reforming of bonds.

Table 6.2 Energy values for the breaking of covalent bonds

Bond	Energy needed to break it, kJ/mol
O=O	496
C–H	412
H–H	436
C=O	743
O–H	463
C–C	348
N≡N	944
C=C	612
N–H	388

H

The **breaking** of a bond **requires** energy. The **formation** of a bond **releases** energy. The difference between the energy needed to break bonds and the energy released when the new bonds are formed determines whether the overall reaction is exothermic or endothermic.

Bond energy data

Table 6.2 gives some energy values for the breaking of some covalent bonds. (There is more about covalent bonds in Chapter 7.)

The complete combustion of methane:

$$CH_4(g) + 2O_2(g) \rightarrow CO_2(g) + 2H_2O(l)$$

The bonds broken are four C–H bonds and two O=O bonds:

$$(4 \times 412) + (2 \times 496) = 2640\,kJ$$

The bonds formed are two C=O bonds and four O–H bonds:

$$(2 \times 743) + (4 \times 463) = 3338\,kJ$$

Therefore more energy is given out than is taken in, and the reaction is exothermic. We know this anyway because methane is a fuel that gives out heat when burnt.

Example

Ammonia is made by the combination of nitrogen gas and hydrogen gas. Find out if the reaction below is exothermic or endothermic.

$$N_2(g) + 3H_2(g) \rightarrow 2NH_3(g)$$

The bonds broken are one N≡N bond and three H–H bonds. The energy needed is:

$$944 + (3 \times 436) = 2252\,kJ$$

The bonds formed are six N–H bonds, so the energy given out is:

$$(6 \times 388) = 2328\,kJ$$

More energy is given out when the ammonia forms than when the nitrogen and hydrogen molecules break. Therefore the reaction is exothermic.

Environmental aspects of burning fossil fuels

Incomplete combustion

If there is insufficient oxygen for the complete combustion of a hydrocarbon, then, as well as carbon dioxide, **carbon** and **carbon monoxide** may be formed. Solid carbon is soot, which can be found in chimneys as a result of burning coal. The inside of the exhaust pipe of a car may contain deposits of soot.

It is carbon particles in a candle flame that make it yellow, and this is why a candle flame leaves a black deposit if it is allowed to touch a cold surface. Soot from chimney smoke can be deposited on buildings, causing them to become dirty. In days gone by when there were many domestic coal fires, fine soot particles contributed to the formation of dense fogs, such as the London 'pea-soupers'. Soot particles in car exhaust fumes could cause health problems if inhaled.

Carbon monoxide is a toxic gas, and it is found in the exhaust fumes from petrol engines. There is insufficient oxygen in the petrol–air mixture for the complete combustion of all the hydrocarbon molecules in petrol.

If gas central-heating boilers are not serviced regularly, they may emit carbon monoxide. As carbon monoxide is colourless and odourless, it may go undetected, and it can lead to death if inhaled over a period of a few hours. The carbon monoxide combines with the haemoglobin of the blood, preventing the blood from carrying oxygen around the body. This can cause breathing difficulties. Many houses have carbon monoxide detectors that warn if carbon monoxide is present.

Did you know?

If you have been inhaling carbon monoxide, you may have a headache and feel fuzzy. **React quickly! Get fresh air immediately**. Open the doors and the windows. Turn off combustion appliances, and leave the building.

Question

3 Explain the meaning of **each** of the following, illustrating your answer with a suitable example:

- **a** fractional distillation
- **b** fossil fuel
- **c** acid rain.

Combustion of sulphur in fossil fuels

Most fossil fuels contain some sulphur. Some natural gas reserves contain so much sulphur that it has to be removed before the gas can be used as a fuel. The sulphur is then used industrially to make sulphuric acid. The sulphur in fossil fuels burns to form sulphur dioxide.

sulphur + oxygen → sulphur dioxide

$$S + O_2 \rightarrow SO_2$$

H

Once in the atmosphere, the sulphur dioxide reacts with water vapour and oxygen to form sulphuric acid. Rainfall containing acids is called **acid rain**:

- Acid rain damages buildings. This is an important issue when historical buildings are being conserved (see Figure 6.18).
- Acid rain damages trees. The pH of the water in the soil falls, and root systems are damaged, causing foliage to yellow (see Figure 6.19).

Figure 6.18 Damage to stonework caused by acid rain

Figure 6.19 Tree damage by acid rain

- Acid rain harms life in lakes and rivers. It leaches harmful metals out of the soil and into the water. Aluminium is one example. If the concentration of aluminium in water is too high, then fish and other creatures die.

Most sulphur dioxide contamination comes from power stations. Acid rain pollution in Scandinavia was found to have originated from power stations in other countries in Europe.

Acid rain in Canada:
www.ec.gc.ca/acidrain

Production of oxides of nitrogen

Nitrogen oxides are produced (see Figure 6.20):

- naturally in thunderstorms
- when fossil fuels are burnt
- extensively in motor car petrol engines when the ignition spark converts nitrogen and oxygen in the air mixed with petrol into oxides of nitrogen.

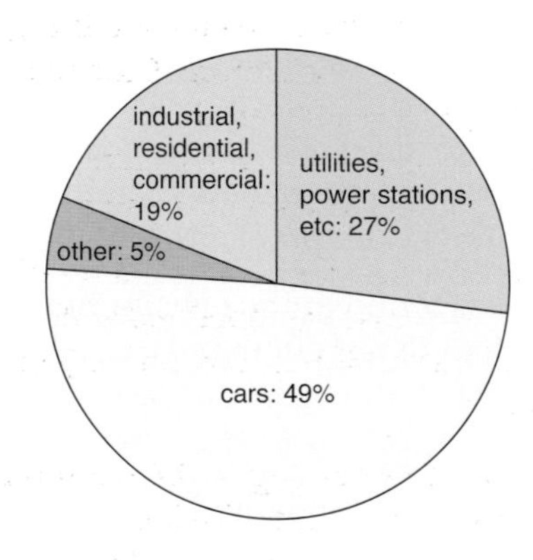

Figure 6.20 Each state in the USA publishes environmental data; this chart is typical for sources of nitrogen oxides in the atmosphere.

Oxides of nitrogen have a detrimental effect on the respiratory tract. When they interact with water in the atmosphere, they form nitric acid, which contributes to acid rain:

Another effect of oxides of nitrogen is low-level ozone pollution and photochemical smog. Photochemical smog is formed when three factors are present:

- oxides of nitrogen
- hydrocarbons
- sunlight in the form of ultraviolet rays.

The result is a build-up of pollutants such as ozone and a brown haze in the atmosphere. This smog may cause eye irritation, damage to the lungs, and damage to the respiratory tract. People in poor health may be even more susceptible to its effects. Smog can also damage plant cells.

In California in the USA, smog is common, as a large number of car users are driving in a sunny climate. The area is also known for temperature inversions, where a layer of warm air sits on top of cooler air at lower levels. This traps the smog.

Some solutions to environmental problems

Changing from a more polluting fuel to a less polluting fuel is an advantage. This has occurred with changes in domestic heating from coal to natural gas.

Fossil fuel power stations produce a significant amount of sulphur dioxide. Although it tends to make electricity generation more expensive, the sulphur dioxide in the flue gases can be removed by reacting it with limestone or by passing it through alkaline solutions such as alkaline seawater.

Nitrogen oxide pollution from car emissions has been reduced by catalytic converters. Catalytic converters have a ceramic or metal honeycomb structure coated with a combination of platinum, rhodium and/or palladium. They can convert up to 90% of hydrocarbons, carbon monoxide and nitrogen oxides from the engine emissions into less harmful carbon dioxide (CO_2), nitrogen and water vapour.

Oxides of nitrogen can be removed from gas streams using ammonia and a catalyst. The method uses ammonia (NH_3) to reduce nitrogen oxides to nitrogen and water in the presence of a catalytic surface. One type of catalyst used is a base metal containing oxides of titanium, molybdenum, tungsten and vanadium. Gaseous ammonia is injected with steam or compressed air into the contaminated gas. The ammonia–gas mixture enters the catalyst, where the oxides of nitrogen are reduced to nitrogen and water.

A new paint has been developed that is claimed to remove oxides of nitrogen from the air. The paint's polymer base is embedded with nano-sized spherical particles (approximately 30 nm in diameter) of titanium dioxide and calcium carbonate.

The nitrogen oxide gases are absorbed into a coating of the paint on an exposed surface, and they adhere to the titanium dioxide particles. The titanium dioxide uses the sun's energy to combine the nitrogen oxide and water to make nitric acid. The acid is then either washed away by rain or neutralised by the calcium carbonate particles to create carbon dioxide, water and calcium nitrate.

Global warming and greenhouse gases

The greenhouse

Figure 6.21 The Palm House at Kew Gardens

The glass of a greenhouse (see Figure 6.21) allows the rays of the Sun to enter the structure and warm up the contents inside. The contents then radiate infrared radiation of a longer wavelength than can pass through the glass, and so heat builds up in the greenhouse.

Some molecules behave in a similar way to a greenhouse, and these are known as greenhouse gas molecules. Radiation from the Sun heats up the Earth's surface, but when the Earth radiates heat energy back into space, it is in the form of infra-red radiation with a longer wavelength. Molecules such as carbon dioxide absorb some of this longer wavelength radiation, and trap the energy within the atmosphere. This is the **greenhouse effect** (see Figure 6.22), which is a natural process.

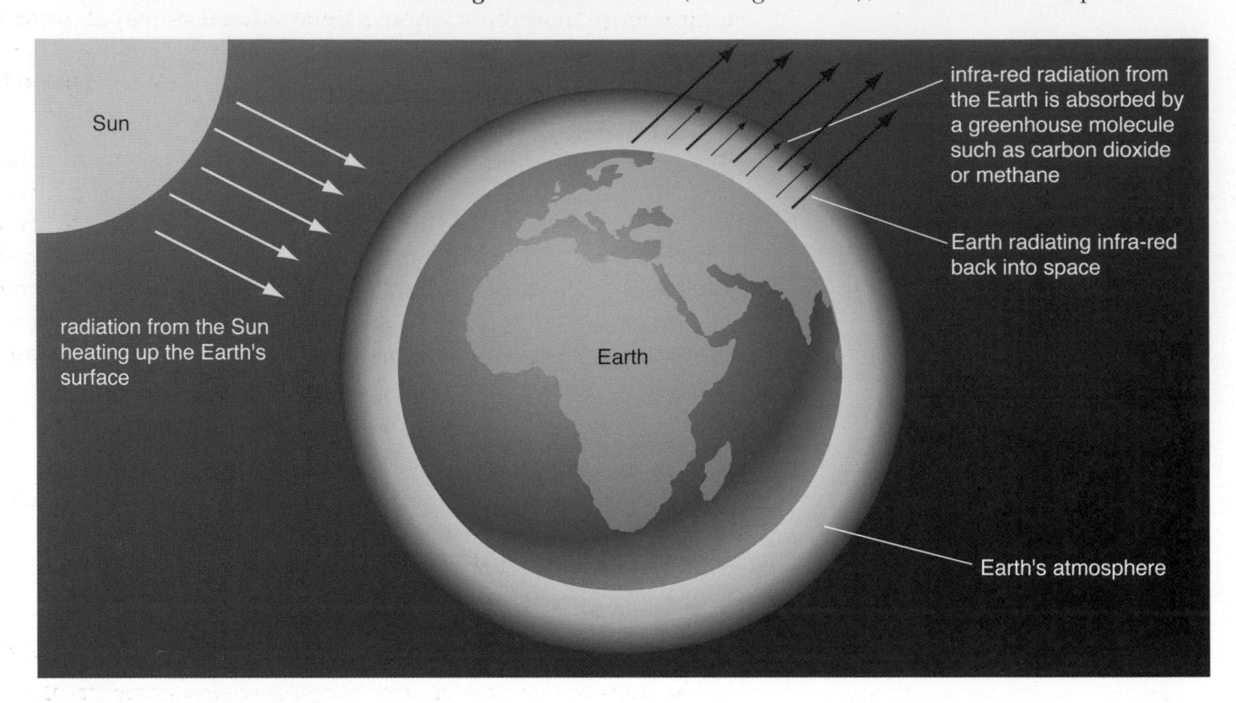

Figure 6.22 Greenhouse effect

Global warming:
www4.nationalacademies.org/onpi/webextra.nsf/web/climate

What is new is that recently more and more fossil fuels have been burnt, adding more carbon dioxide to the atmosphere. The more greenhouse gas molecules there are in the atmosphere, the less heat can be radiated into space, and the more the Earth and its atmosphere heat up. The resulting increase in the Earth's temperature is known as **global warming**.

The increase in carbon dioxide has been made worse by massive deforestation. For example, large areas of the Amazon rain forest have been cleared. Green plants absorb carbon dioxide in photosynthesis. The combustion of fuels and respiration increase the amount of carbon dioxide in the atmosphere, but photosynthesis decreases the amount of carbon dioxide in the atmosphere.

$$\text{carbon dioxide + water} \xrightarrow{\text{sunlight and chlorophyll}} \text{carbohydrates + oxygen}$$

With fewer trees to remove carbon dioxide, the amount in the atmosphere increases. The increase in the amount of carbon dioxide in the atmosphere has been determined by researchers taking ice cores in polar regions and examining the bubbles of air trapped in the ice. Deeper ice is older than ice near the surface, and a timeline can be drawn up that shows that over recent decades, the amount of carbon dioxide in the atmosphere has increased.

H

(a)

(b)

Figure 6.23 NASA pictures of the Arctic sea ice cover in (a) 1979 and (b) 2003

Worldwide, many governments are taking the threat of global warming seriously, and many have signed the Kyoto agreement to reduce carbon dioxide emissions. One fear of global warming is that ice caps on land and glaciers will melt, leading to rising sea levels and flooding. This would also affect wildlife such as polar bears. The US National Aeronautics and Space Administration (NASA) stated in September 2005 that the Arctic summer ice was smaller in surface area than at any time since records began (see Figure 6.23).

Unpredictable weather may arise from climate changes brought about by global warming. Since hurricanes and tornadoes derive their energy from the sea, some believe that they may increase in intensity if the sea temperature increases. These fears have been aggravated by the destruction and flooding of New Orleans in the USA by Hurricane Katrina.

Although most scientists accept the theory that increased carbon dioxide emissions are causing global warming, some scientists are cautious about global warming claims. It has been pointed out that it takes a long time for climate patterns to be established, and that the current data may be just a blip in terms of the geological and atmospheric timescales over hundreds and thousands of years. There is uncertainty about the part that the oceans play in absorbing carbon dioxide.

Preventing global warming

Several remedies have been proposed to offset global warming. They include the following:

- Restrict carbon dioxide emissions using various types of legislation, for example higher taxes on fossil fuels.
- Use renewable energy sources, such as wind and wave power.
- Use fuels that are renewable. Biomass is organic matter that has used the Sun's power to grow. It can be turned into fuels such as methane. Any carbon emissions are carbon-neutral, since the carbon the plants removed from the atmosphere in growing returns to the atmosphere on combustion.
- Use nuclear power.
- Remove carbon dioxide by trapping it in an alkaline medium such as limestone slurry.

Some of the issues about global warming and energy use are described in Chapter 10. Another possible type of atmospheric pollution is from particulate matter, such as dust particles in the atmosphere. Dust sometimes enters the atmosphere through natural events such as volcanic eruptions.

Question

4 a Explain what is meant by **greenhouse gas** and **global warming**.

b State **two** causes of global warming.

c State **two** courses of action that would reduce global warming.

If additional man-made particles from the burning of fuels and mining reached the atmosphere, the dust might prevent sunlight from reaching the Earth. This could cause a fall in temperature, leading to global cooling and adverse effects on both plants and animals.

Geological processes and the atmosphere

Plate tectonics

The surface of the Earth, or lithosphere, is made up of seven large plates and some smaller ones, about 70 km thick, which move a few centimetres per year with respect to one another. This movement is often called **continental drift**.

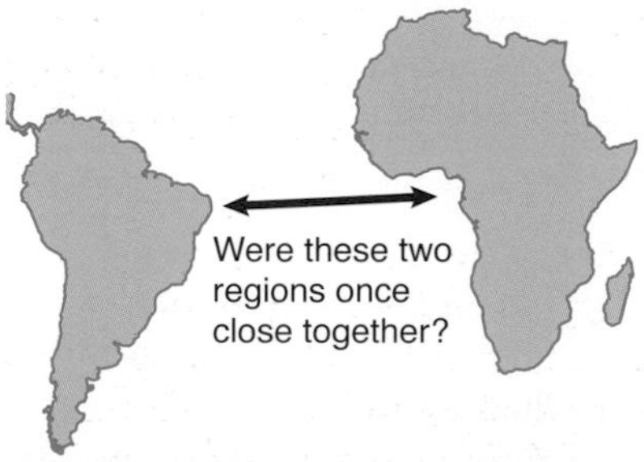

Figure 6.24 The shape of the coastlines of Africa and South America

The idea of continental drift was put forward by Alfred Wegener (1880–1930). He tried to find evidence that would explain the fit between the coastlines of South America and Africa (see Figure 6.24). Wegener studied rock types and fossils that showed the two continents could have once been joined. Wegener's critics demanded to know how continents could move but he was unable to give a convincing explanation. His continental drift theory was not accepted at the time.

As new evidence emerged, other scientists modified the theory. In the 1960s surveys of the ocean floor provided strong geological evidence that new ocean floor is created between the continents over millions of years. Seismologists showed that there is a **mantle** layer beneath the Earth's crust. Seismic studies also showed that earthquakes occurred in a pattern, which is now known to be the boundaries between plates (see Figure 6.25). Since the 1960s scientists have accepted the theory of moving plates. The large moving plates are called **tectonic plates**. The theory of plate tectonics describes movement of the ocean floor as well as the continents, and includes an explanation of how the plates move.

Wegener could not explain the forces that could move large land masses. Today scientists think that **convection currents** in the Earth's mantle cause the plates to move. The mantle can flow slowly, even though it is solid. Hot rocky material rises in the mantle and then cools as it nears the crust. As it cools, the material becomes denser and begins to sink, producing a convection current. Sideways movements at the top of the mantle drag and move the plates above. The main source of heat that drives the convection currents is thought to be radioactivity deep in the Earth.

Plate tectonics:
www.moorlandschool.co.uk/earth

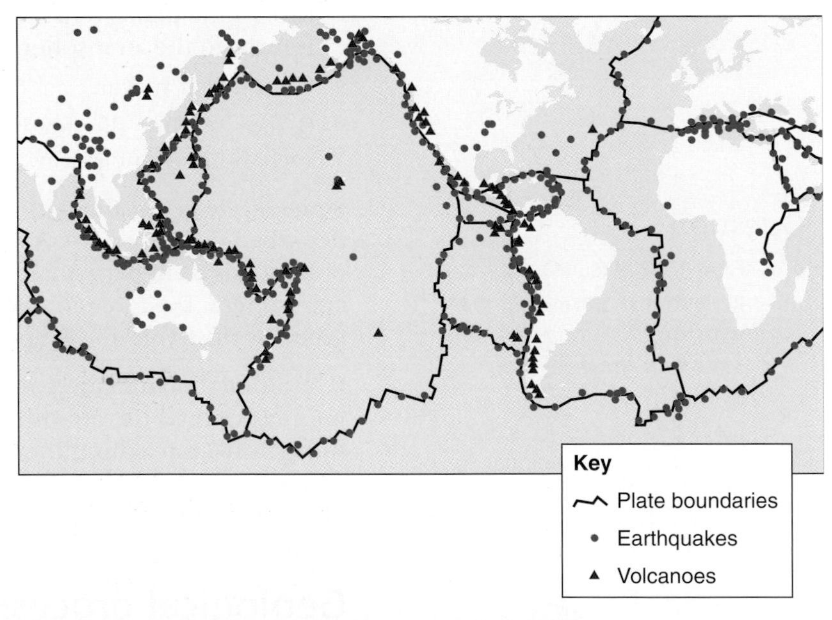

Figure 6.25 The pattern of earthquakes and volcanoes defines plate boundaries. For example, the ring of volcanoes around the Pacific Ocean is the boundary of the Pacific plate.

What happens at plate boundaries?

The boundaries between tectonic plates have high volcanic and earthquake activity (see Figure 6.25). They are also regions where new rocks are formed and where rocks can be deformed and recycled.

When a plate begins to break this is called **rifting**. Rising, hot mantle pushes up the crust and the pressure forces the plate to break and separate. Where plates move apart, molten **magma** (liquid rock beneath the Earth's surface) can rise up in the gap and form new **igneous rocks** as it cools. The mid-Atlantic ridge is a region where two plates are moving apart, and undersea volcanoes form new sea floor. The formation of new rock gradually pushes Africa and South America further apart.

When plates collide, rocks in the crust are deformed and either pushed up or pushed down into the mantle. Where two continental plates collide, the rocks are folded and faulted and pushed up to form high mountain ranges, such as the Himalayas and the Tibetan plateau. The collision may produce large earthquakes.

When an oceanic plate collides with a continental plate, the denser oceanic plate is forced down into the mantle – this is called **subduction**. The rocks begin to heat up as they near the hotter mantle, and they are also under increasing pressure. The subducted plate partially melts and the magma gradually rises through the crust. If the magma cools slowly underground, igneous rocks such as granite are formed. If the magma is forced to the surface, volcanoes erupt molten lava that cools quickly to form igneous rocks such as basalt.

Question

5 **a** What is meant by the term continental drift?
b Name **two** continents that have moved apart over millions of years.
c State **two** things that may occur when two oceanic plates collide.

The rock cycle

Plate tectonics links igneous, metamorphic and sedimentary rocks in the **rock cycle**. Rocks that are formed by volcanoes or are exposed by uplift and deformation are subjected to weathering. Weathered rock particles are transported by water and wind and deposited elsewhere. As the thickness of sediments builds up the particles are compacted and sedimentary rocks can be formed. When plates collide or are subducted, **metamorphic rocks** form as the existing rocks are heated and pressurised but do not melt. Subduction returns rocks from the crust into the mantle, and also recycles crust and mantle material to form new igneous rocks.

The atmosphere

The Earth's atmosphere has the composition by volume shown in Table 6.3.

The air is a source of nitrogen, oxygen and the noble gases that can be separated by the fractional distillation of liquid air (see Figure 6.26).

Table 6.3 Composition by volume of the Earth's atmosphere

	Chemical symbol	Percentage by volume
Nitrogen	N_2	78.08
Oxygen	O_2	20.95
Carbon dioxide	CO_2	0.036 (but variable)
Water	H_2O	Variable
Argon	Ar	0.93
Helium	He	0.000 5
Krypton	Kr	0.000 11
Neon	Ne	0.001 8
Xenon	Xe	9×10^{-6}

During the fractional distillation shown in Figure 6.26, the air is dried, carbon dioxide is removed, and then the air is liquefied. The various gases are then separated according to their boiling points. The rarer noble gases, neon (boiling point: –246 °C) and krypton (boiling point: –152 °C), are extracted by further fractional distillation.

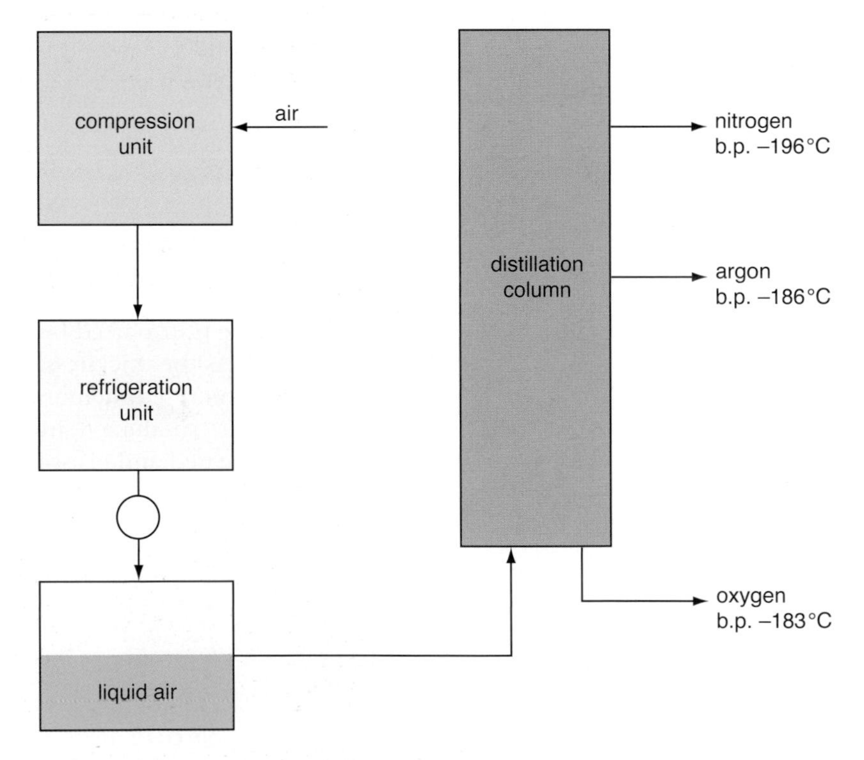

Figure 6.26 Fractional distillation of liquid air

Uses of the gases of the atmosphere

Life on Earth is dependent on oxygen. **Oxygen** is also used:

- in medicine as an aid to breathing
- for producing high temperatures, as in oxyacetylene welding

Question

6 Give **one industrial use** for each of the following gases that are found in the Earth's atmosphere:
- **a** oxygen
- **b** nitrogen
- **c** helium
- **d** argon
- **e** neon.

- in liquid form as the oxidant in some rockets
- for enriching the air supply in some steel-making processes
- in high-altitude aircraft.

Nitrogen is used:

- for making ammonia, but is taken directly from the air and not from the factional distillation of liquid air
- in liquid form to cool things down in low-temperature processes
- as an unreactive gas that provides an inert atmosphere (one that will not react).

Carbon dioxide is used:

- in the manufacture of fizzy drinks
- in fire extinguishers
- as solid carbon dioxide (dry ice), used to keep some things cool (solid carbon dioxide is a white solid that turns directly back to the gas without going through the liquid state).

Argon is used:

- for filling light bulbs
- for providing inert atmospheres.

Helium is used:

- for filling airships and weather balloons (most of the helium used for this does not come from the atmosphere, but is found underground in association with natural gas, where it is thought to originate from radioactivity in rocks).

Neon is used:

- in fluorescent tubes in advertising displays (see Figure 6.27).

Figure 6.27 Neon signs in Piccadilly Circus

Did you know?

Limewater is a saturated solution of calcium hydroxide.

Chemical tests

Oxygen relights a glowing wooden splint.

Carbon dioxide gives a white precipitate when passed through limewater. If the gas is passed for a longer period of time, the white precipitate disappears to leave a colourless solution.

Practical work

Chemical tests

1. Wear eye protection.
2. Place a little manganese(IV) oxide in a test tube, and add 10-volume. hydrogen peroxide solution.
3. Test the gas for oxygen.
4. In another tube place a small piece of calcium carbonate, and add dilute hydrochloric acid.
5. Test the gas for carbon dioxide.

The origin of the atmosphere

The Earth was formed about 5 000 000 000 years ago. Any hydrogen or helium present in the Earth's early atmosphere disappeared, because the molecules of these light gases move so quickly that they can escape the gravitational pull of the Earth. Other gases, which had been expelled from the Earth's interior by volcanic activity, included methane, carbon dioxide, carbon monoxide, water vapour and nitrogen. There was no oxygen. We know this because early rock formations do not contain oxides. As the Earth cooled, water condensed from the atmosphere and formed the rivers and oceans (the hydrosphere).

Changes in the atmosphere

Fossil evidence shows that about 3 billion (3000 million) years ago **cyanobacteria** evolved. These bacteria could use the Sun's energy to produce glucose and oxygen from carbon dioxide and water in the process we call **photosynthesis**. The amount of oxygen increased in the atmosphere and the amount of carbon dioxide decreased. The oxygen produced was a 'polluting' gas and organisms had to evolve that could cope with oxygen in the atmosphere. About one billion years ago simple algae appeared, producing more oxygen and using up carbon dioxide during photosynthesis. Simple plants evolved, then simple animals. These organisms all **respired**, using up oxygen and releasing carbon dioxide. Carbon dioxide was converted into carbonate rocks, sometimes via animals that produced calcium carbonate shells and skeletons (see Figure 6.28).

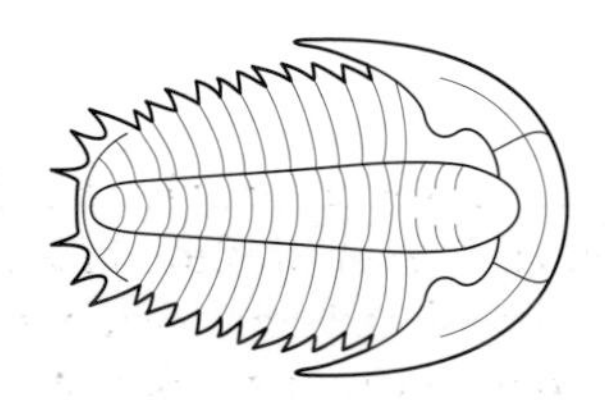

Figure 6.28 Sketch of a trilobite fossil

Eventually about 200 million years ago the levels of carbon dioxide and oxygen in the atmosphere stabilised to their current values (see carbon cycle Figure 6.30 on page 142).

As oxygen levels rose, ammonia and methane reacted with the oxygen, producing more water, more carbon dioxide and nitrogen. The amount of ammonia in the atmosphere decreased.

Did you know?

Many strata of limestone contain the fossil remains of coral and the skeletons of other marine organisms. This is why limestone is mainly calcium carbonate.

Ozone in the atmosphere

With the presence of oxygen in the atmosphere, radiation from the Sun was able to interact with some oxygen molecules (O_2) and split

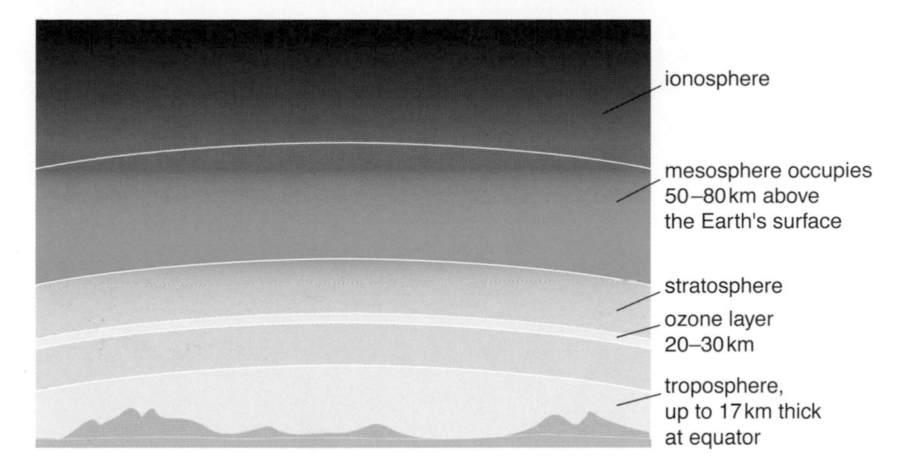

Figure 6.29 The structure of the Earth's atmosphere

them into two oxygen atoms. Oxygen atoms are an example of very reactive particles called free radicals. One oxygen atom can combine with an oxygen molecule to form a molecule of ozone (O_3). Over time, a layer containing ozone molecules built up in the stratosphere. We call this the **ozone layer** (see Figure 6.29).

The ozone layer is very important, because it filters out harmful short-wavelength ultraviolet radiation from the Sun that can cause skin cancer. Since the 1970s, it has been known that chemicals such as **chlorofluorocarbons** (CFCs), once used in aerosols and refrigerators, react with the ozone molecules in the ozone layer, causing it to be less effective in filtering out harmful ultraviolet radiation.

The carbon cycle

The balance of oxygen and carbon dioxide in the atmosphere today is maintained by the carbon cycle. As the carbon cycle shows (see Figure 6.30 overleaf), there is a constant use of carbon dioxide in photosynthesis and release of oxygen, and a constant release of carbon dioxide in respiration and combustion of fossil fuels. The increase in carbon dioxide in the atmosphere over the last 100 years can be attributed to an increase in the burning of fossil fuels and, to some extent, the deforestation of many areas of the world (see section on Global warming and greenhouse gases).

Question

7 a When the atmosphere first formed it contained both hydrogen and helium gas. State why these gases were soon lost by the atmosphere.

b State **two** processes that increase the amount of carbon dioxide in the atmosphere.

c Explain how the carbon cycle maintains a constant amount of oxygen in the atmosphere.

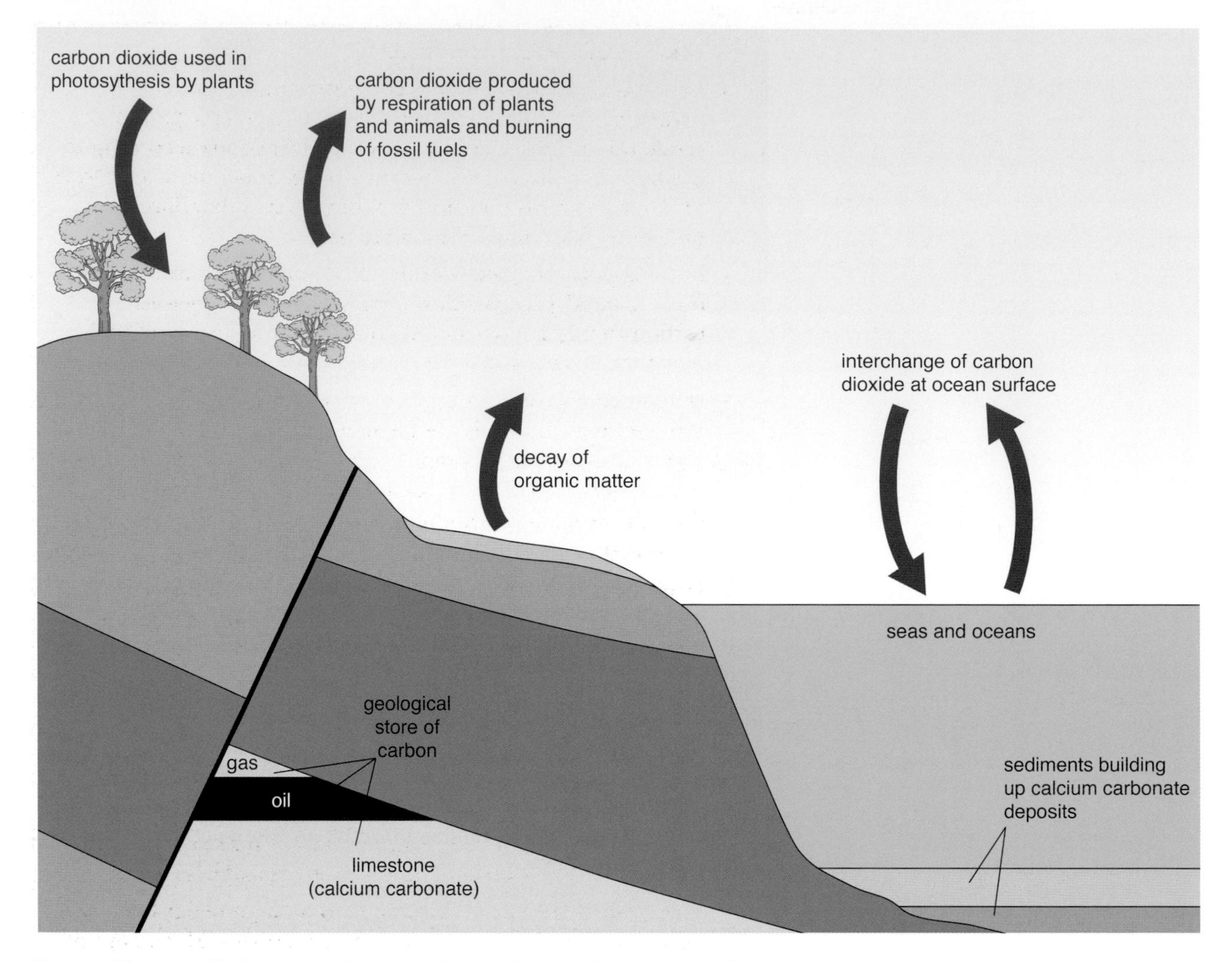

Figure 6.30 A simplified carbon cycle, showing how carbon dioxide is removed from and added to the atmosphere

Summary

1. Using the resources of earth, atmosphere and sea, scientists are able to produce a wide range of useful chemicals by a series of physical and chemical processes. Some processes can cause environmental problems that have to be addressed.
2. Chemical reactions take place over time. Some reactions are very rapid, such as precipitation reactions, and others, such as rusting, are slow.
3. Reactions occur by collisions of particles. Not all collisions are successful. Only a small percentage of collisions lead to reaction. When changes in concentration, temperature or particle size occur, the number of successful collisions changes and therefore the rate changes. Catalysts speed up reactions without changing chemically. They are important in industrial chemistry. Biological catalysts are called enzymes.
4. An increasing number of applications are being found for very small particles called nanoparticles. Very small particles may have quite different properties from the same material in its normal form.

5 Fossil fuels are finite resources.

6 Crude oil or petroleum is a complex mixture of hydrocarbons. Fractional distillation is used to separate crude oil into simpler mixtures called fractions according to the boiling points of the hydrocarbons. Small molecules with few carbon atoms have lower boiling points than large molecules with more carbon atoms.

7 Natural gas and some products from petroleum are used as fuels. When hydrocarbons burn completely, they form carbon dioxide and water. Incomplete combustion of fuels may produce the toxic gas carbon monoxide.

8 When chemical bonds are broken, energy must be supplied. When chemical bonds are formed, energy is released. In chemical reactions, if more energy is released in making new bonds than is used up in breaking bonds in the reactants, then the reaction is exothermic. If the opposite is true, then the reaction is endothermic.

9 Fossil fuels contain sulphur. This forms sulphur dioxide when fuels are burnt. Sulphur dioxide in the atmosphere forms acid rain, which damages buildings and plant life.

10 Burning fossil fuels and using the motor car produce oxides of nitrogen that contribute to acid rain, low-level ozone and photochemical smog.

11 Carbon dioxide is a greenhouse gas. Carbon dioxide molecules absorb infra-red radiation from the Earth, trapping the energy in the atmosphere. This raises the temperature of the Earth and is called global warming.

12 The surface of the Earth is made up of large plates that are slowly moving. They are called tectonic plates. Movement of these plates caused the separation of the South American continent from the African continent millions of years ago. This is called continental drift. Earthquake and volcanic activity is high at plate boundaries. Plate movement drives the rock cycle.

13 The Earth's atmosphere has the following composition: approximately one-fifth oxygen, four-fifths nitrogen, and some carbon dioxide, water vapour and noble gases. The atmosphere has changed over millions of years. Oxygen appeared when cyanobacteria and blue-green algae used photosynthesis. The amount of carbon dioxide was initially very large, but decreased as more plants appeared.

14 There is a balance between carbon dioxide removed from the atmosphere by the photosynthesis of plants, and that introduced into the atmosphere by respiration and the combustion of fossil fuels. This can be shown as a carbon cycle.

15 The balance of carbon dioxide and oxygen is disturbed by continued burning of fossil fuels and by deforestation.

Chapter 7 Chemical bonding, structure and the properties of elements and compounds

By the end of this chapter you should:

- know how atoms are joined in some elements and compounds;
- understand the structures of some elements and compounds;
- understand how the chemical bonding in a substance is linked to its physical and chemical properties (the way it behaves).

Chemical bonding and structure

Electrons in atoms

Before we start to think about the bonding that holds atoms together, we need to remind ourselves of how the electrons in an atom are arranged. This is called the **electron configuration** (see Table 7.1). The electrons occupy **energy levels** that are sometimes called **shells**, or **orbits**.

As we saw in Chapter 5, we sometimes picture these arrangements of electrons in terms of a diagram, as shown in Figure 7.1.

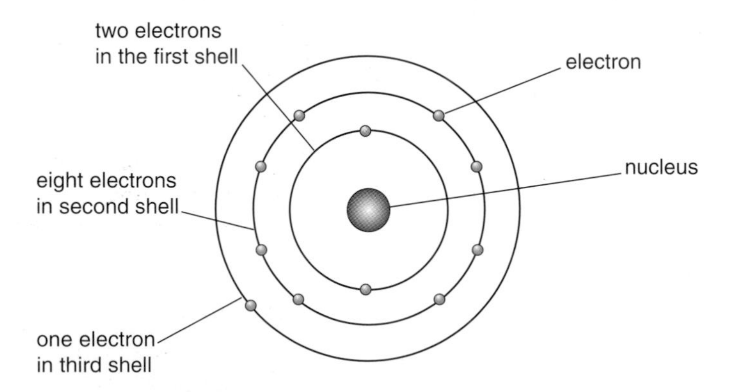

Figure 7.1 The arrangement of electrons in a sodium atom

Did you know?

Scientists are able to design materials with special properties for particular uses, for example:

- alloys such as stainless steel
- polymers such as Kevlar
- textiles such as polyesters
- ceramics such as oven-to-table ware.

Table 7.1 Electron configurations

		Electron configuration			
Atomic number	**Element**	**First shell**	**Second shell**	**Third shell**	**Fourth shell**
1	Hydrogen	1			
2	Helium	2			
3	Lithium	2	1		
4	Beryllium	2	2		
5	Boron	2	3		
6	Carbon	2	4		
7	Nitrogen	2	5		
8	Oxygen	2	6		
9	Fluorine	2	7		
10	Neon	2	8		
11	Sodium	2	8	1	
12	Magnesium	2	8	2	
13	Aluminium	2	8	3	
14	Silicon	2	8	4	
15	Phosphorus	2	8	5	
16	Sulphur	2	8	6	
17	Chlorine	2	8	7	
18	Argon	2	8	8	
19	Potassium	2	8	8	1
20	Calcium	2	8	8	2

Did you know?

The number of Groups 1 to 7 in the Periodic Table of the elements indicates the number of outer electrons in the atoms that make up the group.

When atoms combine chemically, it is the outer electrons that interact. The way atoms behave chemically depends on the arrangement of the electrons within the atoms.

Questions

1 Draw diagrams to show how the electrons are arranged in the following atoms:
a phosphorus **b** calcium **c** chlorine **d** magnesium
e aluminium.

2 Write down the electron configurations of atoms of elements with the following atomic numbers:
a 12 **b** 17 **c** 19 **d** 18 **e** 2.

Hint: The atomic number is equal to the number of electrons in the neutral atom of the element.

								H 1									He 2
Li 3	Be 4											B 5	C 6	N 7	O 8	F 9	Ne 10
Na 11	Mg 12											Al 13	Si 14	P 15	S 16	Cl 17	Ar 18
K 19	Ca 20	Sc 21	Ti 22	V 23	Cr 24	Mn 25	Fe 26	Co 27	Ni 28	Cu 29	Zn 30	Ga 31	Ge 32	As 33	Se 34	Br 35	Kr 36
Rb 37	Sr 38	Y 39	Zr 40	Nb 41	Mo 42	Tc 43	Ru 44	Rh 45	Pd 46	Ag 47	Cd 48	In 49	Sn 50	Sb 51	Te 52	I 53	Xe 54
Cs 55	Ba 56	La 57	Hf 72	Ta 73	W 74	Re 75	Os 76	Ir 77	Pt 78	Au 79	Hg 80	Tl 81	Pb 82	Bi 83	Po 84	At 85	Rn 86

Figure 7.2 Periodic Table

Giant metallic structures

Metals:
www.metals.about.com

Most of the elements in the Periodic Table (see Figure 7.2) are metals (the non-metals are shown in yellow). Metallic elements have well defined properties. The properties of metals and non-metals are compared in Table 7.2.

Table 7.2 Properties of metals and non-metals

Metals	Non-metals
Good conductors of electricity (high electrical conductivity)	Poor conductors of electricity (low electrical conductivity)
Good conductors of heat (high thermal conductivity)	Poor conductors of heat (low thermal conductivity)
Malleable (can be beaten into sheets)	Not malleable (tend to be brittle in the solid state)
Ductile (can be drawn into wires)	Not ductile
Lustrous (freshly exposed surfaces are shiny)	Not usually lustrous
High melting points	Low melting points
High boiling points	Low boiling points

Figure 7.3 The gold mask of Tutankhamen

The structure of metals must explain the behaviour (properties) of metals. Figure 7.3 shows the beautifully worked gold mask of Tutankhamen.

The accepted view of the structure of a metal is as a regular arrangement of positive ions in a 'sea' of free electrons. The electrons behave as a sort of glue that holds the structure together (see Figure 7.4). Regular arrangements of atoms or ions are called **lattices**. The mobile electrons come from the outer electrons of the metal atoms.

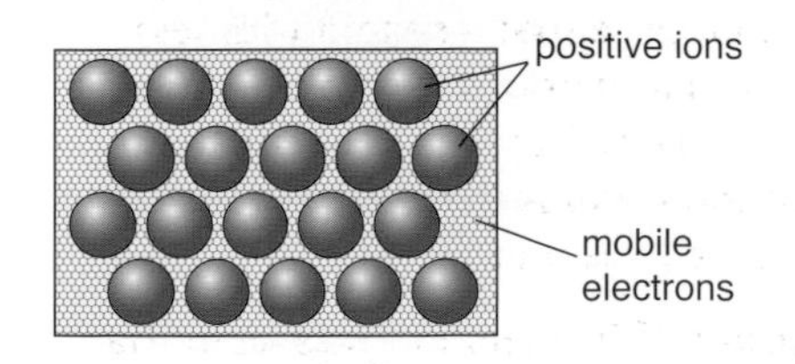

Figure 7.4 Simple representation of a metallic lattice

Using this theory of the structure of metals, we can explain the following:

- **Electrical conductivity:** When a metal is connected to the terminals of a battery, the negative electrons are free to move towards the positive terminal, and an electric current flows.
- **Malleability:** The sea of electrons can act as a lubricant between the layers of positive ions. When a piece of metal is beaten with a hammer, the layers of ions can slide over one another so that the metal can be formed into a sheet. (In the same way, try to explain why metals can be drawn into wires.)
- **The melting points of metals:** In general, the melting points of metals are high, showing that the bonding in metals is strong. The strength of the bonding depends on the number of outer electrons in the atoms. This is demonstrated in Table 7.3, which shows the melting points of the elements in Groups 1 and 2 of the Periodic Table.

Table 7.3 Melting points of the elements in Groups 1 and 2

Group 1		Group 2	
Element	**Melting point (°C)**	**Element**	**Melting point (°C)**
Li	181	Be	1278
Na	98	Mg	650
K	63	Ca	839
Rb	39	Sr	769
Cs	29	Ba	725

Did you know?

The metal with the highest melting point is tungsten, which is used for light-bulb filaments. Its melting point is 3410 °C.

The Group 2 elements have two outer electrons, and the Group 1 elements have only one outer electron. As a result, the corresponding Group 2 elements have a much higher melting point than the Group 1 elements (see Table 7.3).

Magnesium is a much harder metal than sodium, which can easily be cut with a knife. Aluminium has three outer electrons, and so its mechanical strength is very great compared with that of metallic sodium.

H

Alloys

Mixtures of metals are called **alloys**. Stainless steel is used to make saucepans, cutlery and sinks. Metallic spectacle frames are made from a variety of alloys. Nickel alloys and titanium alloys are common, and some modern frames are made from smart alloys that allow the shape to be restored after bending or deformation.

Did you know?

Stainless steel is an alloy of iron, manganese, chromium and nickel, with traces of the non-metals carbon, phosphorus and silicon. Alloys that contain the metal mercury are called **amalgams**. One amalgam containing mercury, silver, tin and copper has been used for dental fillings for the last 150 years, although recently questions have been raised over its toxicity.

Metallic glasses

Metals have a regular lattice in which the atoms have a high degree of order. It has been known for some time that a system with less order will have enhanced mechanical and anti-corrosion properties. In recent years, alloys have been discovered that form solids with a far more random arrangement of their atoms than normal crystalline metallic alloys. These are called metallic glasses.

A number of alloys have been produced in metallic glass form that have been used in electronic components, replacement joints, jewellery, tennis rackets, the aerospace industry, military hardware and golf-club heads. These alloys often contain zirconium (for instance one alloy consists of zirconium, titanium, copper, nickel and aluminium). It was reported in 2004 that pure zirconium had been produced in metallic glass form.

One of the properties of metallic glasses is improved elasticity when hitting an object, which is ideal for a golf club head. Typical metallic glasses are three times stronger than steel and ten times more 'springy'.

Materials science

The study of substances used in building, construction and manufacture is known as materials science. Many benefits of new materials are obvious, but it is important that any drawbacks should be understood. Historically, some materials have been used widely and then found to have negative health risks. One example was the widespread use of asbestos in the construction industry in the twentieth century.

Questions

3 What is meant by the term **alloy**? Name two alloys in everyday use.

4 State the physical properties of a typical metal.

5 Which properties of copper make it suitable for making electrical goods?

Giant ionic structures

Ions are electrically charged atoms or groups of atoms. Positive ions are called **cations**. Negative ions are called **anions** (see Table 7.4). Atoms form ions by losing or gaining electrons.

Table 7.4 Positive ions and negative ions

Positive ions (cations)		
Charge		
+1	+2	+3
Sodium, Na^+	Magnesium, Mg^{2+}	Aluminium, Al^{3+}
Potassium, K^+	Calcium, Ca^{2+}	Iron(III), Fe^{3+}
Lithium, Li^+	Barium, Ba^{2+}	Chromium(III), Cr^{3+}
Ammonium, NH_4^+	Copper(II), Cu^{2+}	
Silver, Ag^+	Lead(II), Pb^{2+}	
	Iron(II), Fe^{2+}	
Negative ions (anions)		
Charge		
−1	−2	−3
Chloride, Cl^-	Oxide, O^{2-}	Phosphate, PO_4^{3-}
Bromide, Br^-	Sulphate, SO_4^{2-}	
Iodide, I^-	Carbonate, CO_3^{2-}	
Hydroxide, OH^-		
Hydrogen carbonate, HCO_3^-		
Nitrate, NO_3^-		

H

Atoms usually lose or gain electrons to reach the electronic arrangement of a noble gas. Apart from helium, the noble gases have eight outer electrons, an octet. Helium has two outer electrons.

Did you know?

The annual production of salt worldwide is over 210 million tonnes.

The compound sodium chloride is common salt, which is used to flavour food and is an essential ingredient of our diet. An atom of sodium has the electron configuration 2.8.1 (see Figure 7.5). An atom of chlorine has the electron configuration 2.8.7 (see Figure 7.6).

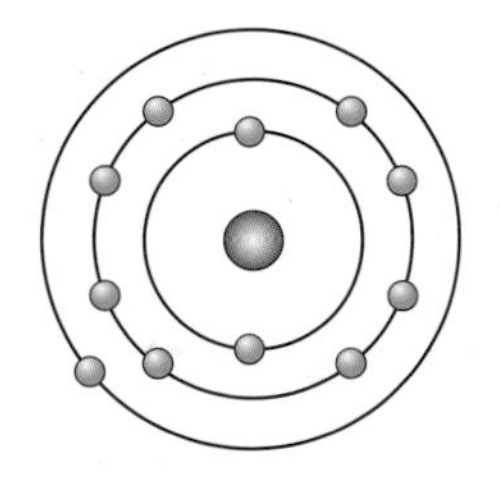

Figure 7.5 Sodium atom

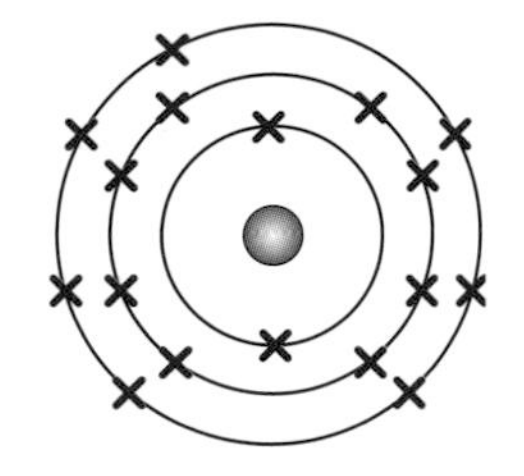

Figure 7.6 Chlorine atom

If an atom of sodium loses its outer electron, its electron configuration is 2.8, like the noble gas neon. It is now no longer a neutral atom but a positively charged sodium ion. This is written as Na^+. In the same way, if an atom of chlorine gains one electron, its electron configuration is like that of the noble gas argon. It is now no longer a neutral atom but a negatively charged chlorine (or chloride) ion. This is written as Cl^-. The process is shown in Figure 7.7. (Diagrams like Figures 7.5–7.7 are called dot-and-cross diagrams.)

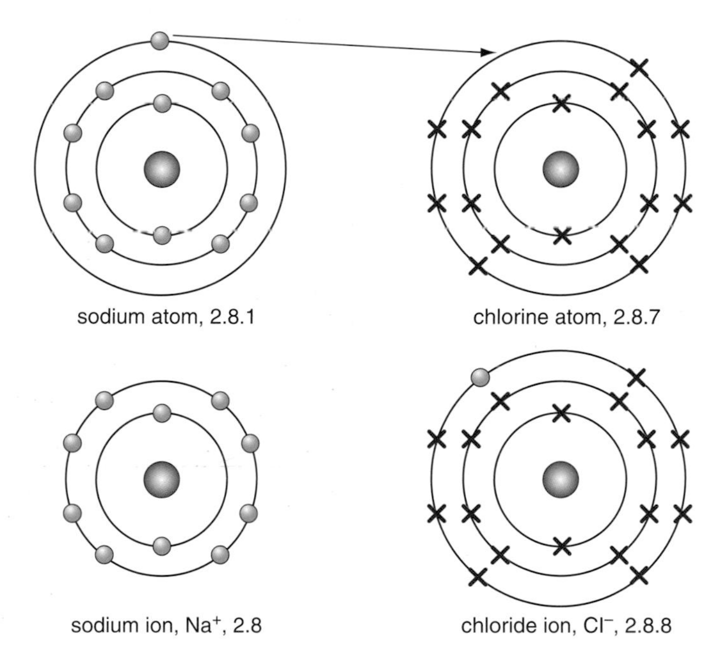

Figure 7.7 Formation of sodium chloride by electron transfer

Sodium chloride contains sodium ions and chloride ions. Since sodium chloride is electrically neutral, there must be equal numbers of sodium and chloride ions. The positive charges cancel out the negative charges. The formula for sodium chloride is NaCl.

Positive charges attract negative charges strongly. The smallest crystal of sodium chloride contains millions of sodium ions and chloride ions held together in a regular arrangement, or lattice, by these strong electrostatic forces (see Figure 7.9).

Figure 7.8 Crystals of impure rock salt; even in this impure sample, you can see the cubic shape of sodium chloride crystals.

H

The ions in sodium chloride are arranged in a cubic lattice, each ion being surrounded by six nearest neighbours of opposite charge. This is shown in Figure 7.9. The simplest way of drawing the arrangement of the ions in the lattice is shown in Figure 7.10.

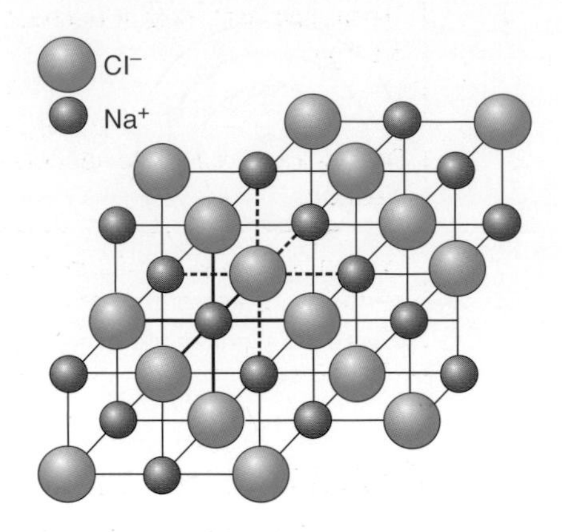

Figure 7.9 Sodium chloride lattice of ions

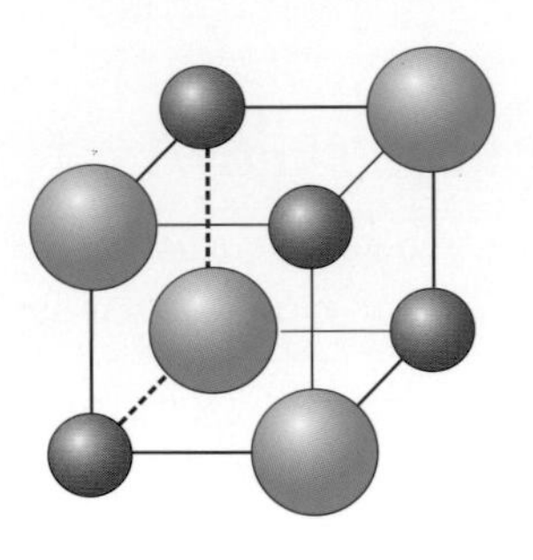

Figure 7.10 Part of the sodium chloride lattice

H

Question

6 How does the shape of the crystals in Figure 7.8 relate to the lattice structure in Figure 7.9?

H

Activity

1 Show the formation of the ions of lithium, potassium, magnesium, calcium, oxygen, sulphur and fluorine.
2 Write the formulae of the oxides, fluorides, chlorides and sulphides of Li, Na, K, Mg and Ca. (Look at the example below.)

H

Example: calcium fluoride

Each calcium atom loses its two outer electrons to reach the electron arrangement of the noble gas argon, and these electrons are taken up by two fluorine atoms to reach the electron arrangement of the noble gas neon.

The result is one calcium ion, Ca^{2+}, and two fluoride ions, F^- (see Figure 7.11). Therefore the formula of calcium fluoride is CaF_2. In a crystal of calcium fluoride, there are twice as many fluoride ions as there are calcium ions.

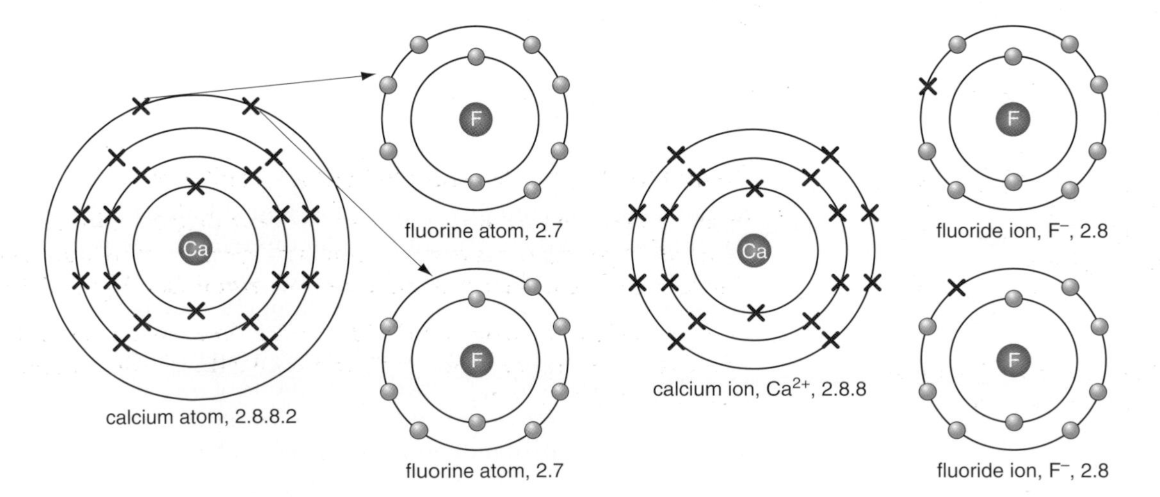

Figure 7.11 Formation of calcium fluoride by electron transfer

Some properties of ionic compounds are listed in Table 7.5.

Table 7.5 Properties of ionic compounds

Property	Cause
The melting points of ionic compounds are high	The forces between the ions are strong electrostatic forces which need a lot of energy to break them
Solid ionic compounds **do not** conduct electricity.	The ions are held in fixed positions and are not free to move.
When molten or dissolved in water, ionic compounds **do** conduct electricity.	The lattice breaks down, and the ions are free to move and conduct an electric current

Covalent molecular substances

Covalent molecular compounds exist as neutral particles called **molecules**. Molecules are formed from atoms through the sharing of electrons. A covalent bond exists when a pair of electrons is shared between two atoms.

Tables 7.6 and 7.7 list some common covalent molecular compounds and elements.

Table 7.6 Some common covalent molecular compounds

Name	Formula
Hydrogen chloride	HCl
Water	H_2O
Carbon monoxide	CO
Carbon dioxide	CO_2
Ammonia	NH_3
Sulphur dioxide	SO_2
Methane	CH_4

Table 7.7 Covalent elements

Name	Formula
Oxygen	O_2
Nitrogen	N_2
Chlorine	Cl_2
Sulphur	S_8

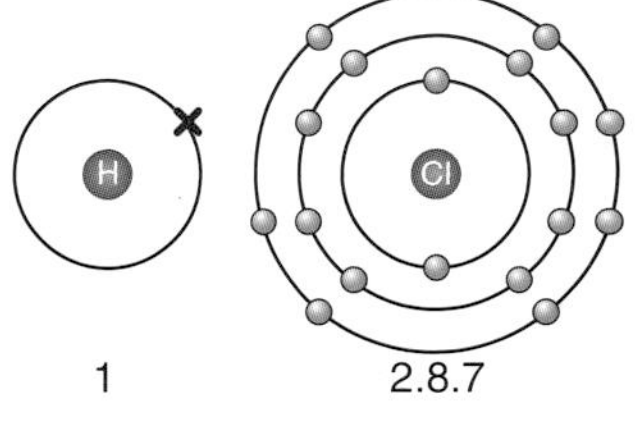

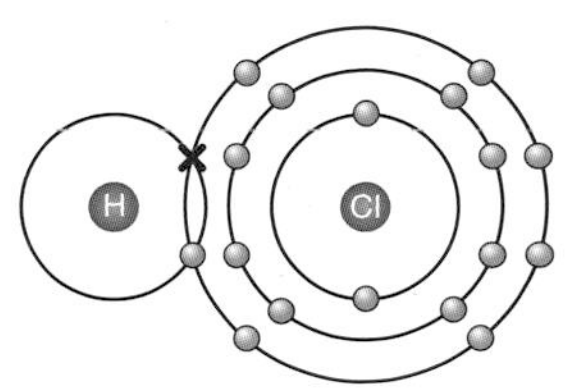

Figure 7.12 Formation of hydrogen chloride molecule by electron sharing

The formation of molecules

The formation of a molecule of hydrogen chloride

Because a pair of electrons is shared in the hydrogen chloride molecule, hydrogen has the electron arrangement of helium, and chlorine has the electron arrangement of argon (see Figure 7.12).

The molecule of hydrogen chloride is sometimes written as H–Cl, where the line between the atoms represents the covalent bond (a shared pair of electrons).

H

Chapter 8 Metals, chemical calculations and green chemistry

By the end of this chapter you should:

- know that metals occur in ores that are naturally occurring minerals;
- know that ores can usually be turned into oxides from which a metal can be produced;
- know that when an oxide is turned into a metal, it has been reduced;
- know that reduction can be described as removal of oxygen, and oxidation can be described as gain of oxygen;
- know that the reactivities of metals vary and that metals can be arranged in a reactivity series;
- know that one metal can displace another from a solution of one of its compounds;
- understand that the method of extraction of a metal depends on its position in the reactivity series (aluminium and iron are used as examples);
- understand the corrosion of metals, rusting, and its prevention;
- understand how to carry out chemical calculations using relative atomic masses;
- understand the meaning of green chemistry;
- understand the importance of water, its purification, the public water supply, and hardness of water.

The production and uses of metals

Metallic ores

Only the least reactive of metals occur uncombined in the Earth's surface. The most obvious one is gold, which has been known since prehistoric times since it is found in nature as the metal. The other metals exist as ores in combination with other elements such as oxygen and sulphur (see Figure 8.1), and some common ores are given in Table 8.1. Metals are produced from their ores by a process called **reduction**.

Metals have different reactivities, and are often placed in a reactivity series.

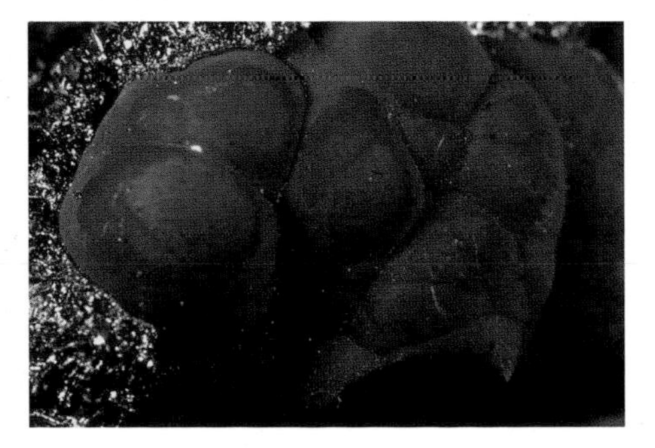

Figure 8.1 Kidney ore is a form of haematite

Table 8.1 Examples of some common ores

Element	Ore	Formula of ore
Iron	Haematite, an oxide ore	Fe_2O_3
Iron	Iron pyrites (fool's gold), a sulphide ore	FeS_2
Iron	Siderite, a carbonate ore	$FeCO_3$
Aluminium	Bauxite, an oxide ore	$Al_2O_3.2H_2O$
Lead	Galena, a sulphide ore	PbS
Magnesium	Epsomite (Epsom salts), a sulphate ore	$MgSO_4.7H_2O$
Titanium	Rutile	TiO_2

The reactivity series of metals

Extraction of metals:
www.gcsechemistry.com

Different metals have different reactivities. This can be shown by looking at how calcium and magnesium react with water.

If a piece of calcium is added to cold water, it immediately gives off bubbles of gas and a white precipitate forms.

calcium + water → calcium hydroxide + hydrogen gas

Magnesium shows little reaction with cold water, but reacts vigorously with steam (see Figure 8.2).

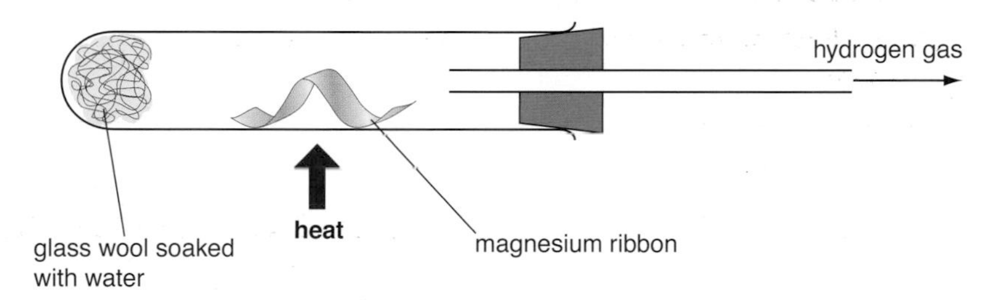

Figure 8.2 The heat applied to the test tube turns the water into steam, which passes over the hot magnesium; vigorous reaction is seen:
magnesium + water → magnesium oxide + hydrogen

A more reactive metal will remove oxygen from the oxide of a less reactive metal when a mixture of the two is heated. One interesting example of this is the **Thermit reaction** (see Figures 8.3 and 8.4).

Figure 8.3 Rail welding using the Thermit reaction

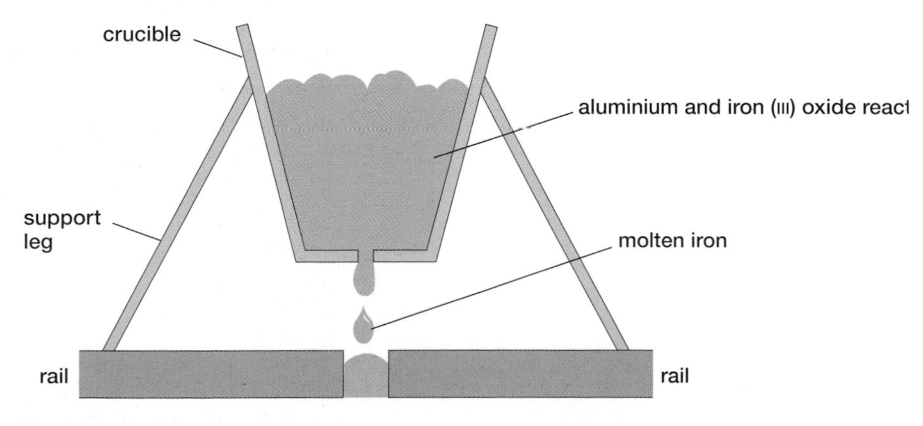

Figure 8.4 The Thermit reaction

Figure 8.5 The Parys Mountain copper mine

When a mixture of powdered aluminium and iron(III) oxide is ignited by a high-temperature fuse, molten iron is formed. This reaction is used to weld railway lines.

An alternative method of constructing a reactivity series is by investigating **displacement reactions**. In a displacement reaction, one metal displaces another from a solution of one of its compounds.

Did you know?

The water running through spoil heaps around old copper mines contains small amounts of dissolved copper. If this water is allowed to run over scrap iron, small amounts of copper metal are formed. This is a displacement reaction.

Did you know?

Copper used to be mined in several regions of Wales. In the eighteenth century, copper from the Parys Mountain mine in Anglesey (see Figure 8.6) was exported all over the world.

Practical work

Reactivities

1 Wear eye protection.
2 Place a few pieces of zinc metal in a beaker of copper(II) sulphate solution, and observe what happens.
3 Stir the solution. The blue colour of the copper(II) sulphate solution fades, and the zinc pieces become coated with brown copper (see Figure 8.6).

Figure 8.6 Zinc displaces copper from copper sulphate solution.

The more reactive zinc displaces the copper from the solution.

zinc + copper(II) sulphate solution → zinc sulphate solution + copper

$$Zn(s) + CuSO_4(aq) \rightarrow ZnSO_4(aq) + Cu(s)$$

4 Repeat the experiment, but replace the zinc with iron and then with magnesium.
5 Place a piece of copper foil in some silver nitrate solution, and leave it to stand.

Experiments of this kind lead to the table of reactivities shown in Table 8.2 overleaf.

Table 8.2 The reactivity series of common metals

Element	Method of extraction	Reaction with acid
K most reactive	Manufactured by electrolysis	*Too dangerous!*
Na		
Ca		Gives hydrogen with dilute hydrochloric acid
Mg		
Zn	Manufactured by chemical reduction	
Fe		
Sn		
Cu		Does not give hydrogen with dilute hydrochloric acid
Ag least reactive		

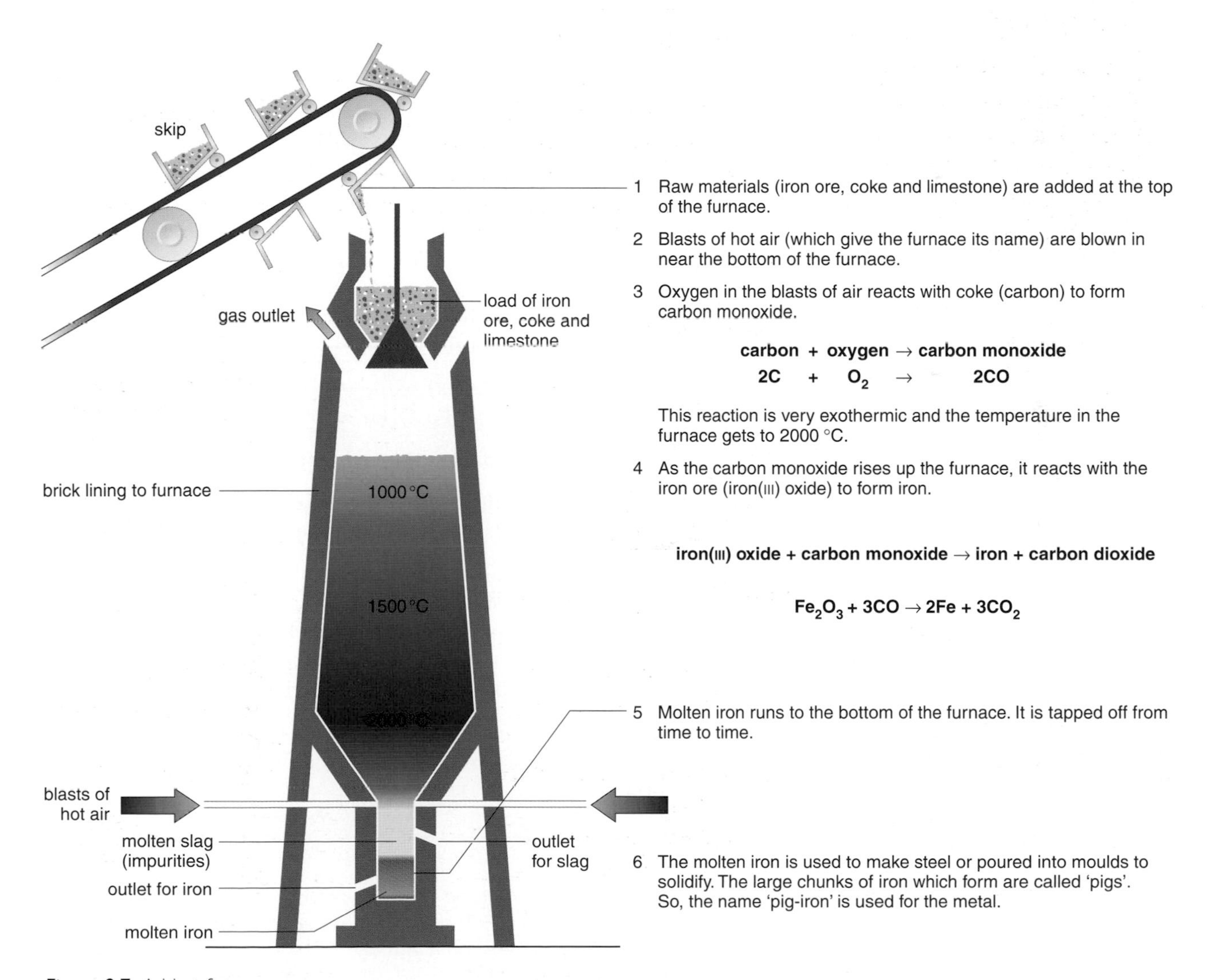

Figure 8.7 A blast furnace

Figure 8.8 Iron and steel industry, South Wales

Iron

The extraction of iron

Iron is a metal that is extracted from its ores by chemical reduction with carbon, which is more reactive than iron (see Figure 8.7 on page 161). Although the iron and steel industry has decreased in size in recent years, especially in Wales (see Figure 8.8), it is still a very important industry worldwide.

The extraction is carried out in a blast furnace, as shown in Figure 8.7. Iron ore, coke and limestone are heated in the blast furnace to make iron. The problem is to turn the iron(III) oxide in the iron ore into iron. This means removing oxygen. Hot air is blown into the furnace, where it combines with the coke (which is mostly carbon) to form carbon monoxide and liberate heat. The carbon monoxide reacts with the iron ore high in the furnace to form molten iron, which collects at the base of the furnace. The reaction is:

carbon monoxide + iron(III) oxide → carbon dioxide + iron

$$3CO(g) + Fe_2O_3(s) \rightarrow 3CO_2(g) + 2Fe(l)$$

H

The coke itself may take part in the reduction of the iron.

carbon + iron(III) oxide → carbon monoxide + iron

$$3C(s) + Fe_2O_3(s) \rightarrow 3CO(g) + 2Fe(l)$$

H

The limestone removes sandy materials from the iron ore to produce a molten slag of calcium silicate that floats on top of the molten iron. The slag is used as hardcore for roads and in building. The waste gases from the top of the furnace are used to preheat the blast of air at the bottom. In some furnaces this air is enriched with oxygen.

The iron that is produced by the blast furnace is often called **pig iron**. It is a brittle metal as it contains a significant amount of carbon, up to 4.5%.

Questions

1 Write word equations for the following reactions in the blast furnace:
  a the conversion of coke (carbon) into carbon monoxide
  b the conversion of iron oxide into iron

2 Why is limestone added to the blast furnace?

Steelmaking:
www.schoolscience.co.uk

Figure 8.9 The five millionth Audi A6 leaves the production line.

Figure 8.10 A team member taking iron samples from a blast furnace at Corus Port Talbot steel works

Properties and uses of steel

Most of the pig iron from blast furnaces is converted to steel, which is far more useful, by the removal of some or most of the carbon. Steel with the lowest percentage of carbon (0.1–0.4%) is called **mild steel**. This is used widely for car bodies (see Figure 8.9), tin plate (used for making cans for food storage), piping, screws, and nuts and bolts. The industrial development of China has brought about an increase in the worldwide demand for mild steel.

Hard steel contains a higher percentage of carbon, up to 1.5%. It is used in machine tools in industry and domestic tools sold in DIY stores.

The properties of steel can be further changed by heat treatment and alloying with other elements. Stainless steel contains chromium and nickel. The addition of tungsten produces extremely hard steel that can be used in high-speed cutting tools. The addition of manganese produces hard, tough steel used in the rims of the wheels of railway rolling stock.

Steel is such an important material that it is recycled on a large scale. Recycling steel:

- saves up to 50% of the energy costs
- helps to conserve iron ore
- cuts down the emission of greenhouse gases.

The iron and steel industry in Wales (see Figure 8.10) initially grew out of the availability of the raw materials coal, iron ore and limestone. However its size has diminished in recent years, with the closure of a number of plants.

Electrolysis

Electrolysis is a chemical reaction that is brought about by an electric current passing through a conducting liquid. The conducting liquid is called the **electrolyte**. The current enters the electrolyte via two solid conductors called **electrodes**. The positive electrode is called the **anode**, and the negative electrode is called the **cathode**. The current is carried in the electrolyte by the movement of ions. Negative ions move towards the anode and are called **anions**. Positive ions are attracted to the cathode and are called **cations**.

A typical set-up is shown in Figure 8.11.

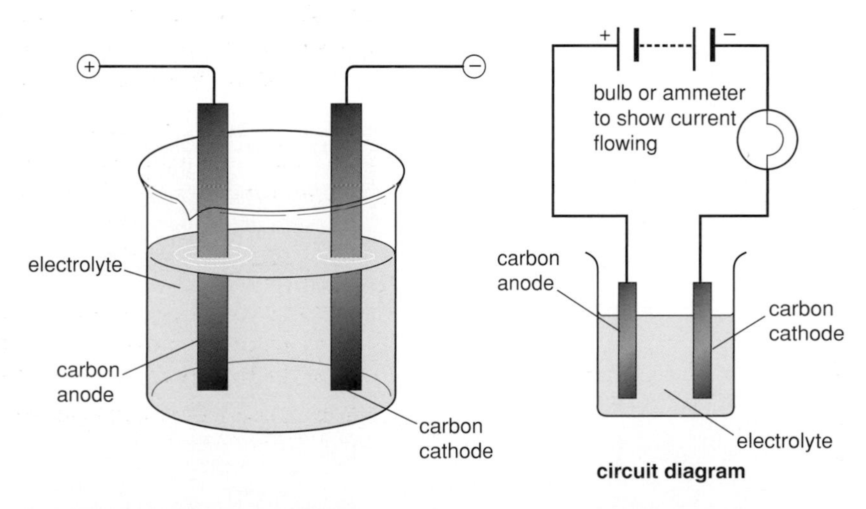

Figure 8.11 Electrolysis apparatus

Did you know?

Sir Humphrey Davy (see Figure 8.12) discovered sodium and potassium using electrolysis.

Did you know?

Chlorine is manufactured on a large scale by the electrolysis of sodium chloride solution. Sodium is also manufactured by electrolysis, and chlorine is a by-product.

Figure 8.12 Sir Humphrey Davy

The simplest case of electrolysis is one where the electrolyte contains only two ions. One example is an electrolyte of molten sodium chloride, NaCl. Sodium chloride melts at quite a high temperature, and so a crucible is required, and care must be taken in setting up the electrolysis. Carbon electrodes are satisfactory. Molten sodium chloride contains only sodium ions (the cations) and chloride ions (the anions) (see Figure 8.13).

When sodium chloride is dissolved in water, the electrolysis products are different because water itself produces small amounts of hydrogen and hydroxide ions. Hydrogen is released at the cathode and chlorine at the anode.

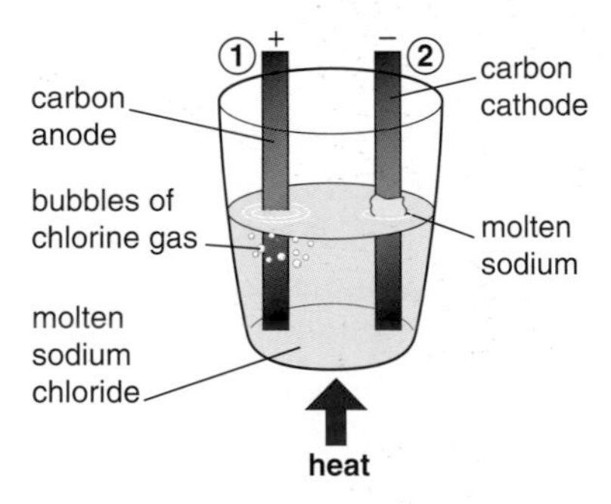

Figure 8.13 The electrolysis of molten sodium chloride

Practical work

Electrolysis

1 Wear eye protection.
2 Remember that chlorine is toxic. If you identify that the substance is present, switch off the circuit. Inhaling the gas can lead to breathing difficulties.
3 Set up electrolysis experiments for the following:
   a dilute sulphuric acid
   b dilute sodium chloride solution
   c dilute copper(II) sulphate solution
   d dilute hydrochloric acid.

All the very reactive metallic elements are usually extracted from ores by electrolysis because they cannot easily be displaced by chemical reduction reactions. Aluminium is a very reactive metal, although it can be used for window frames and for other construction purposes. It does not seem to be reactive because a very thin film of protective aluminium oxide forms on the surface of the metal. This film protects the metal from any further attack.

Did you know?

The thin oxide layer on aluminium can be increased in thickness through a process called *anodising*. The new oxide layer can be treated with dyes to give the metal an attractive colour.

Did you know?

Aluminium is the most abundant metal in the Earth's crust.

Questions

3 Explain what is meant by the words **anode**, **cathode**, **electrolyte**.

4 Draw a simple labelled diagram of the electrolysis cell for the extraction of aluminium, showing the anode, cathode and electrolyte.

Aluminium

The manufacture of aluminium by electrolysis

The usual source of aluminium is the ore bauxite. The bauxite is treated chemically to remove impurities, and is finally turned into the white solid aluminium oxide.

Aluminium oxide is sometimes called **alumina**. It has a very high melting point. For aluminium oxide to be electrolysed (see Figure 8.14), it has to be dissolved in molten cryolite. This brings the working temperature of the electrolyte to about 950 °C. Oxygen gas is formed at the carbon anodes, and at high temperature the anodes react with oxygen, burning away and having to be replaced from time to time. The carbon lining of the cell is also the cathode, and aluminium is formed here as the molten metal. Periodically, aluminium metal is removed, the crust is broken, and more aluminium oxide is added.

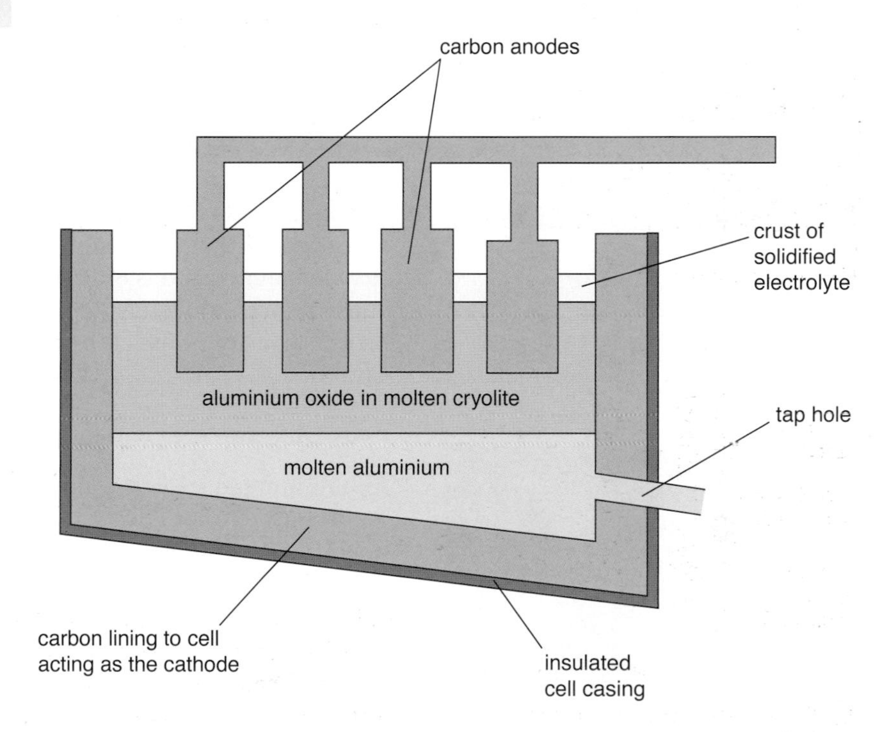

Figure 8.14 The electrolysis of aluminium oxide

The aluminium ions are attracted to the cathode and form aluminium metal.

Al^{3+}	$+ 3e$	$\rightarrow$	Al
aluminium ion			aluminium atom

The oxide ions are attracted to the anode and form oxygen gas.

$2O^{2-}$	$- 4e$	$\rightarrow$	O_2
oxide ions			oxygen molecule

H

The siting of an aluminium plant

Did you know?

Figure 8.15 The Anglesey Aluminium Metal aluminium smelter

The electrolytic production of aluminium requires very large quantities of electricity. The aluminium smelting plant at Anglesey (see Figure 8.15) consumes 250 MW, which represents over 12% of all the electrical power consumed in Wales. Aluminium plants are often sited where there is cheap power, for example near hydroelectric power sources.

The factors to be considered in the siting of an aluminium plant are very important. In Wales, the Anglesey site was chosen because it offered a deep-sea port and good road and rail links to customers in the UK and European Union. Aluminium and coke are imported by sea, and power is supplied by the National Grid from a local nuclear power station.

The high energy costs involved in the production of aluminium make recycling very economical. The energy cost per tonne of recycled aluminium is only about 5% of the energy cost per tonne of aluminium produced from bauxite. In addition, the electrolytic process consumes the carbon anodes and produces oxides of carbon.

Did you know?

An aluminium can is very thin, but the pressure of the fizzy drink inside strengthens the can and its seal.

Figure 8.16 Aluminium cans

Properties and uses of aluminium

Aluminium is a strong, light metal. Its density is 2.7 g/cm^3. It is a good conductor of heat and electricity. Although it is a reactive metal, it is resistant to corrosion because of the protective layer of oxide that builds up on its surface.

Some of the uses of aluminium are the following:

- overhead high-voltage power cables for the National Grid; its lightness enables the pylons to be lightweight structures
- saucepans, aluminium cooking foil (linked to its good conduction of heat and non-toxicity); its strength and lightness make it suitable for window-frames and greenhouse construction
- drinks cans (see Figure 8.16), because of its lightness and non-toxicity
- in the manufacture of aeroplane and car bodies, since it is light and has high tensile strength.

Question

5 Give **three** physical properties of aluminium, and give **three** large-scale uses of aluminium.

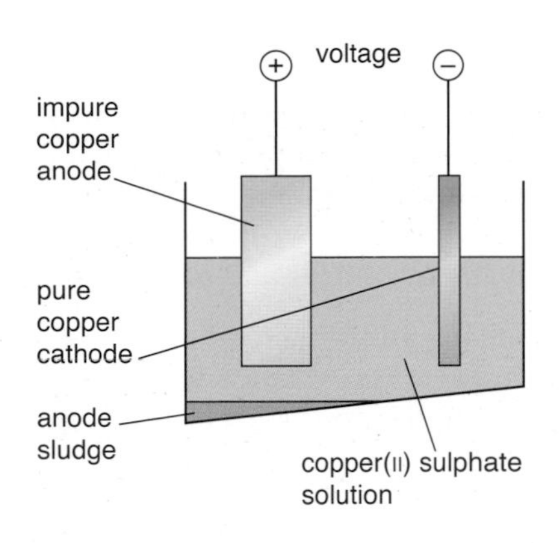

Figure 8.17 Purification of copper by electrolysis

Figure 8.18 Three copper ingots smelted at the Hafod Copper Works, Swansea, Wales, 1890

Copper

For many uses, for example copper piping and copper wire, copper must be very pure. To obtain the level of purity required, impure copper formed by chemical reduction is purified by electrolysis (see Figure 8.17). The anode (positive electrode) is a lump of impure copper. The cathode (negative electrode) is a plate of pure copper.

When a current is passed, copper atoms from the anode become positive copper ions that are deposited as pure copper atoms on the cathode:

- The anode decreases in size.
- The cathode grows in size as pure copper is deposited.
- The impurities from the anode form sludge at the bottom of the cell. This sludge is rich in silver, gold and other elements, and can be treated economically to recover these valuable elements.

At the **anode** (the positive electrode), copper atoms become copper ions by losing two electrons.

H

copper atoms – electrons → copper ions

$$Cu - 2e \rightarrow Cu^{2+}$$

At the **cathode** (the negative electrode), copper ions become copper atoms by gaining two electrons.

H

copper ions + electrons → copper atoms

$$Cu^{2+} + 2e \rightarrow Cu$$

The properties and uses of copper

Copper is an ancient metal that can be found in an uncombined state in nature, native copper, although this is rare. As it is unreactive, it is easily obtained from its ores (see Figure 8.18). When alloyed with tin, it forms bronze. Bronze was an early prehistoric discovery, and was the main alloy in use in the Bronze Age.

The other important alloy of copper is brass, which contains copper and zinc.

Pure copper is an attractive metal in terms of its colour and lustre, and so it is used for jewellery and ornaments. It is also widely used today in plumbing as copper piping, and as copper wire in electrical circuits, motors and other electrical goods.

- It is an excellent conductor of electricity and heat.
- Its malleability and ductility make it easy to form into copper sheets, wires and pipes.

Nineteenth- and early twentieth-century plumbing used lead pipework. Copper or plastic piping is now preferred for the drinking water supply.

Many stainless steel saucepans incorporate a copper bottom for extra heat conduction.

Figure 8.19 Sea King helicopter HU5

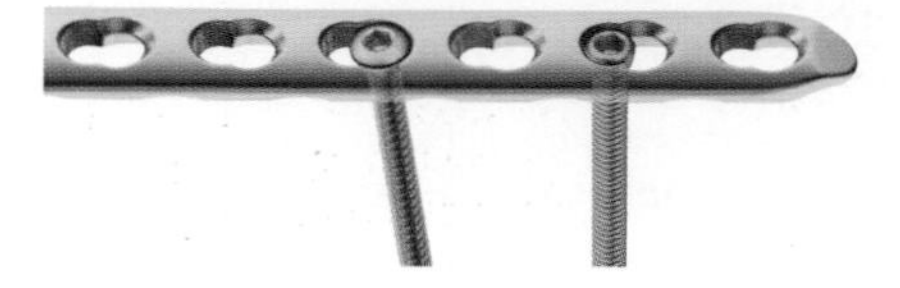

Figure 8.20 Titanium surgical plates

Titanium

Titanium is an important modern metal. For many years it has been made from titanium dioxide. Titanium dioxide is a plentiful ore that is also the basis of a white pigment in paints. Titanium is a hard, strong metal with a high melting point, and it is corrosion-resistant. It has a low density and is a good conductor of heat and electricity.

It is used in the aircraft industry, especially in rotors for helicopters (see Figure 8.19). Its lightness, toughness and resistance to corrosion make it suitable for surgical applications such as hip joints and pins and plates for bone fractures (see Figure 8.20). Other domestic uses are in jewellery and golf clubs.

The consequences of metal extraction

Most metals are extracted from ores that have to be mined. When the ores run out, the mine workings may be abandoned. The mines themselves and the abandoned workings can cause a deterioration of the landscape. There are still old mines in Cornwall and parts of Wales that have not been worked for decades.

On the other hand, the presence of ores that can be extracted is important economically. The existence of ready supplies of iron ore, coal and limestone was essential for the economic growth of Great Britain during the Industrial Revolution. The extraction and processing of metallic ores brings employment. If a country has to import metals, the bill can be very large and cause an impact on the country's balance of payments. The country's finances can also be affected by world prices for the various metals.

Most ores have to be processed, and this may have a detrimental effect on the environment. The production of aluminium, for example, requires large amounts of electricity that in turn may cause greenhouse gases to enter the environment. Many ores are sulphides and have to be smelted, which produces sulphur dioxide (the gas that causes acid rain). Sometimes a particular metal is found with a less useful and more toxic metal. For example, there are recorded cases of land contamination with the toxic metal cadmium near chemical plants producing zinc.

Corrosion of metals

Figure 8.21 Cars and bikes require protection against corrosion otherwise they can rust.

We all know that a metal like gold does not corrode in the air. We also know that motor cars and bridges made from steel show signs of rusting over a period of time (see Figure 8.21). Corrosion can be described as the chemical interaction of a metal with the atmosphere. The basic reaction is oxidation by the oxygen in the atmosphere.

The rusting of iron requires the presence of both oxygen and water. This can be shown by setting up the experiments in Figure 8.22.

Rust is hydrated iron(III) oxide. When rust forms on the surface of an iron or steel object, it flakes away and exposes more of the metal, which in turn forms more rust. This is why rust spots on motor car bodies must be treated as quickly as possible. Aluminium does not corrode as easily as iron, for instance, because the oxide layer is firmly attached to the surface of the metal and does not flake away.

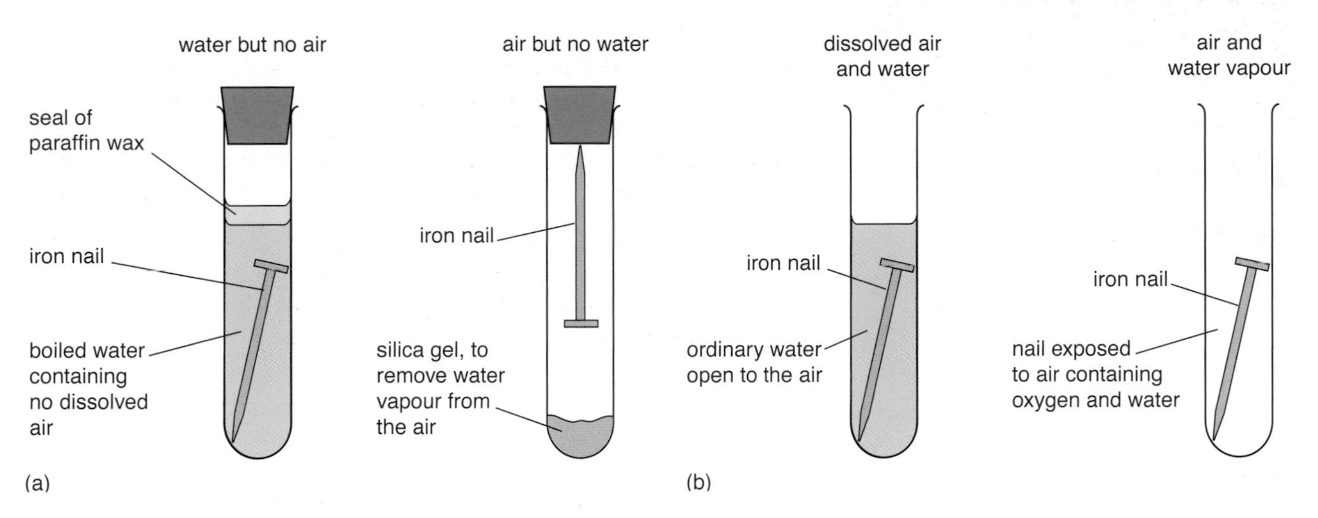

Figure 8.22 The rusting of iron: (a) in each case there is no rusting, (b) rusting occurs slowly.

Figure 8.23 Steel bridges like the Menai Bridge require maintenance to prevent corrosion. Painting a bridge prevents the rusting reaction.

Figure 8.24 Giant oil platform in the Gulf of Mexico

Prevention of rusting

The most obvious way to stop iron and steel from rusting is to prevent moisture and air from reaching the metal surface.

- One way is to paint iron and steel objects (see Figure 8.23). This excludes the atmosphere and prevents rusting.
- If an iron surface is coated with oil, the same effect is produced. If a bicycle chain is well oiled, then it does not rust.
- One can electroplate a steel object with a metal that protects the steel. The object is made the cathode of an electrolysis cell, and the metal is deposited. Chromium plating provides such protection, and the polished chromium layer provides an attractive finish.
- Zinc is a more reactive metal than iron, and it is used to protect steel objects in a process called *galvanising*. The steel object is dipped into molten zinc so that its surface is coated with a thin layer of zinc. Even if the zinc is scratched and the iron surface is exposed, it does not rust. The zinc, being more reactive, is oxidised by the air preferentially.
- Metal dustbins are usually made from galvanised steel. Some motor-car bodies are given extra protection by being galvanised before being painted. In galvanising, the zinc acts as a sacrificial metal. This principle is extended to many other applications. Underground pipes, ships' hulls and oil-rig installations are often protected by a sacrificial metal (see Figure 8.24). The structure is in electrical contact with a piece of magnesium or zinc. The more reactive metal (magnesium or zinc) corrodes preferentially, and this prevents corrosion of the iron.

Practical work

Try an experiment with two identical iron nails in water, but wrap one with a piece of magnesium ribbon, and see which nail rusts more.

The economics of corrosion

Although oil rigs and platforms are manufactured with corrosion-resistant coatings, they may also have additional protection through connection to sacrificial metals underwater to prevent corrosion. The prevention of corrosion, in general, is an economic necessity.

According to a recent report, the annual cost of metallic corrosion in the USA is $300 billion. The 1986 restoration project of the Statue of Liberty cost more than $200 million. Corrosion causes waste, and the Earth's resources are limited. It is important to make manufactured goods, such as motor cars, last and not corrode, so that resources last longer and pollution by greenhouse gases is kept to a minimum.

Chemical calculations

The atoms of the various elements have different masses. The masses of individual atoms are very small, so it is convenient to state the masses of the various atoms relative to each other. The relative atomic mass of an element is the mass of an atom of that element compared with the mass of an atom of carbon with six protons and six neutrons in its nucleus. This atom of carbon is given the value of exactly 12. The symbol for relative atomic mass is A_r (see Table 8.3).

Table 8.3 A_r values for some common elements

Element	A_r	Element	A_r
H	1.0	P	31.0
He	4.0	S	32.0
C	12.0	Cl	35.5
N	14.0	K	39.0
O	16.0	Ca	40.0
F	19.0	Fe	56.0
Na	23.0	Cu	64.0
Mg	24.0	Ag	107.0
Al	27.0	Pb	207.0

If we know the relative atomic masses of the elements, then we can work out the **relative molecular masses** (M_r) of compounds:

- **Water** (H_2O): In this molecule, there are two hydrogen atoms and one oxygen atom. The relative molecular mass is $[(2 \times 1) + 16] = 18$.
- **Carbon dioxide** (CO_2): In this molecule, there are two oxygen atoms and one carbon atom. The relative molecular mass is $[(2 \times 16) + 12] = 44$.

For ionic compounds, it is more correct to use the term *relative formula mass*, as there are no separate molecules in ionic compounds.

- **Magnesium oxide** (MgO): In this compound, there is one magnesium ion for every one oxygen ion. The relative formula mass equals $[24 + 16] = 40$.

Question

6 Find the relative molecular masses of:

a ammonia, NH_3
b methane, CH_4
c hydrogen sulphide, H_2S

and the relative formula masses of:

d calcium chloride, $CaCl_2$
e copper(II) oxide, CuO.

- **Sodium carbonate** (Na_2CO_3): In this compound, there are two sodium ions, and one carbonate ion made up of one carbon and three oxygen atoms. The relative formula mass equals [(2 × 23) + 12 + (3 × 16)] = [46 +12 + 48] = 106.

H

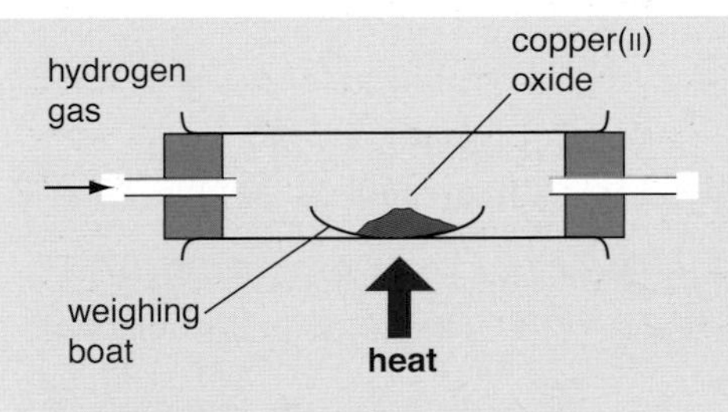

Figure 8.25 Reduction of copper(II) oxide

Practical work

Determining the formula of a compound from experimental data

1 Reduce 4.0 g of copper(II) oxide in a stream of hydrogen (see Figure 8.25).

2 3.2 g of copper remains.

The mass of oxygen in the 4.0 g of copper(II) oxide is (4.0 – 3.2) = 0.8 g.

To find the formula of copper(II) oxide, enter the results into Table 8.4.

Table 8.4 Finding the formula of copper(II) oxide

Element	Mass of element/g	A_r	Mass/A_r	Divide by lowest	Find whole number ratio
Cu	3.2	64.0	3.2/64.0 = 0.05	1	1
O	0.8	16.0	0.8/16.0 = 0.05	1	1

The formula is CuO.

This is called the **empirical formula**, as it shows the simplest ratio of the atoms (or ions) of each element in the compound.

3 Reduce 7.2 g of copper(I) oxide in a stream of hydrogen.

4 6.4 g of copper remains.

To find the formula of copper(I) oxide, enter the results into Table 8.5.

Table 8.5 Finding the formula of copper(I) oxide

Element	Mass of element/g	A_r	Mass/A_r	Divide by lowest	Find whole number ratio
Cu	6.4	64.0	6.4/64.0 = 0.10	2	2
O	0.8	16.0	0.8/16.0 = 0.05	1	1

The formula is Cu_2O.

Sometimes the data is given as a composition expressed as percentage by mass.

Example

Calcium chloride contains 64% chlorine by mass. Find its formula, using Table 8.6.

Table 8.6 Finding the formula of calcium chloride

Element	Percentage of element	A_r	Percentage/A_r	Divide by lowest	Find whole number ratio
Ca	36.0	40.0	36.0/40.0 = 0.9	1	1
Cl	64.0	35.5	64.0/35.5 = 1.8	2	2

The formula is $CaCl_2$.

H

Using equations

A balanced equation is a shorthand way of summarising a chemical reaction. The balanced equation can be used to predict the relationships between the masses of reacting compounds and the products they form.

Example

Find the mass of carbon dioxide formed when 5.3 g of sodium carbonate completely reacts with excess hydrochloric acid.

$$Na_2CO_3(s) + 2HCl(aq) \rightarrow 2NaCl(aq) + H_2O(l) + CO_2(g)$$

sodium carbonate $M_r = 106$; carbon dioxide $M_r = 44$

Having picked out the compounds we are interested in, we use the molecular or formula masses to write a statement:

- 106 g of sodium carbonate form 44 g of carbon dioxide.
- 1 g of sodium carbonate forms (44/106) g of carbon dioxide.
- 5.3 g of sodium carbonate form [(44/106) × 5.3] g of carbon dioxide.

The mass of carbon dioxide formed is 2.2 g.

Example

Calculate the mass of carbon monoxide need to completely reduce 1000 g of iron(III) oxide.

$$Fe_2O_3(s) + 3CO(g) \rightarrow 2Fe(l) + 3CO_2(g)$$

iron(III) oxide $M_r = 160$; carbon monoxide $M_r = 28$

- 160 g of iron(III) oxide is reduced by (3 × 28) g of carbon monoxide.
- 1 g of iron(III) oxide is reduced by (84/160) g of carbon monoxide.
- 1000 g of iron(III) oxide is reduced by [(84/160) × 1000] g = 525 g of carbon monoxide.

Use the equation to calculate the mass of iron formed at the same time.

Questions

7 Calculate the mass of calcium oxide formed by the complete decomposition of 5 kg of calcium carbonate.

$$CaCO_3 \rightarrow CaO + CO_2$$

8 Calculate the mass of sodium chloride that can be formed from the neutralisation of 8 g of sodium hydroxide.

$$NaOH + HCl \rightarrow NaCl + H_2O$$

Atom economy

When a product is formed from reacting compounds, the compounds formed, apart from the product needed, are waste. An indication of the efficiency of a reaction is sometimes given as its **atom economy**, which is defined as follows:

$$\text{atom economy} = \frac{\text{mass of required product}}{\text{total mass of reactants}} \times 100\%$$

Example

Take the reaction between sodium hydroxide and hydrochloric acid to form sodium chloride:

$$NaOH + HCl \rightarrow NaCl + H_2O$$

NaOH	+ HCl	→ NaCl	+ H_2O
sodium hydroxide	hydrochloric acid	sodium chloride	water
$M_r = 40.0$	$M_r = 36.5$	$M_r = 58.5$	$M_r = 18.0$

atom economy = [58.5/(40.0 + 36.5)] × 100 = 76.5%

Not all of the atoms of the reactants are incorporated into the required product – some form a by-product.

H

The traditional manufacture of titanium from titanium dioxide involved the conversion of TiO_2 to $TiCl_4$, followed by reduction to titanium by magnesium.

The overall process is as follows:

TiO_2	+	$2Cl_2$	+	$2Mg$	→	Ti	+	$2MgCl_2$
$Mr = 79.9$		$Mr = 142.0$		$Mr = 48.0$		$Mr = 47.9$		

The atom economy of the process is as follows.

$$\text{atom economy} = \frac{\text{mass of required product}}{\text{total mass of reactants}} \times 100\%$$

$$= \frac{47.9}{79.9 + 142.0 + 48.0} \times 100\%$$

$$= 17.7\%$$

In modern electrolytic extraction, titanium dioxide is electrolysed in molten calcium chloride. This produces the same mass of product as in the above example there are no by-products, so less waste. The reaction is:

$$TiO_2 \rightarrow Ti + O_2$$

$$\text{atom economy} = (47.9/79.9) \times 100\%$$

$$= 59.9\%$$

Green chemistry and water

Green chemistry is the science of carrying out the necessary manufacture of chemicals with minimum effect on the environment. Most chemical manufacturers are responsible and highly regulated by legislation, and it is in the interests of companies and their shareholders to consider green issues:

- Manufacturers should keep waste to a minimum. Often a by-product of a process can be utilised to produce a saleable product rather than being a waste product. In some industries, the toxic gas chlorine is produced for instance, and it can be absorbed into aqueous sodium hydroxide to form bleach.
- Where possible, companies should use renewable resources. For example, aluminium producers, which require large amounts of electricity, try to site their plants near hydroelectric schemes.
- Energy costs money. Wherever possible, energy costs should be cut through insulation and reuse. For instance, the hot gases from the top of a blast furnace are used to preheat the air blown in at the bottom.
- Catalysts are compounds that make reactions proceed more quickly at the same temperature, saving energy and money.
- Companies prefer to use non-toxic reactants wherever possible. Where a product is toxic, research is carried out to find a less toxic one.
- Waste products should be treated to create non-toxic materials that do not harm the environment. In recent times, biological agents have been used to treat hazardous waste. Some hazardous organic materials can be treated with bacteria that degrade the organic molecules to methane, which can be used as a fuel.

Figure 8.26 View from the Afristar satellite, 35 774 km above the Earth

Water

Water is the most abundant substance on the Earth's surface (see Figure 8.26). It is essential to life on our planet. Some predict that there may be a world shortage of drinkable water in the future. In the chemical industries, water is an important raw material that is used as:

- a solvent (solutions of compounds in water occur in many processes); in sulphuric acid manufacture, sulphur trioxide is dissolved in concentrated sulphuric acid to form oleum, which is then diluted with water to form concentrated sulphuric acid
- a coolant (many chemical reactions produce heat that must be taken away from the plant); water cools processes, and then the water is itself cooled before being returned to the environment.

The public water supply and water treatment

Figure 8.27 The Caban Coch Dam

Rainfall is collected in lakes or man-made reservoirs where it is stored. Alternative water supplies are rivers and underground water. The Caban Coch dam in the Elan valley is shown in Figure 8.27; water from the Elan valley is used by the city of Birmingham in England. Before reaching our taps, water has to undergo purification and treatment (see Figure 8.28).

Solids are allowed to settle under gravity in a reservoir. The water is then passed through filter beds that remove smaller particles and some bacteria. For drinking-quality water, chlorine is added, which kills the remaining bacteria. Some water authorities add other chemicals to drinking water. For instance many add fluoride ions, which prevent tooth decay at controlled levels (see page 111).

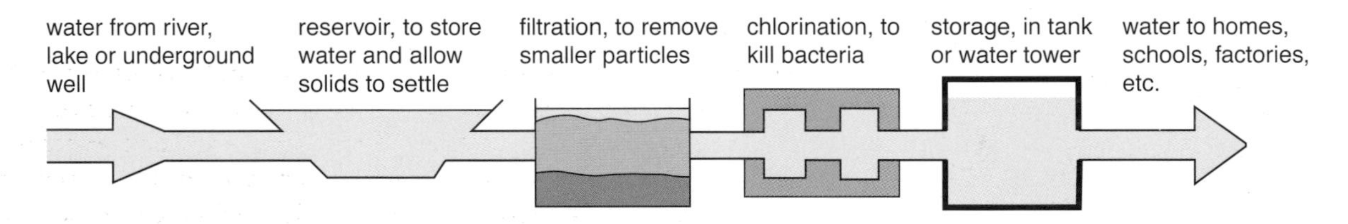

Figure 8.28 The purification of water in the public supply

Figure 8.29 Limestone cavern in the Brecon Beacons in Wales, formed through water dissolving the limestone rocks

Where rainwater flows through rocks (see Figure 8.29), it picks up calcium ions from limestone and calcium sulphate, and magnesium ions from other rocks. Water that contains calcium and magnesium ions is called **hard water**, as it does not readily lather with soap. Soap is the sodium salt of an organic acid derived from oils and fats. A typical soap is sodium stearate. When hard water is shaken with soap, the calcium and magnesium ions react with the soap to form insoluble calcium or magnesium stearate, which we call scum. This does not happen with soft water.

Practical work

Comparing the hardness of water from various sources

1 Make up a solution of soap. (This can be done by dissolving soap flakes in water; commercial solutions are available.)

2 Place the soap solution in a burette, and pipette a known volume (say 25 cm^3) of the water being tested into a conical flask (see Figure 8.30).

3 Run soap solution into the water a few drops at a time while shaking the flask.

4 Stop the titration when a permanent lather (lasting at least 2 min) covers the whole of the surface of the water.

5 Repeat the experiment for the range of water samples available.

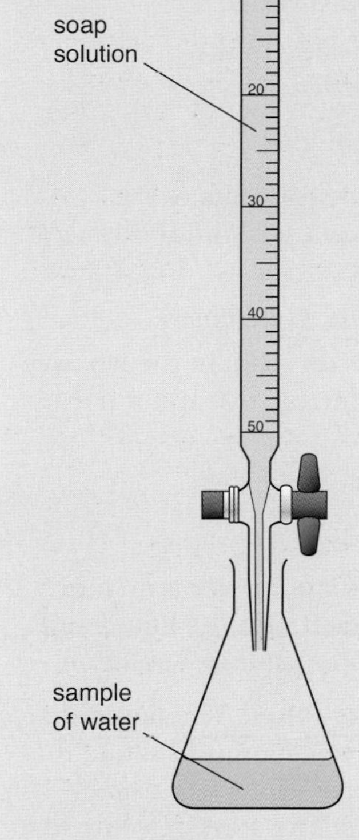

Figure 8.30 Apparatus for comparing the hardness of water samples

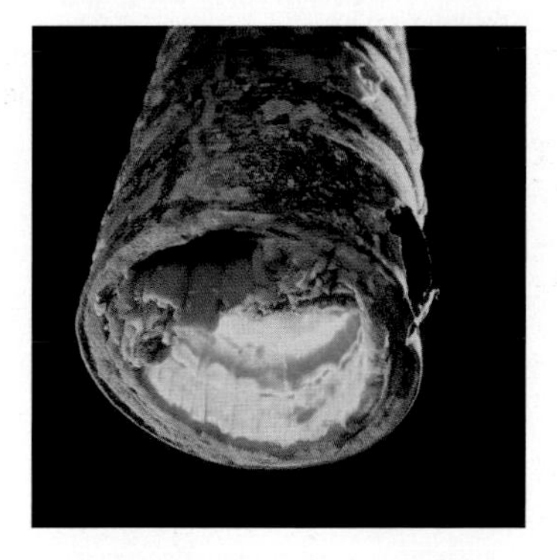

Figure 8.31 The effect of hard water forming scale (calcium carbonate) in a pipe

Types of hardness

Temporary hardness is hardness that can be removed by boiling. It is caused by the presence of calcium and magnesium hydrogencarbonates. When the temporarily hard water is boiled, the hydrogencarbonates are decomposed, and the calcium or magnesium is deposited as insoluble calcium carbonate or magnesium carbonate. This then appears as 'kettle fur' or 'boiler scale' (see Figure 8.31).

H

calcium hydrogencarbonate → calcium carbonate + carbon dioxide + water

$$Ca(HCO_3)_2(aq) \rightarrow CaCO_3(s) + CO_2(g) + H_2O(l)$$

When the hardness of water cannot be removed by boiling, the water is said to be **permanently hard** water. It usually contains calcium and magnesium sulphates. Permanently hard water can be softened by adding sodium carbonate (washing soda) or by ion exchange (see Figure 8.32).

Sodium carbonate simply precipitates the calcium ions as calcium carbonate.

H

sodium carbonate + calcium sulphate → calcium carbonate + sodium sulphate

$$Na_2CO_3(aq) + CaSO_4(aq) \rightarrow CaCO_3(s) + Na_2SO_4(aq)$$

Ion exchange resins replace the calcium ions that cause hardness with sodium ions that do not. The calcium ions remain on the resin.

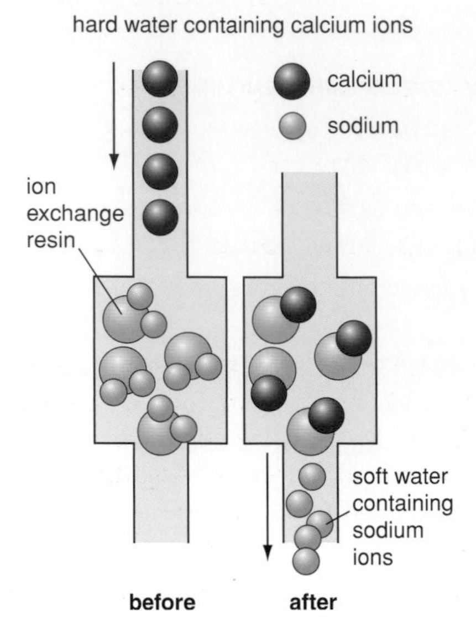

Figure 8.32 How an ion exchange water softener works

Question

9 Explain why two sodium ions are released for every one calcium ion in Figure 8.32.

The advantages and disadvantages of hard water

Hard water usually tastes better than soft water. In the brewing of beer, the water is sometimes artificially hardened.

Disadvantages of hard water are the following:

- The production of boiler scale and kettle fur reduces the efficiency of kettles, boilers, coffee makers, etc. and this causes more energy to be used (see Figure 8.33).
- The removal of boiler scale and kettle fur is expensive.
- Hard water uses up more soap, and creates scum in the laundry. This is why modern detergents are preferred to soap for hard water (see Figure 8.34).

Figure 8.33 A kettle element covered with scale from hard water

Figure 8.34 A water softening product which is added with the detergent to the washing machine

Solubility curves

If we try to dissolve sugar in water, there is a point when no more sugar will dissolve in the water and we are left with the liquid and undissolved sugar. The liquid is then called a **saturated solution**.

The substance that is dissolved is called the **solute**. The liquid that does the dissolving is called the **solvent**. The amount of solute that can be dissolved varies with the temperature. Hot water usually dissolves more solute than cold water does. The mass of solute that dissolves in a given mass of solvent is called the **solubility**. Solubility is usually stated as grams of solute per 100 g of water.

How solubility varies with temperature can be shown as a solubility curve. Curves for sodium chloride and potassium chlorate are shown in Figure 8.35. Notice that the solubility of salt (sodium chloride) does not change very much as temperature rises.

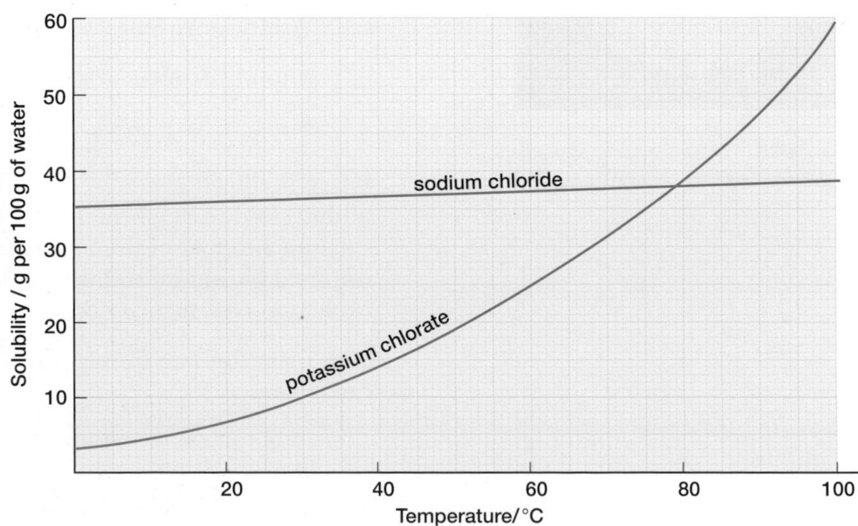

Figure 8.35 Solubility curves for sodium chloride and potassium chlorate

Table 8.7 Solubility data for potassium nitrate

Temperature (°C)	Solubility (g/ 100 g of water)
0	13.9
10	21.2
20	31.6
30	45.3
40	61.4
50	83.5
60	106.0
70	135.0
80	167.0
90	203.0
100	245.0

Activity

Try plotting the solubility curve for potassium nitrate, KNO_3.The data are given in Table 8.7.

Summary

1 Iron is extracted from iron ore by reduction in a blast furnace. Iron ore (iron(III) oxide), coke and limestone are loaded in at the top of the furnace. Hot air is blown in at the bottom of the furnace. Heat is produced by oxygen in the air burning the coke. Carbon monoxide is formed that removes oxygen (reduction) from the iron(III) oxide to form molten iron.

2 Pig iron from a blast furnace contains a lot of carbon (up to 4%). The pig iron can be converted to more useful steel by removal of some of the carbon. Useful alloys can be made by adding other elements to steel. Steel is recycled on a large scale.

3 Electrolysis is chemical action caused by an electric current. The liquid electrolysed is called the electrolyte. Electricity enters the electrolyte via two electrodes. The positive electrode is called the anode and the negative electrode is called the cathode.

4 Aluminium is a reactive metal that is manufactured by electrolysis. The raw materials are bauxite and cryolite for the electrolyte and carbon for the anodes. The process needs large quantities of cheap electricity.

5 Aluminium is a light metal with high tensile strength. Its conducting properties of heat and electricity make it very useful. Recycling aluminium is very economical.

6 Copper is widely used because it is a very good conductor of heat and electricity and can be easily formed into wires and pipes.

7 Titanium is a light, tough metal that is resistant to corrosion and is used in aeronautical industries and medicine.

8 There are often environmental consequences when metals are mined and extracted.

9 It is necessary to be well informed about the economic advantages and disadvantages of such processes.

10 Metals can be arranged in order of their reactivity; this order is called a **reactivity series**. The most reactive metals react readily with the atmosphere.

11 Reaction with the atmosphere is the basis of corrosion. Corrosion costs money. Iron and steel corrode to form rust. Cars, bridges and oil platforms must be protected to prevent corrosion. There are many ways to prevent rusting, including painting, covering with oil, and galvanising. Large structures, such as oil rigs, can be protected by connection to a more reactive metal that sacrificially corrodes, protecting the iron and steel of the rig.

12 Chemical calculations can be carried out by using values of relative atomic masses. Knowledge of the composition of a compound, by mass, allows the empirical formula to be calculated. The **atom economy** of a reaction is important in determining how efficiently a product is formed from the masses of all the reactants.

13 All chemical processes should have minimal effect on the environment.

14 The public water supply collects water from rivers, reservoirs and wells, and purifies it. Drinking water is sterilised by the addition of chlorine. Some water authorities add fluoride ions to prevent dental decay.

15 Water containing calcium and magnesium ions is called **hard water**. Temporary hardness can be removed by boiling. Hardness can be removed by adding washing soda (sodium carbonate) to the water, or by ion exchange. Hard water usually tastes better than soft water, and is used in the brewing industry. Hard water wastes soap, and causes kettle fur and boiler scale.

16 Substances that dissolve in water are said to be soluble. Water is the solvent, and the compound is the solute. **Solubility** is the mass of solute that dissolves in a given mass of solvent at a particular temperature to form a saturated solution. It is usually measured in grams of solute per 100 g of solvent. A solubility curve shows how the solubility of a substance varies with temperature.

Chapter 9 Ammonia and fertilisers, hydrocarbons and polymers, smart materials

By the end of this chapter you should:

- know that nitrogen is essential for healthy plant growth;
- know that nitrogen is an unreactive element that cannot be used directly by plants;
- know that modern agriculture needs nitrogenous fertilisers;
- know that some reactions are reversible;
- know that nitrogen is converted into ammonia by a reversible reaction with hydrogen;
- understand that the yield of ammonia during the manufacturing process depends upon the conditions;
- know that ammonia is very soluble in water and forms a weak alkali;
- know that the reaction between ammonia and sulphuric acid creates the fertiliser ammonium sulphate;
- know that there is a conflict between the use of nitrogenous fertilisers and some aspects of the environment;
- understand some reactions of ammonium compounds and the gas ammonia;
- understand the nature of the alkane and alkene hydrocarbons;
- understand the combustion of alkanes;
- understand the cracking of alkanes to make more useful molecules and alkenes;
- know the nature of some polymers and how they are produced;
- understand the difference between thermoplastic polymers and thermosetting polymers (thermosets);
- know the uses of some common polymers;
- know about some smart materials.

Figure 9.1 Fritz Haber

Ammonia and fertilisers

Nitrogen, ammonia and fertilisers

Did you know?

The process for converting nitrogen into ammonia was developed by the German Fritz Haber (see Figure 9.1). One of the products made from ammonia is nitric acid, which in turn can be made into explosives. This new method for producing nitrogen compounds removed the need to import Chile saltpetre (sodium nitrate) and saltpetre (potassium nitrate).

Figure 9.2 Wheat showing nitrogen deficiency

About four-fifths of the Earth's atmosphere consists of the gas nitrogen. Nitrogen is an unreactive gas, and it cannot be used directly by plants. Plants obtain nitrogen by absorbing nitrates through their roots. Nitrogen enters the soil system naturally via:

- thunderstorms when oxygen and nitrogen combine
- bacteria in leguminous plants such as clover
- rock weathering
- excrement from animals
- decay of plant and animal material.

For a high yield of crops in modern agriculture, nitrogen fertilisers are required, and these must be manufactured (see Figure 9.2). (Organic farmers prefer to use natural fertilisers rather than synthetic ones.) This means that nitrogen from the air must be converted into nitrogen compounds. Converting atmospheric nitrogen into nitrogen compounds by any means is known as **nitrogen fixing**.

Making ammonia

Some reactions can proceed in either direction, depending upon the conditions. If blue copper sulphate crystals are heated gently, they eventually turn white, and water is given off. When water is added to the white residue, it turns back to blue.

H

$$\underset{\text{blue}}{CuSO_4.5H_2O} \rightleftharpoons \underset{\text{white}}{CuSO_4} + 5H_2O$$

(The sign that shows that the reaction is **reversible** is the $\rightleftharpoons$.)

The reaction between nitrogen and hydrogen is a reversible reaction. This reaction is most important, since it means that nitrogen from the air can be converted into ammonia, which is the **Haber process**. Ammonia is the source of many important nitrogen-containing compounds.

$$\text{nitrogen} + \text{hydrogen} \rightleftharpoons \text{ammonia}$$

H

$$N_2(g) + 3H_2(g) \rightleftharpoons 2NH_3(g)$$

The source of the nitrogen for the Haber process is the atmosphere. The source of the hydrogen is usually methane (natural gas) or the naphtha fraction from the fractional distillation of petroleum. The reaction that forms ammonia needs special conditions. A pressure of 200–300 atm (atmospheres) is used in conjunction with a temperature of about 450 °C and an iron catalyst. The ammonia is removed from the system as soon as it forms. There is only a small percentage conversion, and the unused gases must be recycled.

Recently, a new catalyst consisting of ruthenium on a graphite support has been introduced. This is more efficient than an iron catalyst, and enables lower pressures to be used. One US source suggests that a pressure of as little as 40 atm may be possible.

Figure 9.3 Ammonia plant in Trinidad

H

The choice of conditions during the ammonia manufacturing process can be seen by looking at the reaction:

- The production of ammonia (see Figure 9.3) involves a reversible reaction in which there are fewer molecules of gaseous product than of gaseous reactants. In such cases, the yield of product is favoured by a high pressure. However, it must be remembered that high-pressure chemical plant is very expensive.

H

- The combination of hydrogen and ammonia is exothermic. In such reversible exothermic reactions, a high yield is favoured by a low temperature. Low temperatures mean that reactions take place slowly. The temperature used (about 450 °C) is a compromise.
- A catalyst is used so that the reaction proceeds more quickly at a given temperature.

To summarise, the conversion of nitrogen and hydrogen to ammonia needs:

- a high pressure
- as low a temperature as possible with a good rate
- the presence of a catalyst
- removal of the ammonia as it is formed.

Some uses of ammonia

- **Agriculture:** Ammonia is used to make fertilisers such as ammonium sulphate and ammonium nitrate. It may be applied directly to the soil or used to make urea.
- **Manufacture of nitric acid:** Nitric acid is an important intermediate in the production of explosives and of the fertiliser ammonium nitrate.
- **Manufacture of nylon:** Nylon is a polyamide, and a significant proportion of the output of ammonia goes into polyamide manufacture.
- **Manufacture of sodium carbonate:** Sodium carbonate, or soda ash, is used in many industries, including glass manufacture.

Nitrogenous fertilisers: advantages and disadvantages

To increase crop yields, farmers add fertilisers to the soil to replace nitrogen taken out of it by previous crops. Nitrogenous fertilisers produced from ammonia are relatively cheap. Often, the nitrogen compound is mixed with compounds containing phosphorus and compounds containing potassium to form balanced NPK fertilisers. The overuse of nitrogenous fertilisers can lead to detrimental effects on the environment.

Excess fertiliser can by dissolved by rain and washed into streams. Once in streams, rivers and lakes, it causes **eutrophication**. The nitrogen compounds encourage the growth of plants and algae in the water. The algae grow quickly to form a blue-green mass called algae bloom. Although the water plants and algae are eaten by some water creatures, there are too many plants and algae and many die. The dead algae are decomposed by bacteria that use up oxygen in the process. This means that the water loses its oxygen, and fish and other higher animals such as crustaceans die from suffocation.

Nitrogenous fertilisers (overuse): www.bbc.co.uk/schools

The presence of nitrates from nitrogenous fertilisers in drinking water is very dangerous. The nitrates are not removed by the usual water purification. In high concentrations, they form toxic chemicals in the body that can cause cancers. Care has to be taken to prevent the use of excess amounts of fertiliser.

Organic farmers tend to avoid synthetic fertilisers and rely more on natural manure and crop rotation. The planting of fields of clover can restore nitrogen by bacterial fixing (see Nitrogen cycle, Chapter 4).

Do not confuse the words *ammonia* and *ammonium*. Ammonia is the compound with formula NH_3, and is a gas. Ammonium is the ammonium ion, NH_4^+, and it occurs only in compounds or solutions.

Ammonia and water

Practical work

Fountain experiment

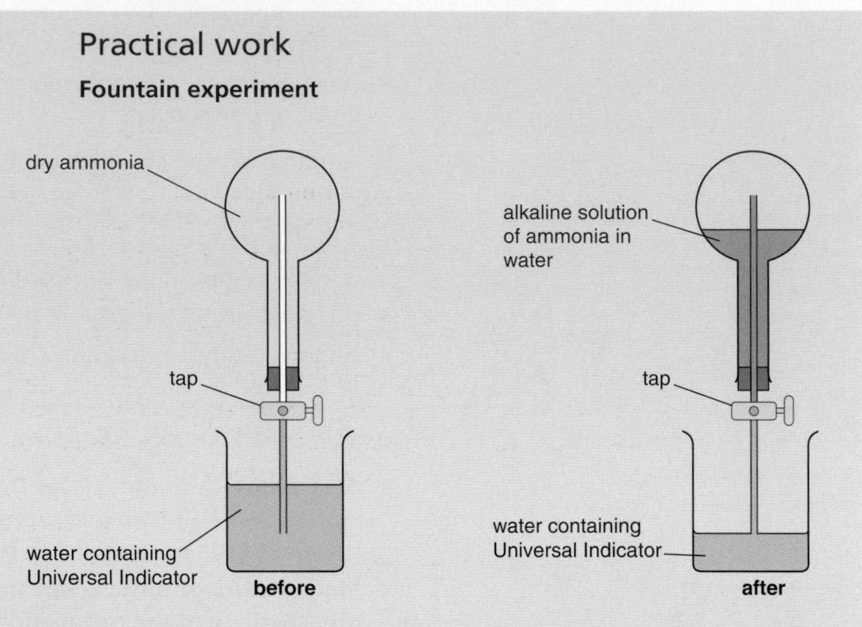

Figure 9.4 Fountain experiment

Ammonia is extremely soluble in water. This can be demonstrated by the so-called fountain experiment (see Figure 9.4).

Method

1 Fill the upper flask with dry ammonia, and close the tap.
2 Immerse the end of the tube in water containing universal indicator, which will have the green colour of a neutral solution.
3 Open the tap.
4 Some of the ammonia will begin to dissolve in the water. As it is very soluble, this will cause the pressure in the flask to fall.
5 Atmospheric pressure will then force liquid from the beaker upwards into the flask in the form of a fountain, dissolving almost all of the ammonia.
6 At the same time, the colour of the liquid in the flask will change to blue-purple, indicating that the solution in the flask is alkaline.

Reaction between ammonia and water

The fountain experiment shows that an aqueous solution of ammonia is alkaline. Alkalis contain hydroxide ions, OH^-. Therefore the ammonia must react with the water.

$$\text{ammonia} + \text{water} \rightleftharpoons \text{ammonium ions} + \text{hydroxide ions}$$
$$NH_3(g) + H_2O(l) \rightleftharpoons NH_4^+(aq) + OH^-(aq)$$

H

We know not all of the ammonia reacts, because a solution of ammonia always smells of ammonia gas. Alkalis of this type are called *weak alkalis.*

Neutralisation

The reaction between an acid and an alkali, or base, is called neutralisation, and the products are salt and water. When ammonia solution is neutralised by an acid, **ammonium** salts are formed.

ammonia solution + sulphuric acid → ammonium sulphate
ammonia solution + nitric acid → ammonium nitrate

Both ammonium sulphate and ammonium nitrate are produced as commercial fertilisers.

Identifying an ammonium compound

All ammonium compounds contain the ammonium ion NH_4^+. When heated with sodium hydroxide solution, ammonium ions form ammonia gas and water.

sodium hydroxide + ammonium chloride → sodium chloride + ammonia + water

Ammonia gas has a characteristic pungent smell. Chemically, ammonia produces an alkali when dissolved in water, so it can be detected by its effect upon **moist** red litmus paper. The red litmus paper turns blue.

Ammonia is formed from urea in urine, and it can be detected when babies' nappies are changed.

Hydrocarbons and polymers

Alkanes

Figure 9.5 Texaco oil refinery in Pembroke, Wales

Figure 9.6 Milford Haven oil refinery, Wales

In everyday life we make use of many materials that are polymers. These are chemicals that are derived from petroleum (crude oil) and are used to make fibres for textiles and a wide range of manufactured items often described loosely as 'plastics'. The basis of polymers is chemicals formed from the components of crude oil called **alkanes**.

The alkanes are hydrocarbons that can be represented by the formula $C_nH_{(2n+2)}$, where n has the value 1, 2, 3, 4, etc. Alkanes are produced by the fractional distillation of crude oil at an oil refinery (see Figures 9.5 and 9.6).

In these compounds, the bonds between carbon atoms and hydrogen atoms and between two carbon atoms are covalent bonds that involve a shared pair of electrons. We can represent each molecule with a **structural formula** that shows the covalent bonds as lines.

The alkanes with 1–5 carbon atoms are shown in Figures 9.7–9.11. Notice that the alkane with five carbon atoms can exist with two

Figure 9.7 Methane, CH_4

Figure 9.8 Ethane, C_2H_6

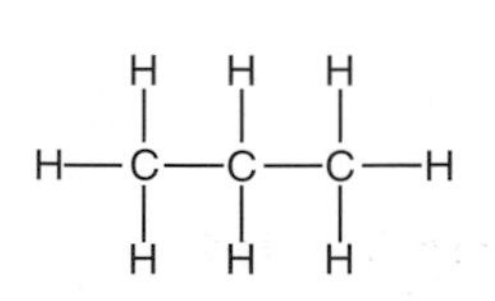

Figure 9.9 Propane, C_3H_8

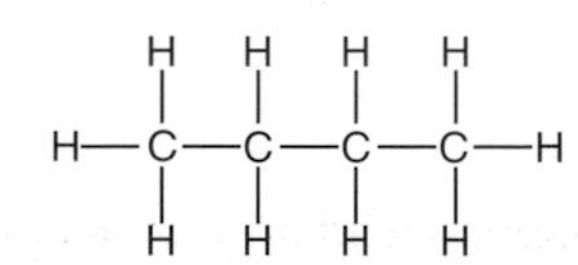

Figure 9.10 Butane, C_4H_{10}

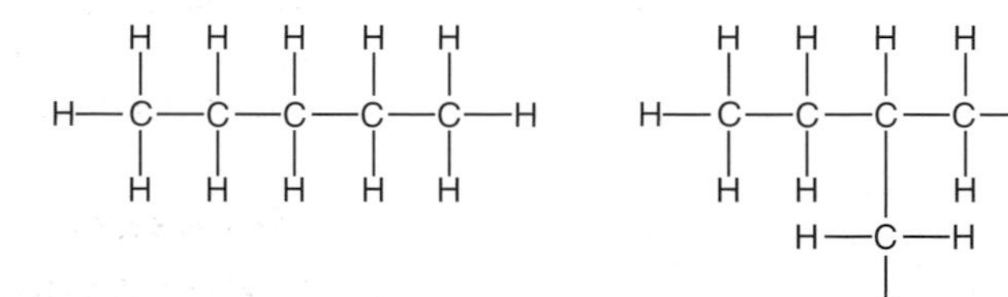

Figure 9.11 Pentane, C_5H_{12} (left) and 2-methylbutane, C_5H_{12} (right)

different structural formulae. These two compounds are called **structural isomers**.

Saturated hydrocarbons

The alkanes that contain one single bond between carbon atoms are called **saturated hydrocarbons**. The alkanes are generally unreactive hydrocarbons. Their most common use is as fuels, and so they undergo combustion.

Methane is the main constituent of natural gas. The complete combustion of methane gives carbon dioxide and water.

$$\text{methane} + \text{oxygen} \rightarrow \text{carbon dioxide} + \text{water}$$
$$CH_4 + 2O_2 \rightarrow CO_2 + 2H_2O$$

Butane is used in camping gas.

$$2C_4H_{10} + 13O_2 \rightarrow 8CO_2 + 10H_2O$$

Making polymers

Many of the larger saturated alkane hydrocarbon molecules obtained by fractional distillation in an oil refinery can be broken down into smaller, more useful molecules. This is done by heating the molecules in the presence of a catalyst, and it is called **cracking**. Among the products of cracking are small molecules (**alkenes**) that are used to make **polymers**.

Decane is an alkane with formula $C_{10}H_{22}$. The cracking of decane (see Figure 9.12) can be shown as an equation:

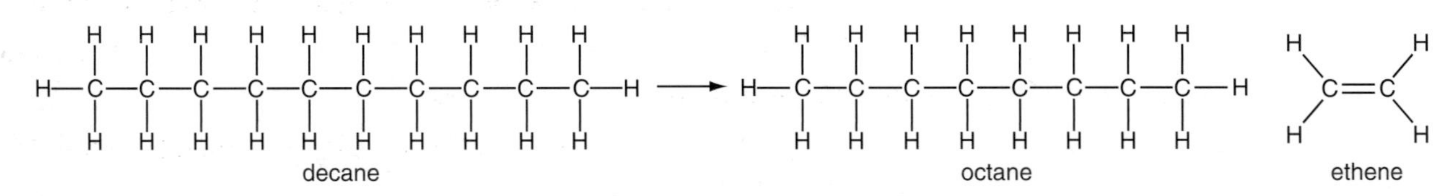

Figure 9.12 Cracking of decane

$$\text{decane} \rightarrow \text{octane} + \text{ethene}$$
$$C_{10}H_{22}(l) \rightarrow C_8H_{18}(l) + C_2H_4(g)$$

Octane can be used in petrol manufacture, and ethene can be used for making a large range of chemicals, including polymers. The industrial demand for ethene and octane is very great, so that cracking is a major component of the petrochemical industry.

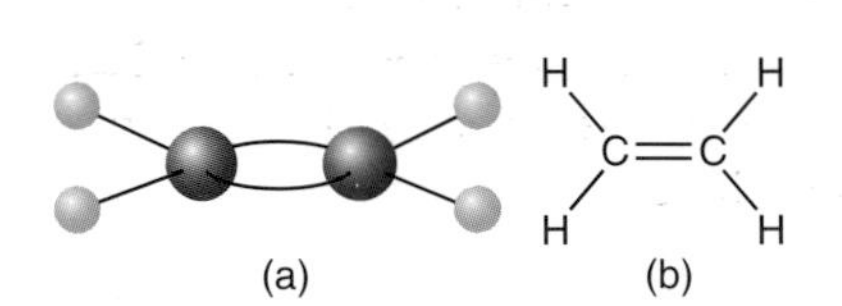

Figure 9.13 Ethene: (a) model of a molecule, (b) structural formula

Ethene is a very reactive molecule because it has a carbon–carbon double bond (see Figure 9.13).

It is said to be **unsaturated**.

Small reactive molecules such as ethene can react to form very large molecules called **polymers**. The small molecules that combine to form a polymer are called **monomers**.

Monomers such as ethene form **addition polymers**. Addition polymers are formed from one type of monomer. The monomer ethene forms the polymer poly(ethene) or polythene (see Figure 9.14).

$$nC_2H_4 \rightarrow (C_2H_4)_n$$

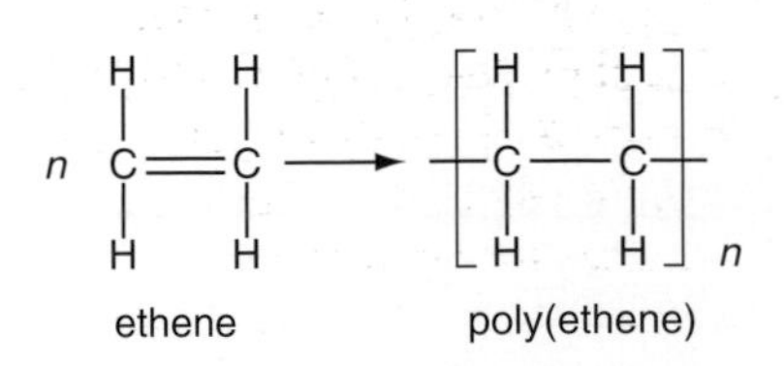

Figure 9.14 Formation of addition polymers (n is a large number)

There are other addition polymers made from compounds that are derived from ethene, all of which are useful everyday polymers.

Poly(tetrafluoroethene) or PTFE is non-flammable, and its most common use is as a non-stick surface for saucepans and frying pans (see Figure 9.15).

Poly(chloroethene) or polyvinylchloride or PVC is strong and light, and it is used to make guttering and window and door frames (see Figures 9.16 and 9.17). When PVC is mixed with other chemicals called plasticisers, it becomes flexible, and can be used for making synthetic leather, cling film, and insulation for electrical cables.

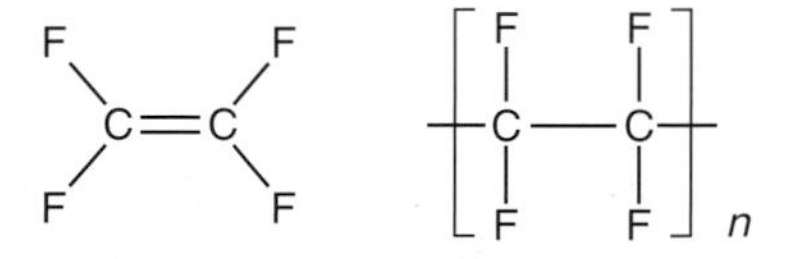

tetrafluoroethene monomer — poly(tetrafluoroethene) polymer (PTFE)

Figure 9.15 Poly(tetrafluoroethene)

chloroethene monomer vinyl chloride — poly(chloroethene) polymer polyvinylchloride (PVC)

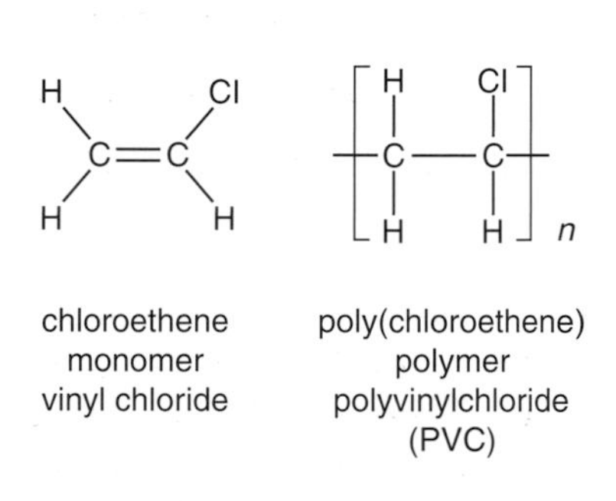

Figure 9.16 Poly(chloroethene) or polyvinylchloride

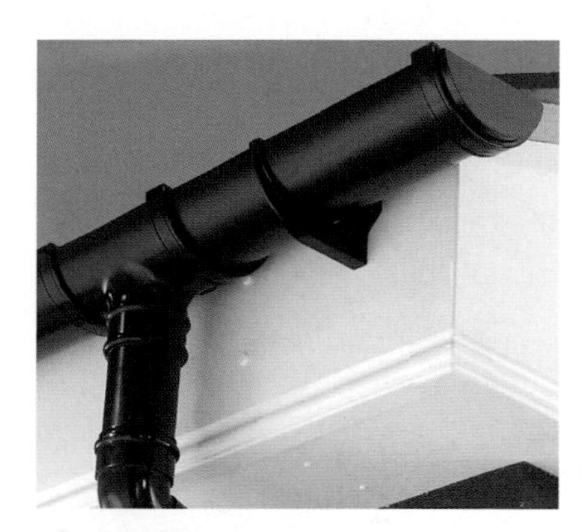

Figure 9.17 PVC guttering

Condensation polymers

Unlike addition polymers, **condensation polymers** are made from two different monomers. This means that their structures are of the following type.

–A–B–A–B–A–B–A–B–A–B–A–B–A–B–A–B–

Addition polymers, which are made from one monomer, have the following structure.

–A–A–A–A–A–A–A–A–A–A–A–A–A–A–A–A–

Among the condensation polymers are polyamides and polyesters, both of which can be made into fibres for textiles. When two monomers combine to form a condensation polymer, the reaction is accompanied by the formation of a small molecule such as water. The water molecule is said to be eliminated.

Practical work

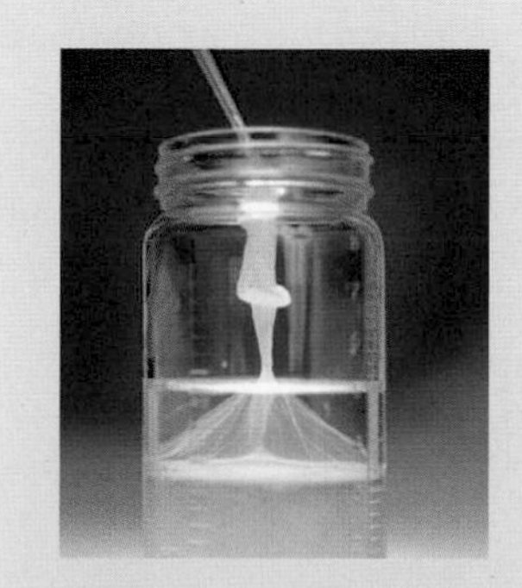

Figure 9.18 Making nylon 6,6 in the laboratory

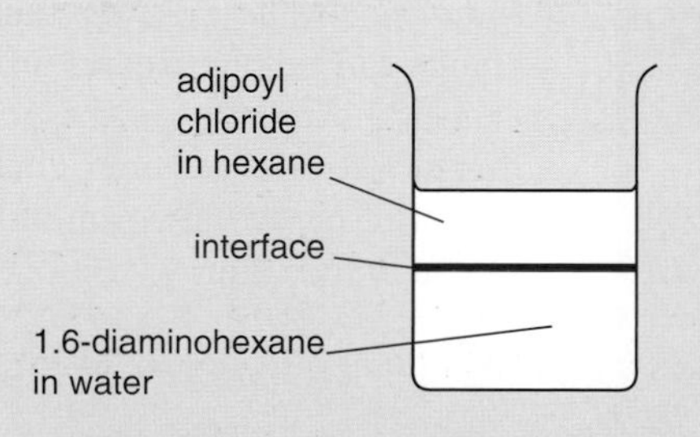

Figure 9.19 The formation of nylon 6,6

Making nylon in the laboratory

This experiment is sometimes called the 'nylon rope trick'. (See Figures 9.18 and 9.19.) It is usually demonstrated by a teacher, for safety reasons.

Method

1 Place in a beaker a solution of 1,6-diaminohexane dissolved in water.

2 Carefully add a solution of adipoyl chloride in hexane. (However, note that this solution is harmful and highly flammable. Cyclohexane is a less hazardous solvent than hexane.) Hexane and water do not mix, so there are two layers. The point where the surfaces of the two layers meet is called the interface.

3 The chemicals react at the interface to form a polymer, nylon-6,6. This can be carefully withdrawn using a glass rod. Then more polymer forms, so that a string of polymer can be wound around the glass rod. This polymer is called nylon-6,6, because both monomers contain six carbon atoms per molecule.

Figure 9.20 1950s radios with Bakelite cases

The nylons and polyesters are synthetic condensation polymers. Similar long-chain molecules exist in nature. Two examples are proteins and starch. Proteins can be considered as condensation polymers of α-amino acids, and starch is a long series of carbohydrate units containing six carbon atoms, $(C_6H_{10}O_5)_n$.

Action of heat on plastics

H

Did you know?

The first commercial thermoset plastic was Bakelite. In the first decade of the twentieth century, Dr Leo Baekeland, a Belgian chemist, discovered a plastic that was resistant to heat and could be moulded into objects. The material was patented under the name Bakelite (see Figure 9.20).

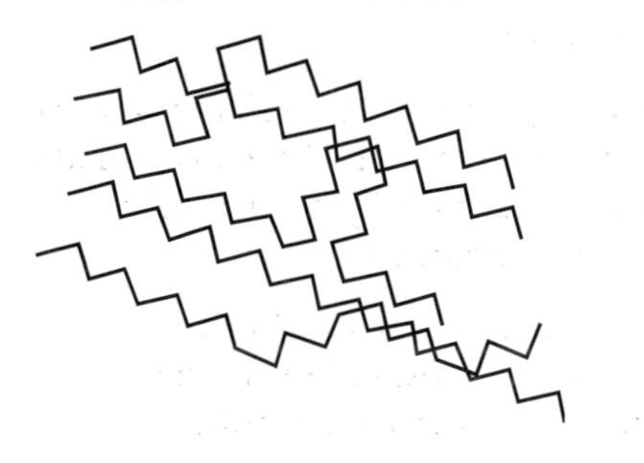

Figure 9.21 Polymer chains in a thermoplastic: there are no strong bonds between the chains.

You may have noticed that some polymers, such as polythene, become soft when heated and go hard again when cooled. Other plastics, such as those used to make saucepan handles, are resistant to heat. Plastics that soften when heated are called thermo-softening plastics, or **thermoplastics**. Those plastics that are resistant to heat are called thermosetting plastics, or **thermosets**. Thermoplastics are used extensively for household containers such as bowls and buckets, and for packaging material. Thermosets are used for electric light fittings, saucepan handles, and other products where heat resistance is important.

Challenge of materials:
www.sciencemuseum.org.uk/on-line/challenge/

The difference in the behaviour of these two types of polymer can be explained in terms of their structures. Thermoplastics are made up of polymer chains that attract one another with forces that are weaker than the covalent bonds that hold the chains of atoms together. The chains can be pictured as independent, and on heating, the chains can slide over one another, causing softening (see Figure 9.21). In thermosets, the polymer chains are linked in a three-dimensional structure by strong covalent bonds. Moderate heat has no effect on this structure, and so the plastic is heat-resistant.

Figure 9.22 Polythene containers

Uses of poly(ethene), known as polythene

Polythene is a very useful material. Its uses include:

- packaging, for example polythene film for sandwich bags and cling film and shrink-wrap
- containers for household liquids, for example squeezy bottles (see Figure 9.22)
- electrical insulation
- moisture barriers in construction.

Advantages and disadvantages of polymers over traditional materials

The widespread use of plastics shows the versatility of polymers. They are cheap, are easy to form into complicated shapes, are unaffected by water, and are light and strong. This makes them suitable for a wide variety of uses.

Figure 9.23 A hedgehog in danger from plastic debris

Unlike metals such as iron, plastics are not subject to corrosion, and are non-conductors of electricity. Transparent plastics such as Perspex are less brittle than glass. Toys and outdoor play equipment are often manufactured from plastics instead of wood and metal.

Although plastics are derived from petroleum compounds, which are a non-renewable resource, the total energy consumption when manufacturing a plastic item may be less than the energy requirement when using a traditional material. This is especially true with respect to glass and plastic bottles.

Up until now, the main drawback with polymeric materials has been their persistence in the environment. Some plastics have been made that degrade under the influence of ultraviolet light in sunlight. However, polymers are usually nonbiodegradable.

Almost all shorelines are littered with plastic refuse. This is a worldwide problem, and it causes extensive loss of life among animals (see Figure 9.23). Wood and cardboard are rapidly degraded by microorganisms. Recycling of polymers is essential to maintain a clean environment.

Smart materials

Smart materials are new materials with properties that change reversibly with a change in the material's surroundings, for example mechanical deformation and changes in temperature, light and pH.

- **Shape-memory polymers**: These polymers are somewhere between thermoplastics and thermosets. When heated, the polymer softens, and it can be stretched or deformed. On cooling, it remains in the deformed state. On being reheated, it 'remembers' its original shape, to which it returns. This property is called *shape retention.* Potential applications are plastic car bodies from which a dent could be removed by heating, and medical sutures that will automatically adjust to the correct tension and be biodegradable, and so will not need to be surgically removed.

- **Thermochromic paints and colorants**: Invisible ink made from cobalt(II) chloride solution has been known for a long time. The pink solution used on paper is almost invisible when dry, but turns blue on heating. More recently, complicated organic molecules have been made that can change colour over a specified temperature range. Applications already in production are T-shirts that change colour at body temperature, and coffee mugs that can indicate the temperature of the drink they contain.
- **Photochromic paints and colorants**: These contain organic molecules that when exposed to light, particularly ultraviolet light, change colour. The light breaks a bond in the molecule that then rearranges itself into a molecule with a different colour. When the light source is removed, the molecule returns to its original form. Manufacturers usually offer four basic colours, violet, blue, yellow and red, from which other colours can be made by mixing.
- **Shape-memory alloys**: Some alloys, in particular some nickel/titanium alloys (often called NiTi or nitinol) and copper/aluminium/nickel alloys, have two remarkable properties: pseudo-elasticity (they appear to be elastic), and shape-retention memory (when deformed, they return to their original shape after heating). Possible applications include deformable spectacle frames, surgical plates for joining bone fractures (as the body warms the plates, they put more tension on the bone fracture than conventional plates), surgical wires that replace tendons, thermostats for electrical devices such as coffee pots, and the aeronautical industry (shape-memory alloy wires can be heated by an electric current and made to operate wing-flaps as an alternative to the conventional hydraulic systems).

Shape-memory alloys: www.cs.ualberta.ca

- **Hydrogels**: These are cross-linked polymers that have the ability to absorb or expel water when subjected to certain stimuli, for example temperature change, exposure to infra-red radiation, or change in pH. The open nature of the cross-linked structure enables water (or some aqueous solutions) to be absorbed within the structure causing the structure to swell. Small changes in the stimuli control the amount of swelling or shrinking. Possible applications include artificial muscles, underground water cut-off in the oil industry (the volume of gel can be pH-controlled), robotic actuators (found in some cases to be more effective than shape-memory alloys), and houses threatened by forest fires (hydrogels can be more effective than fire-fighting foam).

Questions

1 Name the process below.

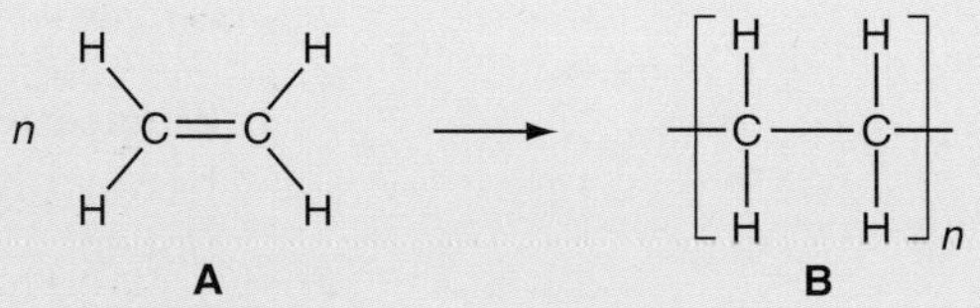

$$n\,\mathrm{H_2C{=}CH_2} \;\longrightarrow\; \mathrm{\left[\!-CH_2-CH_2-\!\right]_{\mathit{n}}}$$

A B

State the names of **A** and **B**.

2 **Read the following**. 'Ammonia is manufactured from nitrogen and hydrogen. The reaction is a reversible one and not all the nitrogen and hydrogen react. Unused gases are recycled. The reaction requires a temperature of about 400 °C and a pressure of 200 atmospheres. A catalyst of iron is used. The ammonia is removed as it forms.'

a State the sources of the hydrogen and the nitrogen used in the manufacture of ammonia.
b Give the meanings of the following:
- **i** 'reversible'
- **ii** 'recycled'
- **iii** 'catalyst'.

c Give **two** large-scale uses for ammonia.

3 Name **three** types of smart material and give **one** use for each type of material you choose.

Summary

1. Healthy plants need nitrogen.
2. Nitrogen is an unreactive gas, and must be 'fixed' before it can be used by plants.
3. Nitrogen can be fixed by a number of natural processes and by the Haber process.
4. Some chemical reactions are reversible.
5. The reaction between nitrogen and hydrogen is reversible.
6. Obtaining the maximum amount of ammonia in the Haber process needs carefully controlled conditions.
7. Ammonia is very soluble in water and forms an alkali.
8. When alkalis react with acids, neutralisation occurs.
9. Ammonium salts can be identified by heating them with sodium hydroxide solution.
10. Ammonia gas turns moist red litmus paper blue.
11. Alkanes are unreactive saturated hydrocarbons that contain only single bonds.
12. Alkenes are reactive unsaturated hydrocarbons that contain a carbon—carbon double bond.
13. Large alkane molecules can be broken into smaller, more useful molecules by cracking.
14. Cracking produces alkenes that can be made into a wide variety of chemicals, including polymers.
15. Poly(ethene), PVC and PTFE are addition polymers that have many everyday uses.
16. Addition polymers are formed from a single monomer, but condensation polymers are formed from two different monomers with the elimination of a simple, small molecule such as water.
17. Some polymers soften when they are heated and go hard again when cooled. These are called thermoplastics.
18. Polymers that are not softened by heat are called thermosets. These polymers are used in electrical fittings.
19. There are modern materials called smart materials that have a variety of uses. They include thermochromic paints, photochromic paints, shape-memory alloys, shape-memory polymers, and hydrogels.

Chapter 10 – Energy

By the end of this chapter you should:

- understand what things influence the types of power stations we build;
- understand who decides whether to build new power stations such as wind farms;
- know why we have pylons and overhead power lines;
- know why power is transmitted at high voltages but used at low voltages;
- know that some of the energy we use is wasted;
- know that we can use less energy by doing things in different ways;
- know how much electrical energy we use in the home;
- know how much it costs;
- know what type of heating is the most economic to use;
- know whether it is worth installing alternative energy sources for home use;
- know that heat can flow from place to place;
- know that we can help heat to flow and can use materials to keep the heat in;
- know that it may be cost-effective to fit double glazing and loft insulation.

The generation of electricity

Energy: using too much of it can be a problem

Scientists and engineers have given us many things that we enjoy so much and make our life comfortable. We are warm and have plenty of food. Travel by car, bus, train or plane is easy and cheap. For these things we have used energy, lots of it, by burning oil, coal and gas. Burning these fuels is polluting the environment and changing the climate.

What is happening to the Earth?

The Earth is getting warmer – this is called **global warming**.

Global warming of even just 2 °C could have serious consequences:

- The deserts will get bigger and there will be less land to grow food: most of southern Spain is turning to desert.
- We shall have more violent and extreme weather: floods and droughts are becoming common.
- Winters will become warmer. Alpine ski resorts are using snow machines much more than before. Soon there will no snow for winter sports.

Figure 10.1 How the UK will look if sea levels rise by 10 m

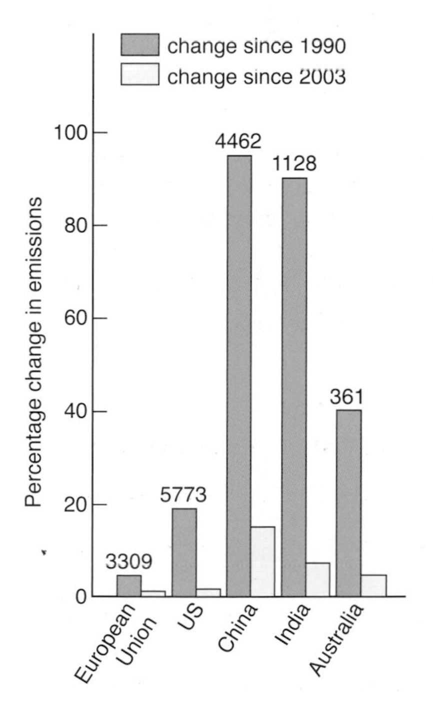

Figure 10.2 Carbon dioxide (CO_2) emissions are increasing. The number for each country is the amount of carbon dioxide released in 2004, in millions of tonnes.

- The ice caps in Greenland, the Arctic and the Antarctic are melting and sea levels are rising. As a result, many low-lying parts of the world including areas in the UK will be flooded (Figure 10.1).
- It could get really cold in the UK. The Gulf Stream is an enormous ocean convection current which keeps the UK 5–8 °C warmer than other countries at the same latitude. It carries 27 000 times more heat than all the UK energy sources put together. If the ice sheets melt they will dilute the seawater and may stop the Gulf Stream. If that happens we could have winters as cold as Siberia.

Did you know?

In the last ten years, more than one million square kilometres of Arctic ice have melted. That is an area more than five times the size of the UK.

What is the cause of global warming?

Oil, coal and gas are fossil fuels. Burning fossil fuels produces carbon dioxide and increases the natural 'greenhouse effect', making the Earth hotter (see the section on global warming and greenhouse gases in Chapter 6).

Natural processes have previously kept carbon dioxide levels in balance. The problem is that coal and oil that took a few million years to form are being burnt in a few decades. We are also using more and more fossil fuels in power stations to produce more electricity.

To reduce global warming, industrial countries have agreed to cut their emissions of carbon dioxide to 5% below 1990 levels by 2010. But that is proving difficult. Total emissions in 2004 were a massive 26% higher than 1990 levels.

Did you know?

How big-screen TVs might cost the Earth!

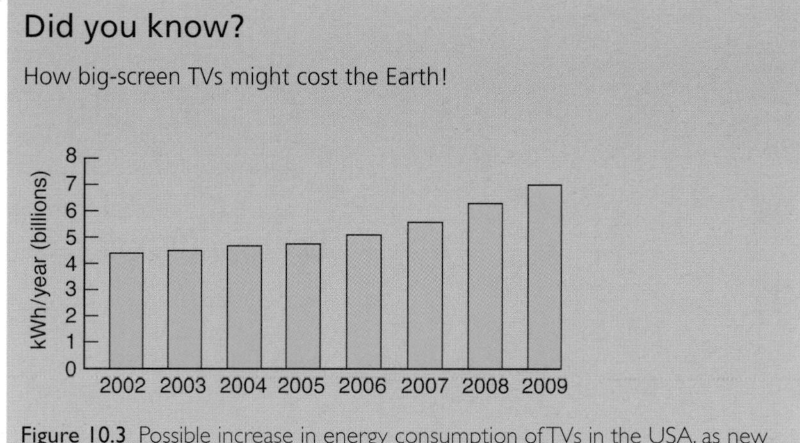

Figure 10.3 Possible increase in energy consumption of TVs in the USA, as new types of sets become more popular (according to the National Resources Defense Council, a US environmental action organisation)

Activity

1 Some people argue that the Earth's temperature change is just part of a natural temperature variation. Divide your class into two. One group should carry out research to support the idea that the change is natural. The other group should perform research to argue that the cause is overuse of fossil fuels. Both groups can use books and the internet for their research. Prepare and deliver your cases using an overhead projector or as PowerPoint presentations. Limit your reports to a maximum of 400 words.

2 Just suppose the warm Gulf Stream was to disappear. What might happen to the climate of the UK as a result? Will it get warmer or cooler? What will we need to do to keep warm in winter? In your group, give an account of what might happen to life in the UK if global warming continued to have an increasing effect on our climate.

Questions

1 What causes global warming?

2 Why is the world's energy use increasing?

3 Look at the graph in Figure 10.2 on page 191 showing carbon dioxide emissions.
   - **a** Which country released most carbon dioxide in 2004?
   - **b** Which countries show the biggest increase in carbon dioxide emissions?
   - **c** Which countries might have increased their industrial production the most?
   - **d** By what percentage have:
     - **i** Australia
     - **ii** the USA

     increased their output of carbon dioxide since 1990?

An energy crisis is coming

We are using more and more fossil fuels every year but eventually they will run out. They are **non-renewable resources**. If we don't develop alternative ways of producing electricity we will have an energy crisis.

- By 2020, many of Britain's power stations will be worn out and have to close. We will lose 45% of our present capacity to produce electricity.
- European Union (EU) plans to cut carbon dioxide emissions mean that by 2012 two-thirds of our coal-fired power stations would have to close.
- Nuclear power stations currently produce 22% of our electricity. They do not emit carbon dioxide, but produce radioactive waste. All but one will close by 2023. No new nuclear power stations are planned at present (2006).
- Gas is expected to provide 75% of the UK's electricity by 2020, but North Sea gas is running out. Eventually 90% of our supplies will have to come from Russia and Algeria. The supply pipeline is long and vulnerable to terrorist attack.
- Wind farms could produce 20% of our electricity needs, but the wind is unreliable and we would have to keep other power stations running all the time just in case.

Energy efficiency

One of the simplest things we can do to reduce the amount of fossil fuels we burn is to use less energy and also to produce electricity more efficiently. We know that for all energy transfers:

- The total energy after the transfer is equal to the total energy before the transfer. This is known as the **law of conservation of energy.**
- Usually we find that some energy is transferred into a form we do not want.
- When we use the term **energy efficiency** we mean the efficiency of the transfer to the sort of energy we want – **useful energy**.

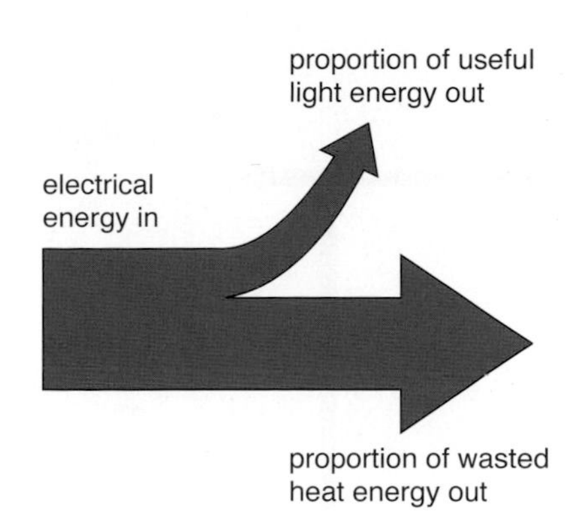

Figure 10.4 An energy transfer (Sankey) diagram for an ordinary electric light bulb

Take an ordinary lamp. We want it to provide us with light, but the lamp gets very hot. A lot of the energy is transferred as heat. We do not want the heat. We say the heat energy is **wasted energy** (Figure 10.4).

We can measure the efficiency of a device using the following formula.

$$\text{energy efficiency} = \frac{\text{useful energy transfer}}{\text{total energy input}} \times 100\%$$

Thermal power stations

Thermal power stations (Figures 10.5–10.7) provide us with most of our electricity. They burn non-renewable high energy density fuels such as coal, gas and oil (or use a nuclear fuel). They boil water to produce steam. The steam drives **turbines** and **generators** which produce the electricity. In a nuclear power station the heat to make the steam comes from a nuclear reactor (Figure 10.6). In most power stations a lot of the energy is wasted and goes up the chimney as heat.

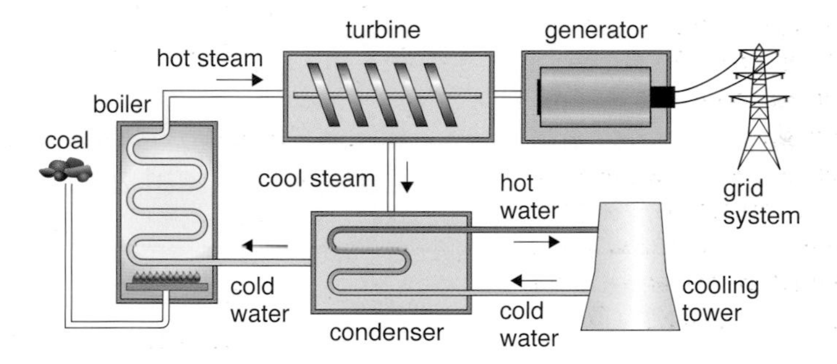

Figure 10.5 A fossil fuel power station. People often think it is smoke coming out of the cooling towers; in fact it's steam.

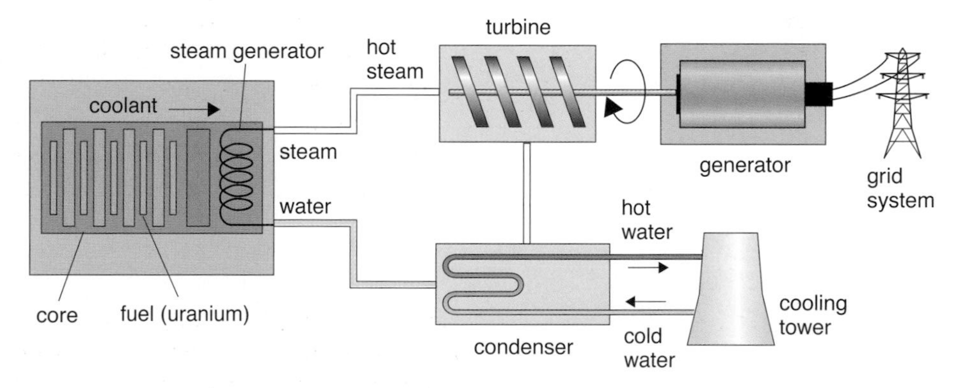

Figure 10.6 A nuclear power station

A hand-driven dynamo can be used to show how a power station produces electricity. Connect a lamp to the dynamo to represent the lights in your house.

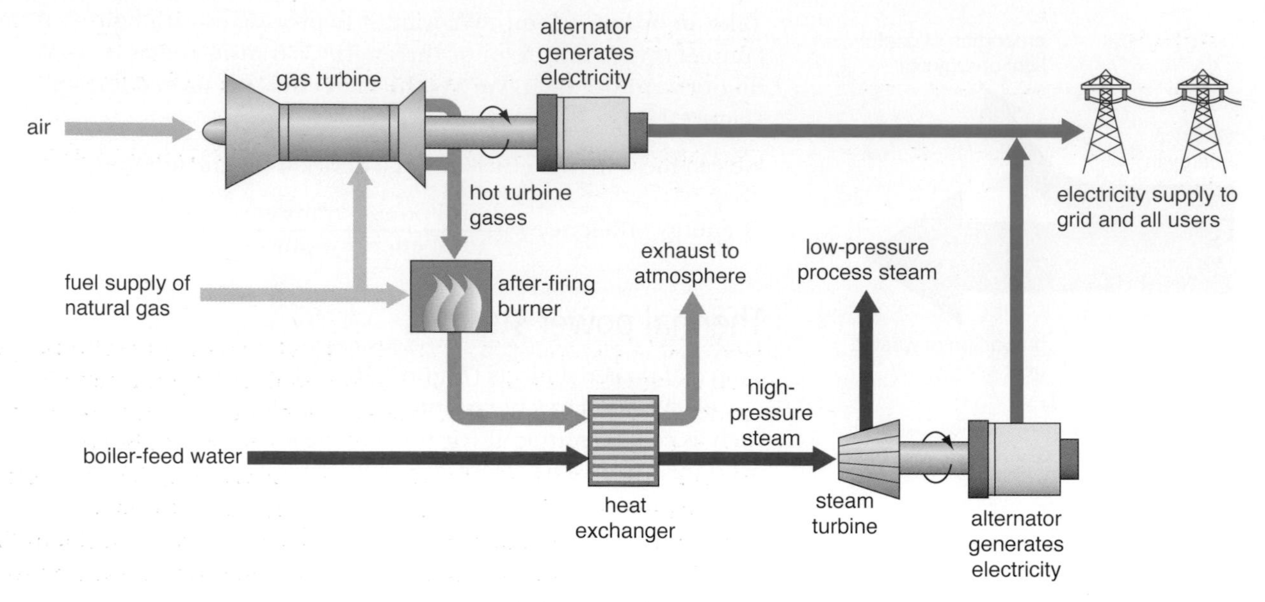

Figure 10.7 Some new gas turbine generators are much more efficient than old power stations. This combined cycle generator is 60% efficient. It burns natural gas (methane) in a gas turbine similar to those used in jet aeroplanes. It generates electricity from the rotation of the gas turbine and then goes on to use the heat from the hot gas to heat water to make steam as in a conventional power station.

Combined heat and power (CHP) units

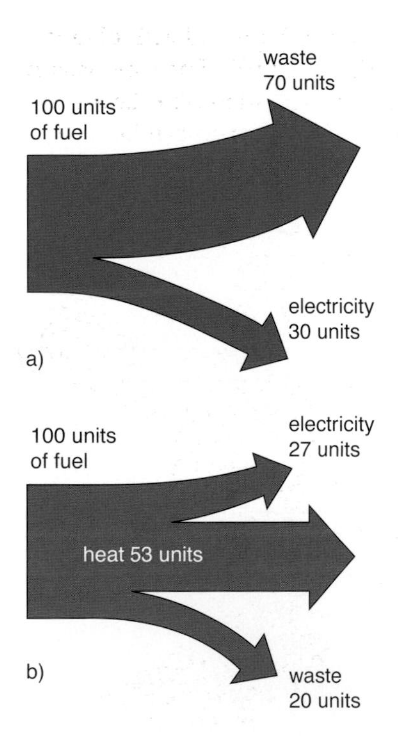

Figure 10.9 Energy transfer efficiency for (a) a conventional electricity generation system, and (b) a combined heat and power unit

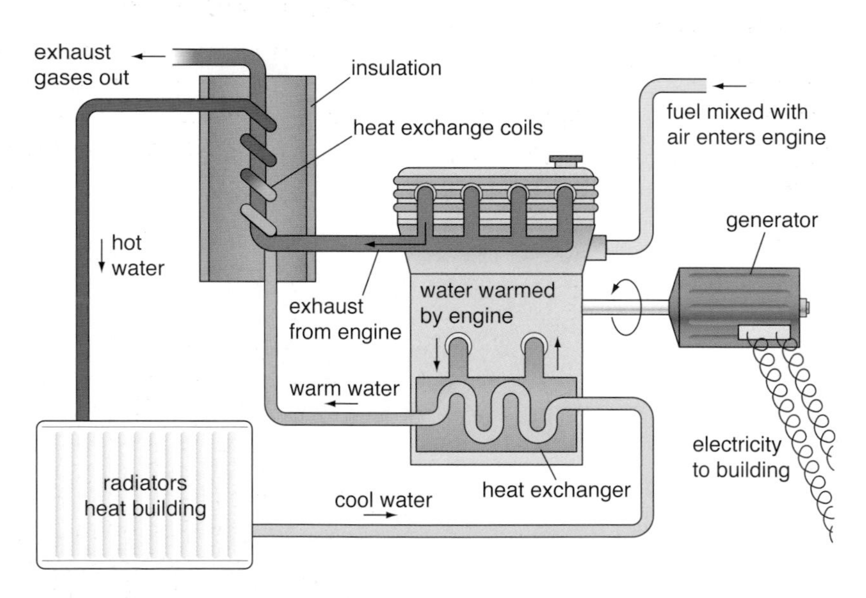

Figure 10.8 A combined heat and power unit like this heavy goods vehicle engine could be used in a large hotel.

Some large users like factories, hospitals and big hotels have their own small power stations. These generate electricity and also use the heat produced to warm the building and to heat water. If the hotel has a swimming pool, the hot exhaust gas is passed through another heat exchanger to warm the pool. These power stations are no better at generating electricity. They are 80% efficient overall because of the way the heat is used.

Activity

Use reference books or the internet to find out more about combined heat and power units and how much power they could provide in the future.

Questions

4 Explain, with the help of a block diagram, how a coal-fired power station works.

5 Explain what is meant by the statement 'a gas-fired power station is 60% efficient'.

The alternatives to fossil fuel power stations

Generating electricity from fossil fuels releases the greenhouse gas carbon dioxide and also causes acid rain from sulphur dioxide emissions. Nuclear power stations do not emit carbon dioxide and other polluting gases. However they are expensive to build (**commission**) and dismantle safely (**decommission**). They produce radioactive waste that needs to be carefully stored for many thousands of years. There is also the possibility that an accident like the Chernobyl disaster could release huge amounts of radioactive material into the atmosphere.

The Sun, wind and waves give us renewable non-polluting energy sources. We can also burn things that we grow (biomass fuel).

Let's look at factors that affect decisions about what types of power station should be built in the future.

Did you know?

You need 2400 windmills to replace just one nuclear power station.

Electricity from wind

Giant windmills (wind turbines) up to 100 m high and with blades up to 30 m long generate electricity (Figure 10.10). They are usually linked together in groups of ten or more and called wind farms. Some are sited on hills, others out at sea. Wind farms can be controversial.

Figure 10.10 A wind turbine can generate 2 MW of electricity on a windy day.

Figure 10.11 Use a hairdryer and this model wind turbine to investigate the effect of wind speed on the generation of electricity.

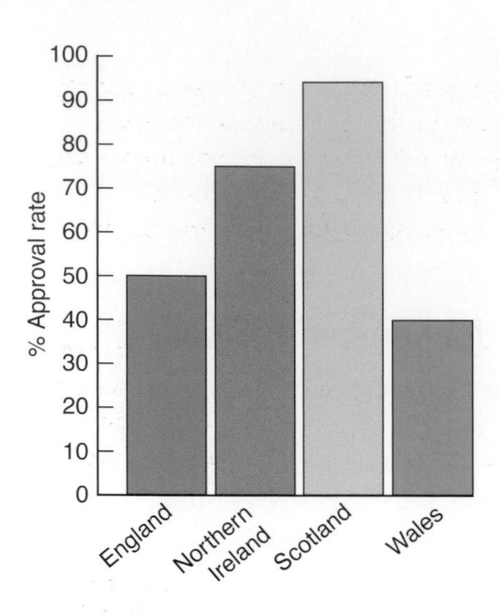

Figure 10.12 Approval rates in the UK, by country, for wind farms producing 1 MW or more of electricity

Who decides whether a wind farm can be built near where I live?

There have been mixed reactions to wind farms. Some people approve; others do not. Country Guardian is a group that campaigns against wind farms.

If the proposal is for generating capacity of less than 50 megawatts (MW) then the local council for the region will make the decision. For generating capacity above 50 MW, the decisions are taken by the Welsh National Assembly or the national government in England.

The Scottish Executive appears to be committed to wind farms as shown in Figure 10.12.

In England there are big differences in approval rates between local authorities.

Activity

Make a list of people's concerns about wind farms. Some of these concerns can be scientifically tested. Others are non-scientific or are the opinion of a group of people.

1 Sort the list you have made into those concerns that can be scientifically tested and those that cannot.
2 For each one outline how you would test to see if the statement is true.

Activity

A wind farm has been proposed for your area. The local council has organised a meeting with all interested parties. In your group decide who will represent the following stakeholders and put forward their policy:

- the wind-farm company
- the locals who are worried about the effects on their local environment
- the local council who are under pressure from the regional assembly to pass the proposal (they are also aware that there is an election next month)
- the farmer whose land will be used for the development.

Use local newspapers and internet search engines to find out as much as you can about the concerns of local residents in places where this has actually happened. Those representing the company and the council will need to prepare answers to the residents' concerns. When you have done your research, set up the debate in your classroom.

Electricity from hydroelectric power and pumped storage

Water flows through pipes from a reservoir at a high level. It turns the turbine which is connected to a generator. In areas with a plentiful supply of water, the water is returned to its natural course.

The pumped storage scheme at Dinorwig in North Wales has two reservoirs. When water flows from the top to the bottom, it generates electricity. At night when there is a low demand for electricity, the water is pumped back up to the top reservoir, using electricity generated by nuclear and coal-fired power stations as these cannot easily be switched on and off.

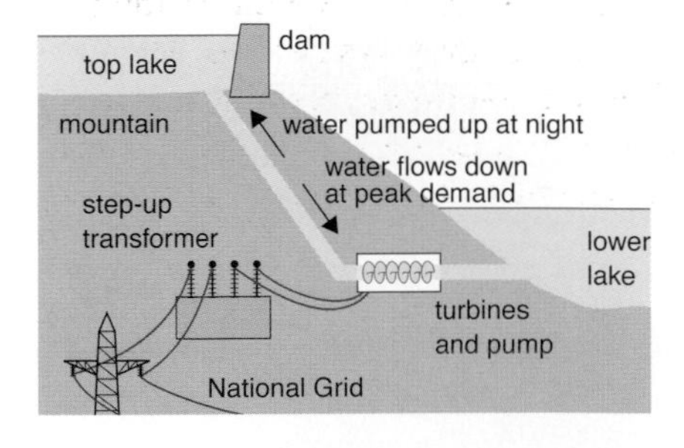

Figure 10.13 A pumped storage scheme

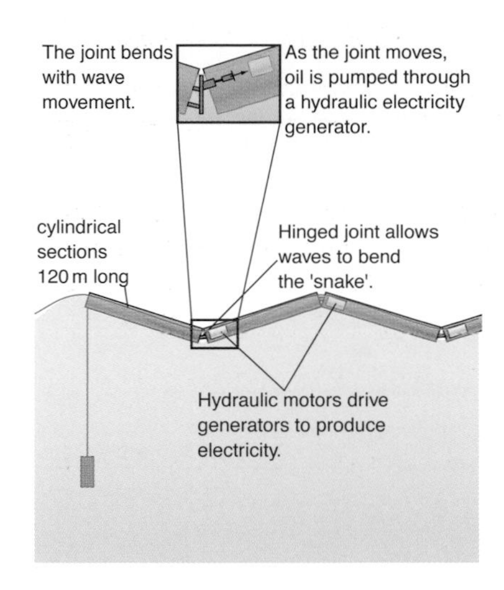

Figure 10.14 The snake, an experimental wave machine

Electricity from waves

A wave machine transfers the energy of waves into electrical energy (Figure 10.14). It consists of a string of five cylinders, each about 3.5 m in diameter. As each one is moved up and down by the waves, it pumps oil through a **hydraulic** generator. The machine is anchored to the seabed and generates up to 750 kW of electricity.

Electricity from tides

France and Canada have tidal power stations. A dam is built across a river estuary. As the tide flows in, the water passes over underwater turbines like those used in hydroelectric schemes, and electricity is generated (Figure 10.15). More electricity is generated when the tide flows out.

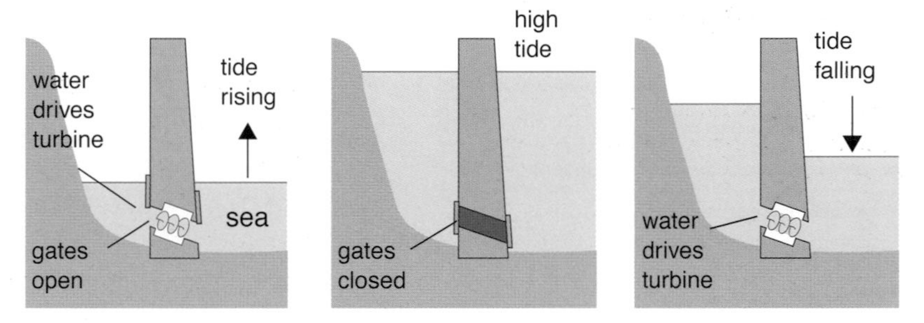

Figure 10.15 A tidal barrage like this was proposed for the Severn Estuary. It could have produced a large amount of the UK's energy needs, but it was never built.

Electricity from biomass

Biofuel is the name given to a fuel that has been grown. Burning biofuels is carbon-neutral (it does not increase carbon dioxide emissions), because when we burn them we replace the carbon dioxide the plant took out of the air when it grew. Many farm waste products can be burnt as a fuel. One power station in England burns chicken droppings. In other regions there is a lot of straw left over after the cereal harvest. A straw-burning power station ensures that this energy source is not wasted. However the energy in biofuels is not very concentrated so you need a huge amount. They could be put to good use in small local power stations.

If we plant as many fast-growing trees as we cut down to burn, we can carry on using this resource. Fast-growing trees like willow and poplar are ideal. After they have been cut down they will grow again from the same roots. This technique is called coppicing and needs three forests. Each one is cut down every third year. One is ready to cut this year, one was cut last year, and the third was cut two years ago. The wood is burnt in a small power station or mixed with coal as at Aberthaw power station in the Vale of Glamorgan. The wood is blown into the furnace in the form of sawdust. It replaces up to 5% by weight of coal.

Did you know?

In 2002, police arrested some drivers in South Wales. They were using cheap vegetable oil from chip shops to replace diesel in their cars. They were in trouble because they were not paying the 45p per litre tax on fuel. Their exhaust fumes smelled of chips. Now some police forces are using a legal mixture of vegetable oil and diesel, called biodiesel. They have been nicknamed 'the frying squad'.

> **Activity**
>
> Collect information and articles from newspapers and the internet. Find out how global warming is affecting the lives of people all over the world, with floods, hurricanes, droughts, etc. Find out about the effects of past nuclear accidents. Then, in an extended report, consider the following.
>
> **1** **a** Should we build more nuclear power stations in Britain?
>
> **b** Should we be influenced by the unsafe operation of Chernobyl, an outdated nuclear power station?
>
> **c** Are the consequences of global warming easier to bear?
>
> **2** Prepare cases for and against these proposals.

Environmental costs

Table 10.1 and Figure 10.16 summarise the environmental impact of building different types of power station. There are also social impacts no matter how we generate electricity. Employment and tourism are affected by where power stations are built, and health could be affected by air pollution.

coal (150)
peat
oil (109)
gas
nuclear
biomass
hydro
solar
wind
best case
worst case
0 10 20 30 40 50 60 70 80 90 100
Thousandths of a euro/kWh

Figure 10.16 The cost to the environment of generating electricity from various energy resources

Can carbon dioxide emissions be reduced?

Power stations could trap the CO_2 before it is released into the atmosphere. They could then pump the gas along a pipeline and into old oil or gas wells. These oil and gas fields stored the original gas safely for millions of years. The estimated cost of disposal is about 30 £/t (£30 per tonne) of gas.

The costs of generating electricity

In Figure 10.17 you can see that renewable resources look rather expensive ways of generating electricity when the cost of CO_2 emissions is not taken into account.

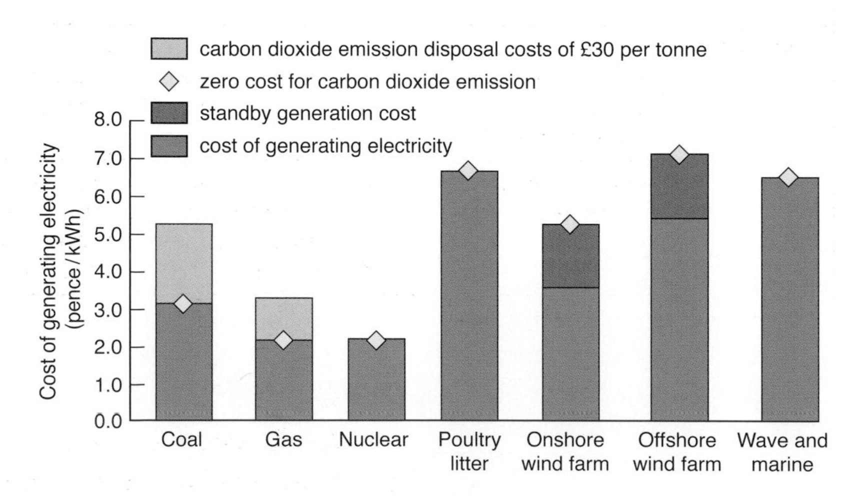

Figure 10.17 Cost of different ways of generating electricity

However, add on the costs of dealing with CO_2 of 30 £/tonne, as shown by the grey bars in Figure 10.17, and renewable energy resources look competitive.

Table 10.1 Advantages and disadvantages of using various energy sources to generate electricity

Fuel	Advantages	Disadvantages	Environmental considerations	Cost to build new
Coal and oil	Relatively cheap fuel; UK has coal and oil reserves; large world reserves, but extraction getting more difficult; high power output	Non-renewable source; much heat wasted in hot flue gases; overall, only 38% efficient; slow to start up, best when running continually	Burning releases carbon dioxide, increases global warming; regulations mean new build must capture and remove carbon dioxide, which is very expensive; releases sulphur dioxide, which produces acid rain and is expensive to remove	£0.8 million per megawatt (MW) produced
Natural gas	Relatively cheap fuel; UK has reserves in North Sea; large reserves in Russia and Middle East; high power output, 60% efficient	Non-renewable resource, UK North Sea reserves now running low; supply will be dependent on long vulnerable pipelines; some resources are in politically unstable regions	Burning releases carbon dioxide, increases global warming; no sulphur content; gas power stations can be switched off and on quickly so used only when necessary	£0.5 million / MW
Nuclear fuels	Very high power output from a small amount of fuel; no increase in global warming; when stations are operated correctly, little or no radiation released	Radioactive waste produced that remains radioactive for a long time and must be stored safely for many years, possibly thousands; non-renewable resource; must be kept running all the time; only 38% efficient	In an accident, radioactive materials could be spread over a large area; vulnerable to terrorist attack and waste could be used to make nuclear weapons	£1.7 million / MW; high cost to build plants and shut them down safely; Dounreay will cost a total of £30 million and 30 years to shut down
Wind turbines	Renewable resource; low running costs	Low power, need many turbines to achieve a useful output; some think wind farms unsightly; can cause noise pollution; only 40% efficient in converting kinetic energy of wind to electrical energy	No air pollution; electricity generated depends on weather, output not constant; other power stations must always be on standby	Depends whether onshore or offshore; Wales to spend £700 million on 200 MW offshore and 800 MW onshore generating capacity
Hydroelectricity	Renewable resource; low running costs; pumped storage schemes not as dependent on rainfall	Many schemes entail flooding river valleys and losing farmland and wildlife habitat	No air pollution; reliable; suitable only for hilly areas; can be switched on instantly	New builds in the UK unlikely because of shortage of suitable sites
Tides	Renewable resource; low running costs	Tidal dams have to be built across river estuaries; problems for shipping and marine wildlife	No air pollution; tides change daily and monthly; not a constant source of electrical energy	Severn Barrage last costed in 1981 at £7500 million for 7000 MW

Questions

6 What are the advantages of fossil fuel power stations?

7 Why does planting and burning wood not increase the amount of CO_2 in the atmosphere?

8 Explain the following.

a Mixing wood with coal at Aberthaw power station is reducing CO_2 levels.

b A combined heat and power unit is more efficient overall than a conventional power station.

Question

9 Why do wind farms need to have other generators on standby?

Wind farm, Scarweather Sands, Swansea Bay:
www.wwf.org.uk, search for Scarweather

Electricity generation FAQ:
www.aepuk.com

Compares energy resources, has an online calculator:
about-gas-electricity.co.uk

Access to national government:
www.direct.gov.uk

Wind energy (go to *Wind farms of the UK* and then *Dynamic map of operational projects*):
www.bwea.com

Centre for Alternative Technology, Wales:
www.cat.org.uk

Activity

You have just been elected president of an island in the North Atlantic. Its weather and resources are similar to those found in the UK. You discover that all the power stations will be worn out in a few years' time. Your government (working group) must now decide on an energy policy: what type of power stations should you build, and how many?. You must keep carbon dioxide emissions, set up and running costs to a minimum and ensure a year-round supply of up to 80 GW (1 GW = 1000 MW). Use the data in this book and any other sources for costs. Each group is to present its energy policy to the class together with all costings. Remember that if you upset the electorate, you will be voted out at the next election.

Transmission of electricity

The National Grid and transformers

The **National Grid** is a network of power lines. It connects all the power stations together and supplies electricity to all our factories and towns. Having such a network means that if something goes wrong with a power station, your electricity supply can be switched to another power station. It also means that some stations can be shut down at off-peak times.

Power stations generate electricity as an alternating current (a.c.). This cannot be stored for use later. The supply always has to match the demand from our homes and industry (Figure 10.18).

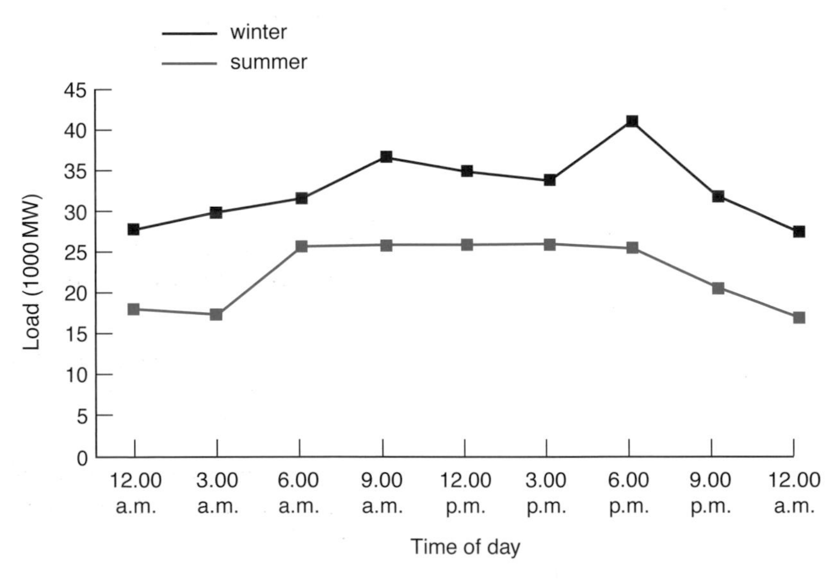

Figure 10.18 Energy demand curves for summer and winter (note the evening peak)

Engineers at the National Grid rely on computers to help them monitor power use and predict demand. Remote sensors in many parts of the country provide data about the weather. The Meteorological Office sends information about predicted weather trends, temperature, wind speed, etc. A network of 1500 free-floating buoys in the Atlantic provides long-term weather forecasting. Television companies pass on details about their schedules and predicted audience. All this, and information about the output power available from the generating stations, is fed into the computer. The supply is then changed to meet the expected demand.

Power, voltage and current

Electrical power is how quickly electrical energy is transferred to other forms of energy.

The unit of power is the watt (symbol W). This is a small unit, so we often use kilowatts (symbol kW) (1 kW = 1000 W). The output from power stations is often measured in megawatts (symbol MW) (1 MW = 1 000 000 W).

Power depends on the supply voltage and the current flowing. Current and voltage are linked to power by the following formula.

power (W) = voltage (V) × current (A)

Example
An electric heater has a current of 10 A flowing through it when a voltage of 230 V is connected to it. What is its power?

Formula first:

power (W) = voltage (V) × current (A)

Put the numbers in:

power (W) = 230 V × 10 A

The answer is

power = 2300 W
= 2.3 kW

Why do we generate and transmit electricity at high voltage?

Power lines carry electricity at very high voltages, up to 400 000 V (400 kV). Homes and offices use electricity at only 230 V. High voltages can be lethal, so why are they used to transmit electricity?

When an electric current flows through a wire some of the energy is transferred into heat. The bigger the current, the more heat is produced.

Remember that power = current × voltage. A small 250 kW power station could transmit electricity at 25 000 V and 10 A, or at 250 000 V and 1 A.

When energy is transmitted at low current, the use of a high voltage means that a lot less energy is lost as heat. The National Grid (Figure 10.19) has to distribute electricity over great distances. If it transmits at 25 kV from the power station, about 40% of the electricity is lost as heat. Electricity costs would be very high. Transmission at 400 kV reduces the heat loss to about 1%.

Tall pylons and large insulators are used to carry the bare cables. With a low current the cables can be thinner, lighter in weight and the pylons do not need to be so strong.

Copper's role in electricity (go to *Site map* and then *Copper and electricity*):
www.schoolscience.co.uk

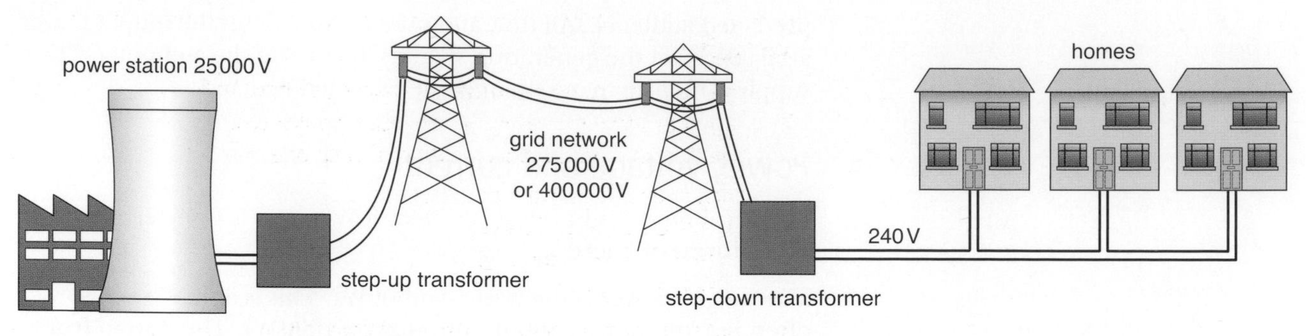

Figure 10.19 The National Grid transmission system

A **transformer** is used to increase or decrease the voltage. A step-up transformer at the power station is used to increase the voltage and decrease the current. At electricity substations, a step-down transformer reduces the voltage to a safe level. The overall efficiency of the grid is about 92%.

Questions

10 Name the units of voltage, current and power.

11 How is power linked to current and voltage?

12 A power station delivers a current of 10 000 A, at a voltage of 25 000 V. What is its power output in megawatts?

13 Why is electricity transmitted around the country at a very high voltage?

14 Why do we not use high-voltage electricity at home?

15 What is the name of the device that changes the voltage?

Energy efficiency, heating and the home

Energy and power

We use a vast range of electrical appliances in our homes. Appliances have a power rating on them (Figure 10.20). If the power isn't shown you can work out the power from voltage × current.

Power is a measure of the rate at which energy is transferred. A high-watt bulb is brighter because it transfers electrical energy to heat and light energy more quickly. Electric cookers usually have the highest rating, up to 10 kW. Electric fires are limited to a maximum of 3 kW; a small radio or lamp will be rated at only a few watts.

1W is the transfer of 1 joule per second (J/s) of energy. A 60 W lamp transfers 60 J/s of energy. The energy transferred by an appliance (and so the electrical energy that you pay for) depends on two things: its power and the time for which the power is used.

energy transferred (J) = power (W) × time (s)

Figure 10.20 This label tells you the rate at which the appliance transfers electrical energy.

Example

A 100 W lamp is left on for 10 minutes. How much electrical energy is transferred?

Formula first:

energy transferred (J) = power (W) × time (s)

Put the numbers in:

Convert the minutes to seconds.

energy transferred (J) = 100 × (10 × 60)

The answer is:

energy transferred (J) = 6000 J
= 6 kJ

> **Activity**
>
> **1 a** Make a list of the electrical appliances you have at home. (Do not try to move heavy or delicate appliances.)
> **b** With care, find the power rating of each one.
>
> **2** Which appliances have a power rating of:
> **a** between 1 kW and 3 kW
> **b** between 100 W and 500 W
> **c** less than 100 W?
>
> **3** What sort of appliances fall into each category?

Cost of energy use in the home

We have to pay for the electrical energy we use. The joule is a very small unit, so a larger unit of energy called the kilowatt-hour (kWh) is used for electricity bills. A kWh is called a unit on your electricity bill. Power is measured in kilowatts and the time is measured in hours (h).

The electricity company fixes a price for each kilowatt-hour or unit of energy that we use. They usually read domestic electricity meters every three months, and then send a bill or a statement showing the amount owing (Figure 10.21).

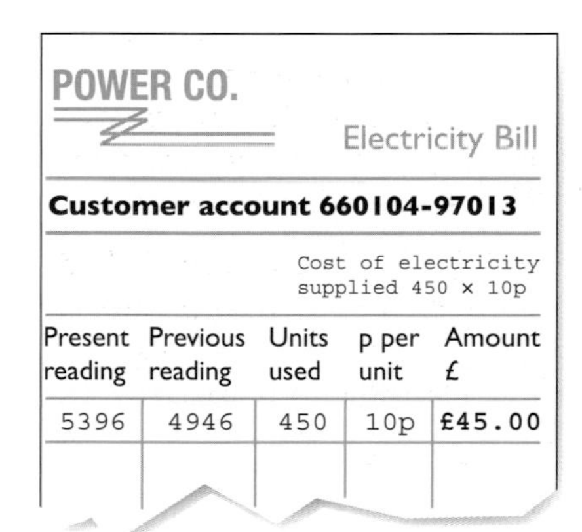

POWER CO.

Electricity Bill

Customer account 660104-97013

Cost of electricity supplied 450 × 10p

Present reading	Previous reading	Units used	p per unit	Amount £
5396	4946	450	10p	**£45.00**

Figure 10.21 Typical electricity bill

Paying the electricity bill

How much does it cost to use my appliance?

Formula first:

number of units used (kWh) = power (kW) × time (h)

cost = number of units × cost per unit

(Each unit typically costs about 10p. You may have to convert the total cost from pence to pounds.)

Example

Suppose that you leave a 3 kW heater on in your room. You put it on at 8 a.m. and forget about it until you get home at 4 p.m. If a unit (1 kWh) costs 10p, how much will it have cost to leave the heater on?

Formula first:

number of units used (kWh) = power (kW) × time (h)

Put the numbers in:

number of units = 3 kW × 8 h
= 24

Formula first:

cost = number of units × cost per unit

Put the numbers in:

cost = 24 × 10p
= £2.40

Questions

16 Beth is worried. She left the heater in her room on, from 7.00 a.m. until 5.00 p.m. It is a 3 kW heater.
- **a** How many hours was it on for?
- **b** How many units of electricity did it use?
- **c** If the electricity cost 10p per unit, how much did her mistake add to the family electricity bill?

17 Aled was last out of the house when they went on holiday for a whole week. He left the hall light on, which had a 60 W light bulb in it. How much did that mistake cost?

18 Aled's dad gets an electricity bill. The current reading is 34 231, and the previous reading was 33 571.
- **a** How many units have been used?
- **b** The electricity company charges 10p per unit for the first 250 units used, and 7p per unit after that.
  - **i** On the electricity bill, how much have the first 250 units cost in total?
  - **ii** How much have the remaining units used cost?
  - **iii** What is the total bill?

Paying for energy (go to *Environment* and then *Energy*) (in English and Welsh): www.consumereducation.org.uk

Different energy resources to heat the home

Heating our homes uses a lot of energy, and costs of electricity, gas, oil and coal are rising. We must use an efficient system. Remember that we measure the efficiency of an energy transfer using the following formula.

$$\text{energy efficiency} = \frac{\text{useful energy transfer}}{\text{total energy input}} \times 100\%$$

Table 10.2 Efficiencies and running costs for a range of heating systems. Running costs are high when the appliance is inefficient. Running costs also include cylinders for bottled gas and service and replacement costs for central-heating systems.

Heating appliance	Space-heating efficiency (%)	Running costs
Open fire burning coal	28 (poor)	High
Closed fire burning anthracite	60 (medium)	Low
Electric radiant fire	100 (good)	High
Gas radiant fire	60 (medium)	Medium
Gas wall convector heater	73 (good)	Low / medium
Bottled gas heater	60 (medium)	High
Paraffin heater	15 (poor)	High
Central-heating systems		
Central-heating system burning anthracite	70 (good)	Low / medium
Electric storage heaters	90 (good)	Medium
Gas central-heating standard system	70 (good)	Low
Gas central-heating condensing boiler	85 (good)	Lower
Oil central-heating system	70 (good)	Medium
Liquified petroleum gas central-heating system	70 (good)	Medium / high

Note Condensing boilers add an extra 15% to the efficiency of a system.

Activity

1. Use an internet search engine to find the latest prices for gas, electricity, oil, and coal prices in your area.
2. Find out how much it costs to heat your home and the type of fuel used.
3. Collate the results (try to make it anonymous) of the whole class. See if there is a pattern to the expenditure.
4. What other factors influence the heating costs?

Did you know?

For each 1 kW h of electricity used 0.51 kg (kilograms) of carbon dioxide are produced.

Figure 10.22 A compact fluorescent lamp

Table 10.2 shows efficiencies and costs for various heating systems. For example, with an open coal fire, for every 100 units of energy in the coal, only about 28 units heat the room. The energy wasted by hot air going up the chimney is about 72 units. Therefore the efficiency of the fire is about 28%.

Bottled-gas and paraffin heaters have two serious disadvantages. First, they do not have a flue to the outside. They discharge a lot of moisture into the room, which condenses and makes things damp. Second, they are a fire risk.

World energy prices are changing rapidly.

Save money and be green

Replace your traditional light bulbs with compact fluorescent lamps (Figure 10.22). They use just a quarter of the energy of ordinary light bulbs.

Leaving televisions and videos on standby costs UK consumers a total of £153 M (£153 million) per year and adds to carbon dioxide emissions.

Do not waste energy by filling the kettle when you need only one cup of tea. If everyone boiled just enough water, we could shut down a power station.

Activity

You could heat water in a pan on the ring of an electric cooker, or you could use an electric kettle. Some electric kettles are tall and thin; others are short and fat. Some have the heating element ring inside the kettle; others have it built into the base.

- Which are the most efficient at heating a large quantity of water?
- Which can heat enough water for a single cupful?

Your task is to design and carry out an experiment to find the answers to the questions above. Alternatively, you may explore secondary sources such as books or the internet to provide the answers.

In designing the experiment think about:

- how you will carry out the investigation
- what things you will keep constant and what you will measure as a result
- what equipment you will need
- what safety precautions you will have to take
- what measurements you will take and how you will record them
- how you will present your findings.

At the end of your work, suggest modifications to your experimental technique.

Figure 10.23 Solar cells are now sometimes used to power parking meters.

Is it worthwhile installing renewable energy systems in our homes?

Photovoltaic solar panels

A **photovoltaic cell** (PV cell) or solar cell (Figure 10.23) converts sunlight into electricity. Small cells can be connected together to make a solar panel. They can save you about 50% of your electricity. PV solar panels are best fixed on a south-facing roof. They will produce electricity on a cloudy day and even in winter. You heat your home and hot water with another fuel.

Solar PV systems cost £8000–£18 000. Grants are currently available from the UK government that reduce this cost by 50%.

How long will it take to recover the initial cost?

Suppose that you fit a solar PV system that costs £4000 (after a 50% grant). Say your annual electricity bill is £300. If a PV system saves 50% of this, the **fuel cost savings** are £150 per year. Therefore it will take you 4000/150 years to recover the cost of the system, which is about 27 years. This is the **payback time**.

After that, you will save £150 per year at today's prices. However electricity prices will have gone up considerably by then. You may well save more.

Did you know?

Several solar cars have been built. The Honda solar car raced over 3000 km across Australia, between Darwin and Adelaide, in 35 h 28 min. It can cruise at 90 km/h, and has a top speed of 140 km/h.

Figure 10.24 The Honda solar car

Questions

19 Make a list of appliances that use solar cells.

20 List places where you have seen PV solar panels installed. This could include photographs, on television, or on the internet.

Thermal solar panels

This is a different type of solar panel that is warmed by the Sun's rays. Over 96% of the light is converted to heat. The heat is absorbed by the water in the copper pipes, and passes through the heat exchanger to warm the water in the hot water system (Figure 10.25). This pre-heated water uses less electricity to get to working temperature, and the panels can produce up to 50% of the average household's water-heating requirements. They cost between £2000 and £4500 to install. They are very popular in Japan, with Tokyo alone having 1.5 million systems.

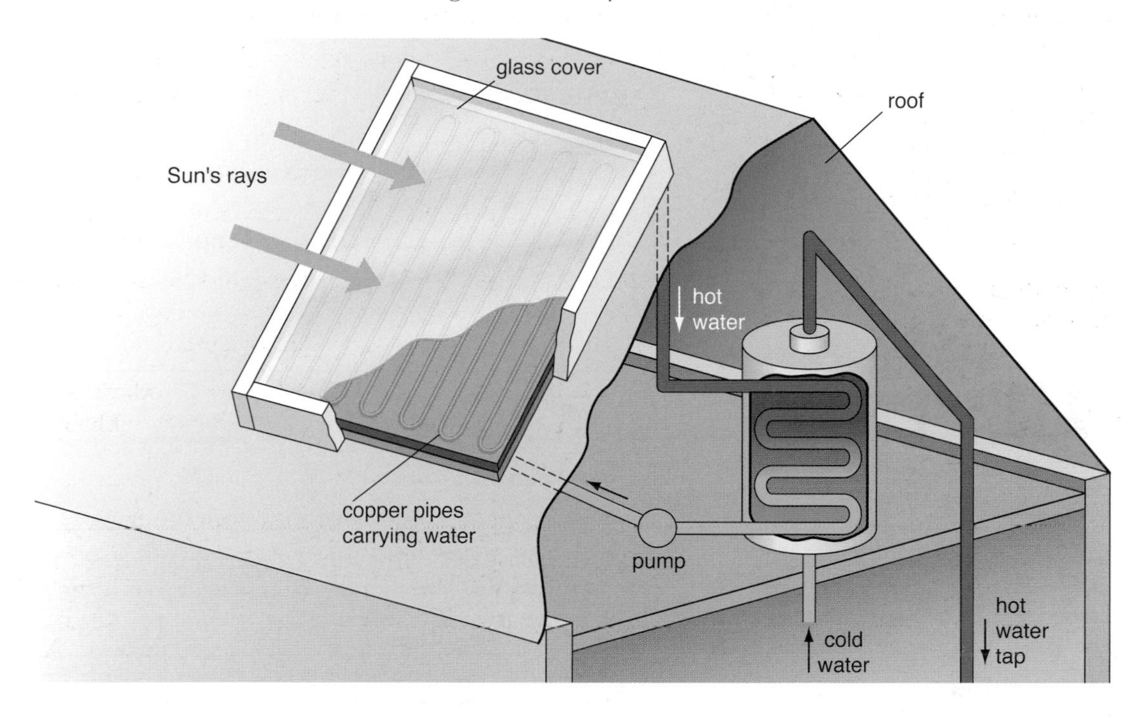

Figure 10.25 Flat plate thermal collector on house roof

Figure 10.26 Home wind-power system

Wind generators

Individual turbines (Figure 10.26) vary in output from a few hundred watts to 2–3 kW. You need planning permission to build one. A system costs £2500–£5000 to install.

How long will it take to recover the initial cost?

You have a 2 kW generator, which, if the wind blew at full strength for 12 h per day all year, would produce the following amount of energy.

$$2 \times 12 \times 365 = 8760\,\text{kWh}$$

This output is very unlikely. Even very tall commercial units rarely reach one-sixth of their best output. Your generator might generate 1500 kWh. At 10p per unit of electricity, the fuel cost savings are £150 per year. If the cost of installing your system was £2500, then the payback time or number of years taken to recover your money would be 2500/150, which is about 17 years.

Every home should have one

High initial costs and relatively cheap electricity from the National Grid make home-based solar and wind installations very expensive. But there is no doubt that energy costs will rise dramatically in the near future. As more PV solar panels and wind turbines are produced their, costs will fall – look at how the price of computers and mobile phones has tumbled.

A micro CHP unit is under development. It is small enough to fit under a kitchen worktop. When you switch on your gas central-heating system in winter, it will generate electricity at no extra cost.

Questions

21 You put a solar hot water system on your roof. It costs £2000 to install. It saves you £200 a year. How long will it take to recover the cost of installation?

22 What are the dangers of heating your home with paraffin heaters?

23 Which is the most economical central-heating system for your home?

Activity

Take a reading of your electricity meter tonight and again at the same time tomorrow.

1 Work out how many units of electricity were used during that day.

2 If one unit (kilowatt-hour) of electricity produces 0.5 kg of carbon dioxide, how much carbon dioxide does your house produce in:
- **a** one day
- **b** one week
- **c** one year?

3 How could you get a more accurate result for one year?

Fuel costs and efficiencies of old and new boilers:
www.boilers.org.uk

Generating your own energy:
www.est.org.uk

Energy, temperature and the transfer of heat energy

Transferring heat energy

Heat energy is always transferred from a hot body to one at a lower temperature. The bigger the temperature difference, the more heat energy is transferred. Heat can be transferred by conduction, convection and radiation.

Transferring heat energy by conduction

Practical work

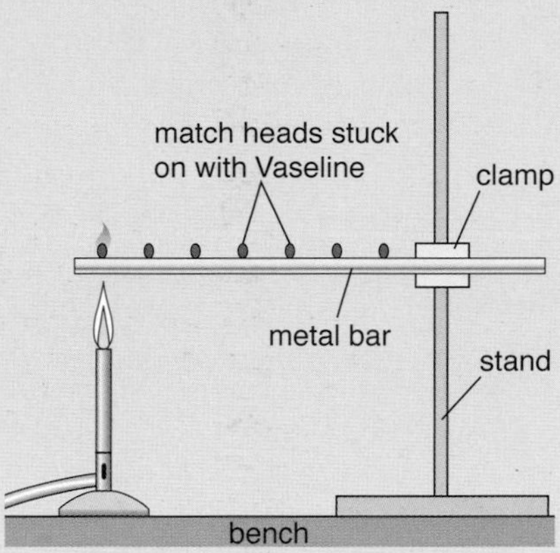

Apparatus to show that heat energy flows from a high temperature to a low temperature (Figure 10.27)

The end nearest the Bunsen burner is hotter than the rest of the bar. This temperature difference means that heat energy moves along the bar. The bigger the temperature difference, the greater is the heat transfer. The process is called **conduction**. Metals are good conductors of heat.

Figure 10.27 The matches burst into flame as the heat travels along the bar.

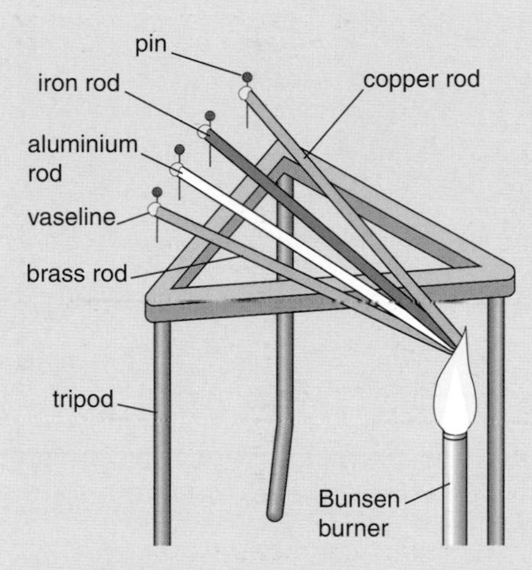

Figure 10.28 Comparing conduction in different metals

Some metals are better conductors than others (Figure 10.28)

- Get some metal rods all the same size but made from different metals.
- Put them on a tripod so that the ends are close together.
- Use Vaseline to stick pins on the other end of each rod.
- Heat the other ends equally with a Bunsen burner.
- Time how long it takes for each pin to drop off.

The hot ends of the bars were all at the same high temperature and the cold ends were all at room temperature, so the temperature difference was the same for all the bars.

1 Which metal bar lost its pin first, and which was last?

2 Which metal was the best conductor, and which was the worst?

Activity

1 Use books, the internet or other people to help you make a list of things that use good conductors to make them work efficiently.

2 Why are radiators made from steel? Would copper radiators be more efficient? Would they be more expensive?

Using good conductors

The pans we use for cooking are made from metals. Heat energy is transferred from the hot surface through the pan so that it can cook our food.

The radiators we use to heat our rooms are made from metal. The heat energy from the hot water is transferred to the outside by conduction. The hotter the water, the more heat energy is transferred.

Conduction in non-metals

Non-metals such as plastic and wood are poor conductors of heat. Plastic and wood are often used for handles on pans and other things that get hot.

Insulators

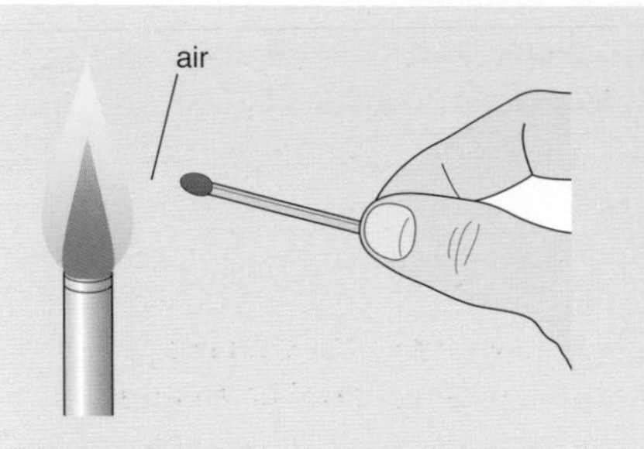

Figure 10.29 Heat conduction through air

Practical work

Showing that air is a poor conductor

1. Hold a live match about 1 cm away from a very hot Bunsen burner (Figure 10.29).
2. The match does not light.

This shows that air is a very poor conductor. Some materials have air trapped inside them. They are very poor conductors. They are called **insulators**.

Figure 10.30 The filling in the duvet traps air; air is an insulator so not much heat escapes, and the duvet keeps you warm.

Did you know?

The US National Aeronautics and Space Administration (NASA) has used space-shuttle technology to design a lightweight insulation material. Its drying time is five times faster than wool and its heat insulation four times greater. It can be used for clothing for firefighters and for insulation for refrigerators, freezers and ovens.

Transferring heat energy by convection

Open the door of a hot oven and you can feel the hot air rushing up past your face. This process is called **convection**. Convection occurs when liquids or gases are heated.

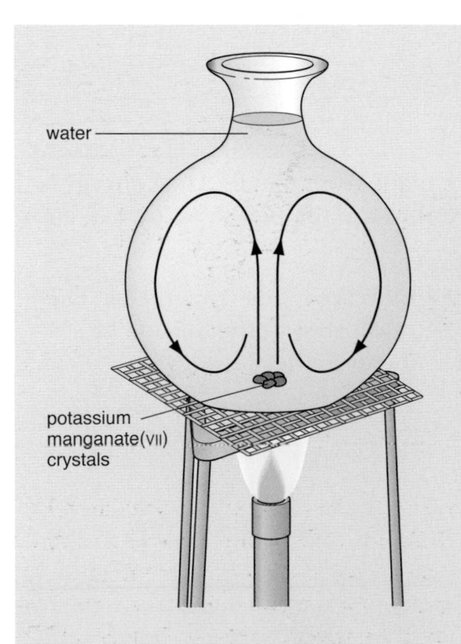

Figure 10.31 Apparatus to show convection currents

Practical work

Showing the path of a convection current

1. Wear eye protection.
2. Your teacher will provide you with a small crystal of potassium manganate(VII) (permanganate). Drop it carefully into a beaker of cold water.
3. Use a very low flame to heat the water near the crystal.

The coloured water shows the path of a **convection current**. It carries the heat energy along with it (Figure 10.31).

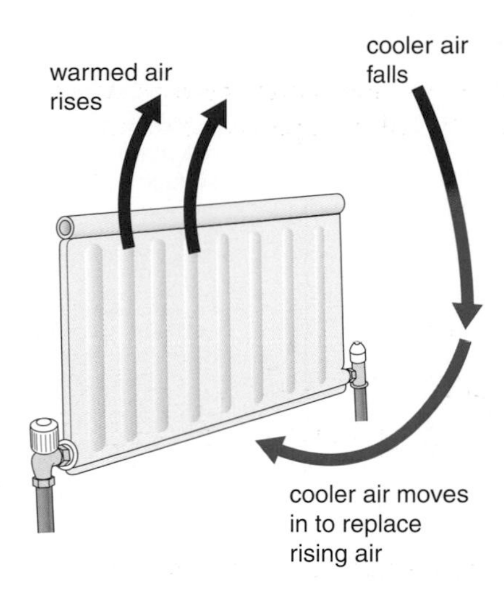

Figure 10.32 Convection currents transfer heat from the radiator to your room.

The hotter a radiator, the more heat is transferred to the room.

Convection currents need a fluid (gas or liquid) to travel through.

Questions

24 Name two good conductors of heat.

25 Name two poor conductors of heat.

26 Name two insulating materials.

27 What do most insulators contain?

Activity

Aled's dad says that copper is a better conductor of heat than iron. His mate Joe does not agree. He says iron is stronger so it will be a better conductor.

Design (on paper) an experiment to find out who is right. Say:

- how you will make it a fair test
- what measurements you will take
- how you will record your results.

Transferring heat by radiation

Heat radiation is a type of electromagnetic radiation. We call it **infra-red radiation**. It travels at the speed of light. All bodies that are at a temperature above absolute zero emit radiation. The higher the temperature, the more thermal (heat) energy is transferred by radiation.

Which surfaces are better radiation emitters?

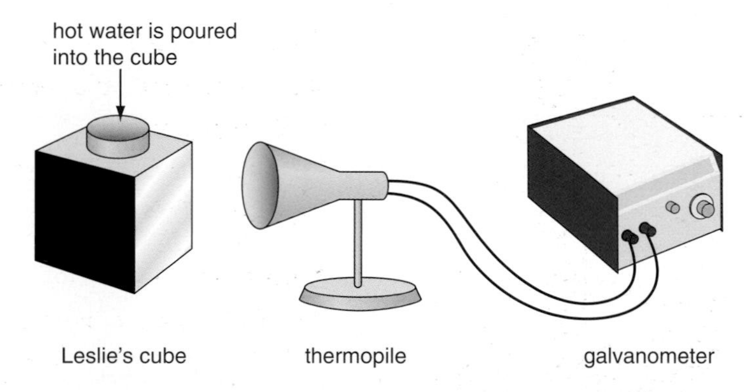

Figure 10.33 Comparing radiation emission from different surfaces

The thermopile shown in Figure 10.33 gives a small electric current when infra-red (heat) radiation falls on it. Each surface of the cube is turned to face the thermopile. The meter shows the greatest reading when it faces the matt (dull) black surface. It shows the smallest reading when it faces the shiny silvery surface. Dull black surfaces are the best emitters of radiation, and shiny metallic surfaces the worst.

Which surfaces absorb most radiation?

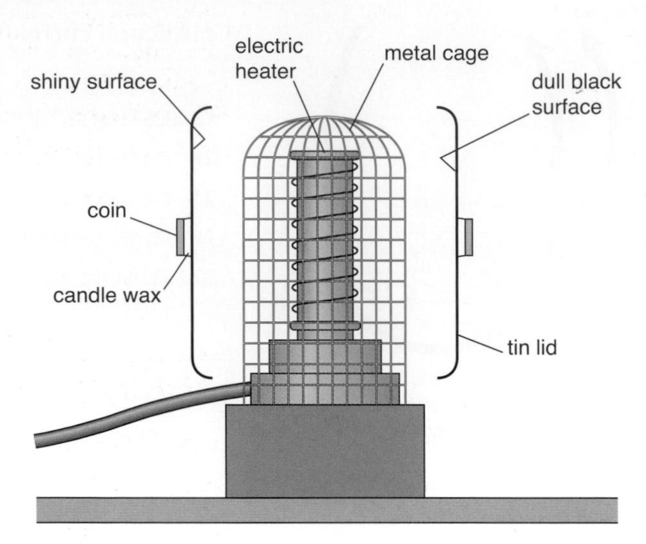

Figure 10.34 Comparing absorbers of radiation

In Figure 10.34, the coin on the matt black surface drops off first. The matt black surface is the best absorber of heat.

This is why solar heating panels that heat the water for a house have dark surfaces to absorb the greatest amount of energy from the Sun. They will work in the UK: on a dull day the water will be heated a little, so less electricity is needed to get it really hot.

Practical work

Testing simple solar water heaters made from black plastic tubing

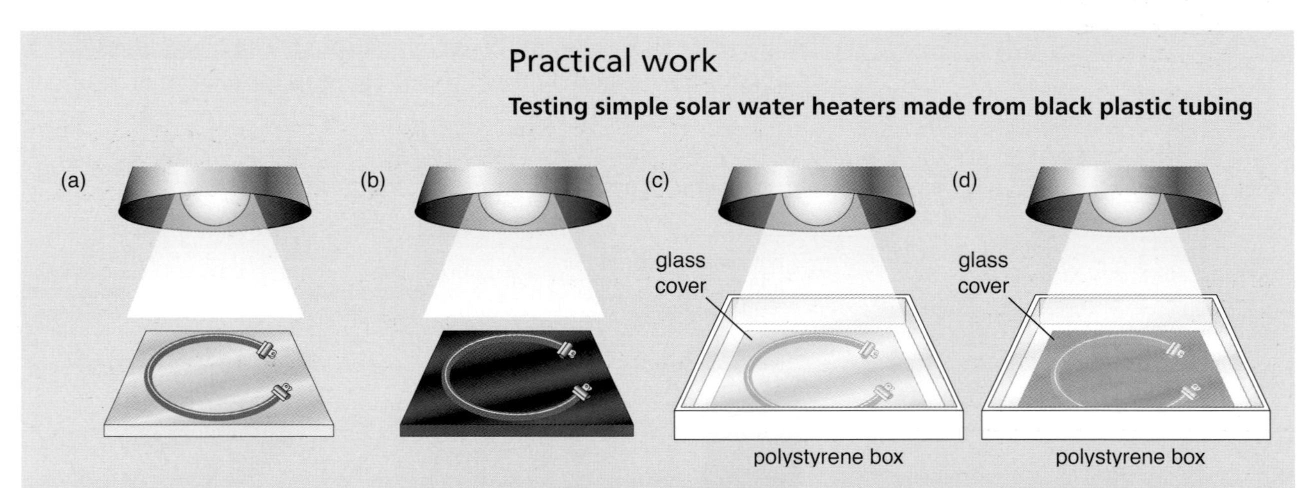

Figure 10.35 Four simple solar heaters: (a) tube A on a shiny metal plate; (b) tube B on a blackened metal plate; (c) tube C on a shiny metal plate with glass cover; (d) tube D on a blackened metal plate with glass cover

1. Take four equal-sized pieces of black plastic tubing, fill them with water and close the ends with clips.
2. Set them up as shown.
3. Measure the temperature of the cold water and again after each one has been under the heater for a set time.
4. Say how you will make it a fair test and how you will record the results.

Simple aluminised survival blankets can save lives (Figure 10.36). They reduce a person's heat loss by radiation, because shiny surfaces give out less infra-red radiation. The foil also reflects back heat produced by the body, because shiny surfaces do not absorb infra-red radiation. You can also reduce the heat loss from houses by putting reflective foil on the walls behind radiators.

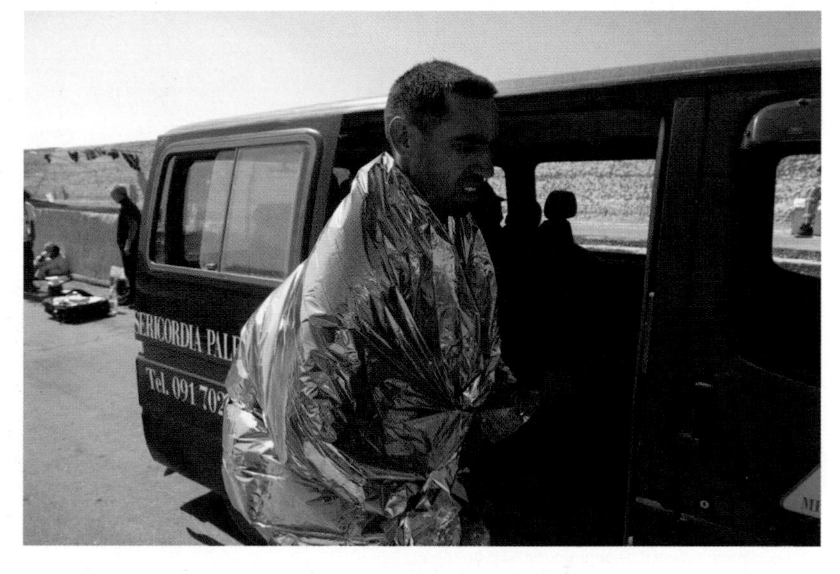

Figure 10.36 Aluminised survival blankets are often used in emergency situations, to prevent victims from getting cold.

Questions

28 Beti has a bungalow in the hills. It is made of wood and the walls are held together with steel bolts. In winter the bolts inside the house have icicles hanging from them, but the wood feels warm. Explain these effects.

29 Glass has a higher conductivity than brick. Why is more heat lost from the walls of a house than the windows?

Table 10.3 The best insulators have the lowest conductivity values

Material	Relative conductivity
Aluminium	8800.0
Steel	3100.0
Concrete	175.0
Glass	35.0
Water	25.0
Brick	23.0
Breeze block	9.0
Wood	6.0
Felt	1.7
Wool	1.2
Fibreglass	1.2
Air	1.0

Activity

Aled's dad doesn't like change. He says his old black teapot keeps tea hotter than that new-fangled shiny one he was given for a Christmas present. Design (on paper) an experiment to find out if he is right. Explain:

1 what measurements you would make
2 how you would make it a fair test
3 how you would present your findings
4 how you would decide whether or not Aled's dad was right.

Insulating our homes

Energy costs are always going up. We lose heat from our homes by conduction, convection and radiation. All new homes must be energy-efficient. Building regulations make sure that they are built with insulating materials. The materials must be poor conductors of heat, or in other words, have a low **conductivity** value (Table 10.3).

We can also improve the insulation of existing homes. Study Figure 10.37 on page 214 to see how much money can be saved. It also shows how much your carbon dioxide emissions can be reduced. High-efficiency condensing boilers also reduce energy costs.

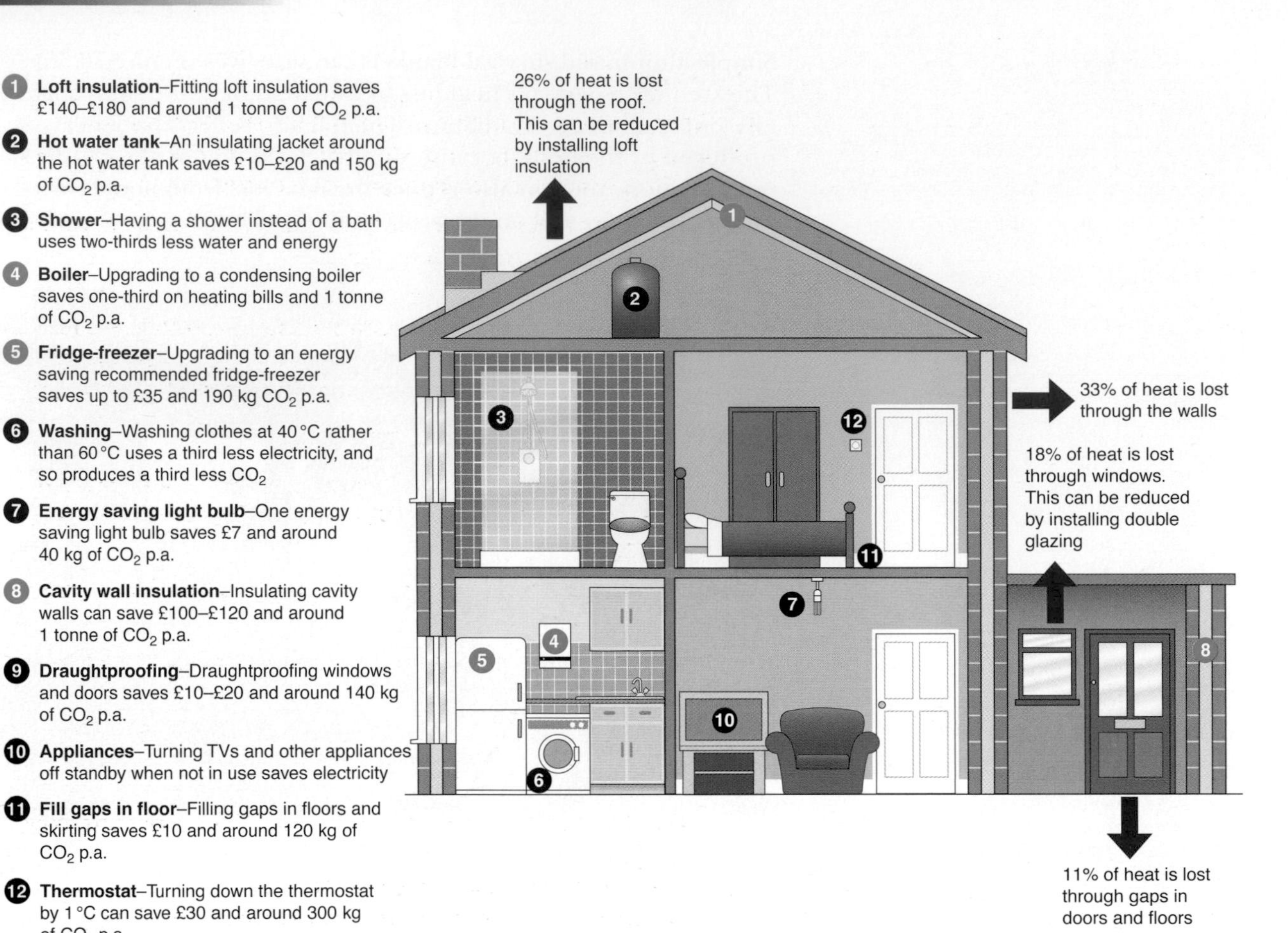

Figure 10.37 How to save energy in your house (the blue circles show that grants and advice are available through the Energy Saving Trust, and the black circles show no-cost or low-cost measures)

Is it worth it?

Houses can be insulated quite easily. Table 10.4 shows the savings for a typical modern house.

Table 10.4 Costs and savings for insulation in a modern semi-detached house

Type of insulation	Cost of fitting (£)	Annual saving on fuel bills (£)
Cavity wall insulation	260–380	100–120
250 mm loft insulation where none at present	220–250	140–180
Fit a jacket to hot water tank (DIY)	10+	10–20
Draughtproofing (DIY)	40+	10–20
Floor insulation (DIY)	100+	15–25
Filling gaps between skirting board and floor	25	5–10
Double glazing	3000+	40

Activity

Use DIY books or the internet to find out how to:

1 draughtproof your home
2 fit a jacket to a hot water tank
3 fill the gaps between a skirting board and a floor.

Saving money on your home heating bills: www.est.org.uk

DIY stores often give away energy saving leaflets.

Questions

30 When you insulate your house, how long does it take you to start saving money? Use Table 10.4 to calculate the payback time for each type of insulation. Which kind is the least cost-effective?

31 Triple glazing is used in cold countries. Explain the advantages of triple glazing compared with double glazing.

32 Heat transfers from one place to another by the processes of **conduction**, **convection** and **radiation**. Copy and complete the sentences below by choosing the suitable method of heat transfer.

a Polystyrene cups are good insulators. They do not let much heat pass through them by _________.

b Heat passes through metals by _________.

c The air near the ceiling of a room is hotter than the air near the floor because of _________. Black surfaces are good at giving out heat by _________.

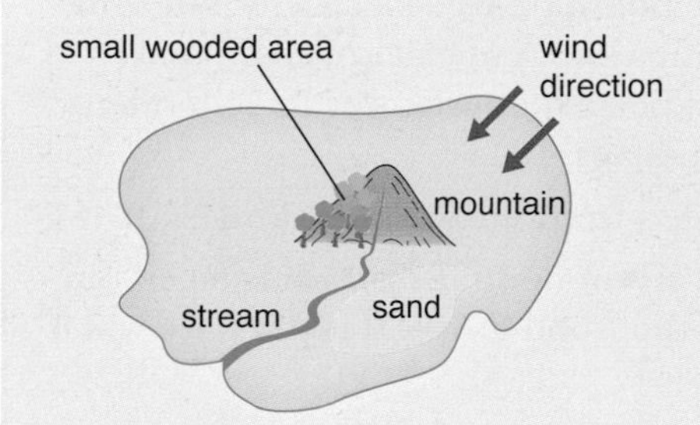

33 Some scientists are going to visit a remote island for a few months (see diagram left).

They will need a supply of electricity. The scientists could produce electricity using:

i solar power
ii wind energy
iii wood.

a Give **two** disadvantages of using wood from the island to produce electricity.

b Give **one** disadvantage of using solar power.

c Which would be the most suitable place for a windmill?

d Copy and complete the energy change for a windmill:
_________ energy → electrical energy.

34 Electricity is sent through the National Grid at high voltages. Transformers play an important part in the National Grid.

a What is the National Grid?

b Give a reason why electricity is carried at high voltages.

c Explain why **two** types of transformers are needed in the National Grid system.

35 The table below shows information about two kettles. Each kettle contains the **same** mass of water.

	Input energy (kJ)	**Energy absorbed by water (kJ)**	**Wasted energy (kJ)**	**Temperature rise (°C)**
Kettle A	600	460		25
Kettle B		460	160	25

a Copy and complete the table.

b At the start the temperature of the water of the kettle was 17 °C. What was the temperature of the water at the end?

c Use the table to explain why kettle A is more efficient than kettle B.

Summary

1. Burning fossil fuel puts a lot of carbon dioxide in the air.
2. The carbon dioxide is causing a rise in global temperatures.
3. Coal, oil and gas power stations burn fossil fuels.
4. Nuclear power plants do not emit carbon dioxide, but they have their own disadvantages.
5. A combined heat and power unit uses its fuel more efficiently.
6. Costs of building power stations (in descending order) are: nuclear, coal and oil, gas.
7. Photovoltaic cells convert sunlight into electricity and can be used to power buildings or, with battery backup, to power motorway signs.
8. Wind farms can generate pollution-free electricity, but some people argue they spoil the view. A nuclear or fossil fuel power station must be on permanent standby to generate electricity when the wind drops.
9. Electricity can be generated from the waves and tides.
10. Burning biomass fuel is carbon-neutral. Carbon is taken out of the atmosphere when the wood grows. It is put back when it burns.
11. The unit of power is the watt (W).
12. 1 kW = 1000 W and 1 MW = 1 000 000 W.
13. Power (W) = voltage (V) × current (A).
14. A system of overhead cables called the National Grid connects all the power stations and our homes and factories together.
15. The National Grid carries electricity at a very high voltage. This reduces the energy lost as heat.
16. Power stations generate alternating-current (a.c.) electricity.
17. 1 W of power is the transfer of 1 J/s of energy. This is a small unit.
18. Energy transferred (J) = power (W) × time (s).
19. Energy use in the home is measured by electricity and gas meters.
20. The unit of energy is the kilowatt-hour (kWh). The number of units used (kWh) = power (kW) × time (h).
21. Our homes can be heated by a variety of energy sources. Some are more expensive to run than others. The installation costs vary.
22. Using renewable energy in the home is an option, but installation costs can be high and the payback time long.
23. The greater the temperature difference, the greater is the transfer of heat energy. This applies to conduction, convection and radiation.
24. Heat energy is transferred through solids by conduction.
25. Metals are good conductors and non-metals are poor conductors of heat.

26 Materials which contain trapped air are very poor conductors and are called insulators.

27 For conduction there has to be a material for the heat energy to travel through.

28 Convection is the main process in which heat is transferred in fluids (gases and liquids). As the hotter fluid rises it takes heat energy with it.

29 For convection there has to be a fluid to carry the heat energy.

30 When heat is transferred by radiation, it travels as electromagnetic radiation. It can travel across a vacuum at the speed of light.

31 Matt black surfaces are the best emitters and absorbers of radiation. Silvered surfaces are the worst.

32 A thermal solar panel can pre-heat domestic hot water.

33 Insulating a house can save money and energy.

34 Many simple measures like draughtproofing, and sealing gaps in floors can save money.

35 With some measures like fitting expensive double glazing, it can take years for payback of their installation costs.

Chapter 11 Waves and electromagnetic radiation

By the end of this chapter you should:

- know how to describe the shape and measure the speed of waves;
- know about the different types of electromagnetic waves;
- know how we use electromagnetic waves and which types are dangerous;
- know about the use of optical fibres, microwaves and satellites for communications;
- understand the perceived dangers of mobile phone masts.

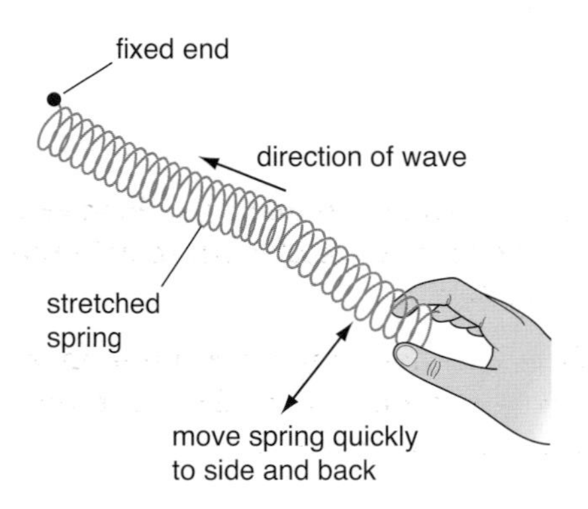

Figure 11.1 Making a wave in a slinky spring

The characteristics of waves

Wave speed, frequency and wavelength

Waves carry energy and information. A tsunami wave carries an awesome amount of energy. Light and sound waves reaching your eyes and ears carry energy and information about the world around you.

To make a wave in a stretched slinky spring, flick one end of the spring from side to side. The wave travels along the spring (Figure 11.1).

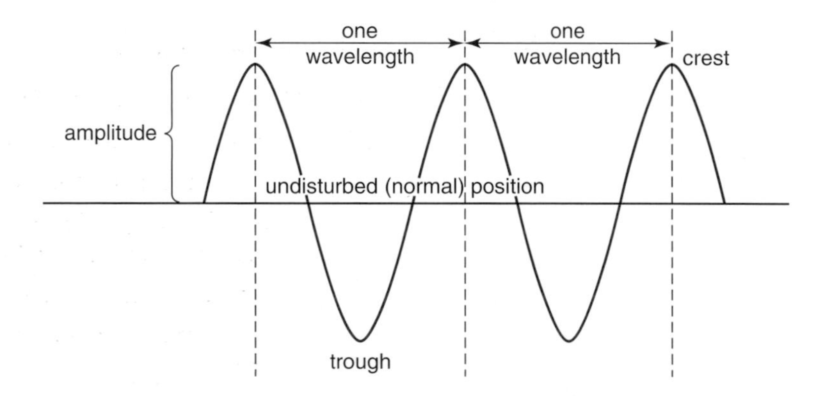

Figure 11.2 Wave measurements

Flick the spring continually and you will get a succession of waves. Look at their **wavelength** (λ); this is the distance from start to finish of one complete cycle. Wavelength is measured in metres (m).

Flick the spring at a faster rate and you will send out more waves per second. You have increased the **frequency** of the waves. Frequency is the number of waves which pass a point in one second. Frequency is measured in units of number per second. This unit is called the hertz (Hz).

Wave speed is the distance a wave travels in 1 second (s) (Figure 11.3). It is measured in metres per second (m/s).

$$\text{wave speed (m/s)} = \frac{\text{distance (m)}}{\text{time (s)}}$$

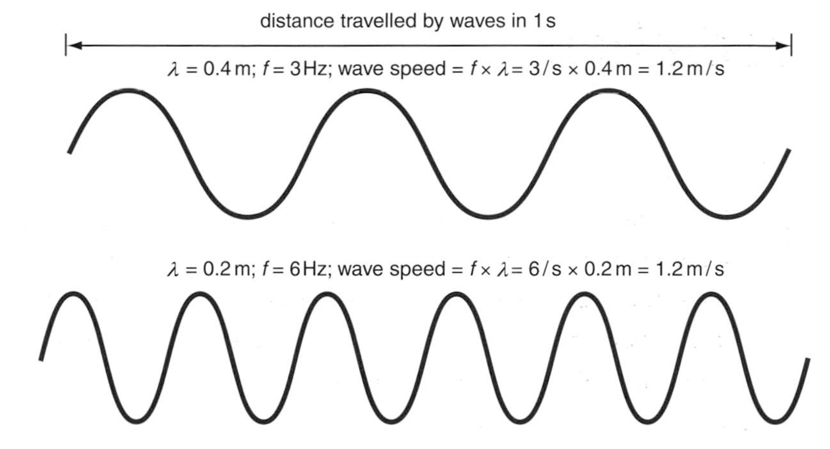

Figure 11.3 These two waves are travelling at the same speed, so the one with a higher frequency has a shorter wavelength.

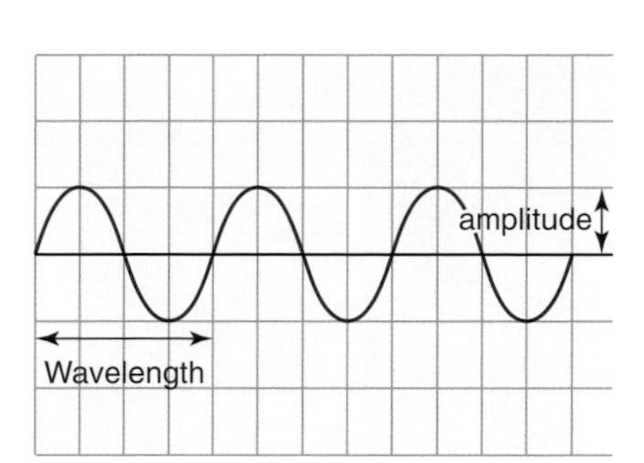

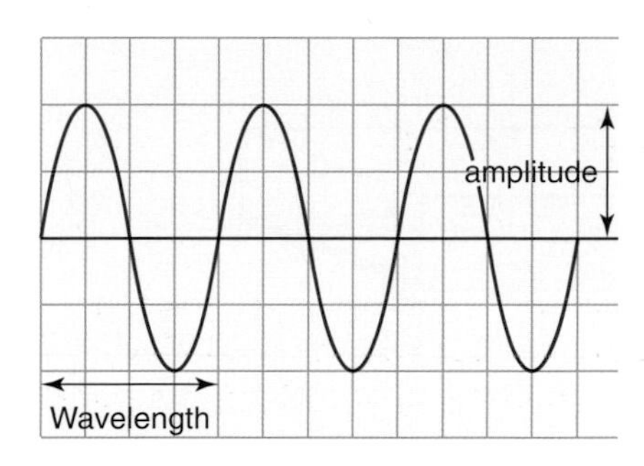

Figure 11.4 These two waves have the same frequency and wavelength but different amplitudes.

Wave amplitude

Keep the frequency of the waves the same and make gentle waves. Then make stronger waves. Frequency and wavelength remain the same, but the **amplitude** increases (Figure 11.4).

The greatest displacement of the material from its rest position is called the amplitude of the wave. Note that amplitude is half the total distance between the crest and trough of a wave.

Practical work

Measuring the speed of waves along a spring

1 Work in a group of three or more.
2 Stretch the spring between you and your partner.
3 Give the spring a single flick to one side to check that the wave can reach the fixed end and be reflected back in a reasonable time.
4 Measure the extended length of the spring.
5 On a given signal, start the wave and time it over as long a path as possible.
6 Calculate the wave speed from the following formula.

$$\text{speed (m/s)} = \frac{\text{distance (m)}}{\text{time (s)}}$$

7 It's better to time the wave over several reflected paths. Multiply the distance by the number of times the wave has travelled this distance.
8 Now find out how, if at all, the wave speed changes when you stretch the spring.

9 Decide how you are going to present your results.
10 Plot a graph to find out if there is a relationship between speed and tension (amount of stretch).

What do you think would happen to the wave if you put the slinky on a carpet? Make a prediction, and if you have time, test your prediction.

Finding what affects the speed of waves in water

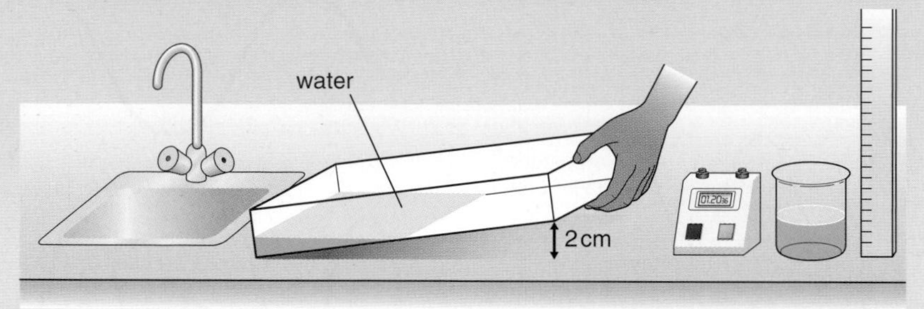

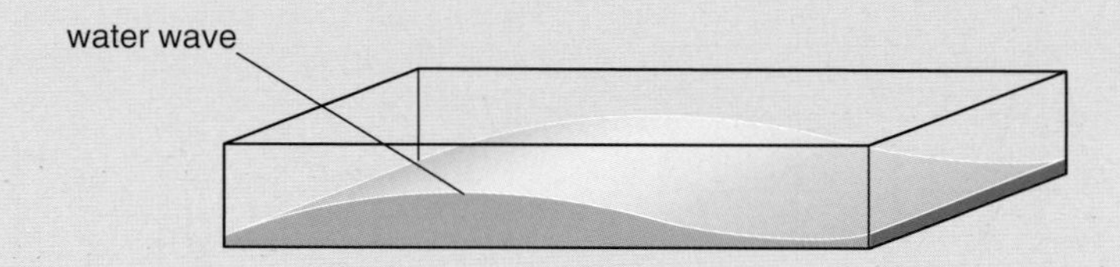

Figure 11.5 Apparatus for measuring waves

1 In your group think about the results of your slinky wave speeds and make a list of things that might affect the speed of waves on water.
2 Decide what apparatus to use.
3 Think about:
   a the sort of container you need to hold the water (the water in your container must not be deeper than 1 cm)
   b how you will make the waves
   c what sort of measurements you will make on the waves
   d how you will make the measurements
   e what apparatus you will need to make the measurements
   f what things you are going to change or keep constant
   g what things you are going to measure as a result of your changes
   h how you are going to record your measurements
   i how you will display and present your findings.

   Write a report on how you carried out the investigation. What were your conclusions?

The wave equation

The speed of a wave and its frequency and wavelength are related by the following formula.

wave speed (m/s) = wavelength (m) × frequency (Hz)

This formula can be rearranged as follows.

$$\text{frequency (Hz)} = \frac{\text{wave speed (m/s)}}{\text{wavelength (m)}}$$

and also:

$$\text{wavelength (m)} = \frac{\text{wave speed (m/s)}}{\text{frequency (Hz)}}$$

Another formula we need is the following.

$$\text{speed (m/s)} = \frac{\text{distance (m)}}{\text{time (s)}}$$

Very large and very small numbers can result in an awkwardly large number of zeros. We use a shorthand way of writing them in which we count the number of zeros. For example, one thousand = 1000 = 1×10^3 and one million = 1 000 000 = 1×10^6.
One thousandth = $\frac{1}{1000}$ = 0.001 = 1×10^{-3} and one millionth = $\frac{1}{1\,000\,000}$ = 0.000 001 = 1×10^{-6}.

Example
A wave has a frequency of 3 Hz and a wavelength of 10 cm. What is its speed?

Formula first:

wave speed = wavelength × frequency

Change the centimetres to metres (10 cm = 0.1 m).

Put the numbers in:

wave speed = 0.1 m × 3 Hz
= 0.3 m/s

Waves:
www.gcse.com

Questions

1 a Draw a diagram of a wave. On it mark:
i the wavelength
ii the amplitude.
b On the same diagram, using a different colour, draw a wave with a higher frequency and smaller amplitude.

2 A wave has a frequency of 50 Hz and a wavelength of 20 cm. Calculate the speed of the wave in metres per second.

3 A sound wave travelling at 330 m/s has a frequency of 220 Hz. What is its wavelength?

4 Copy and complete the table below.

	Speed (m/s)	Frequency (Hz)	Wavelength (m)
Sound waves in air	340	170	
Sound waves in water	1500		3
Sound waves in steel		2000	3

The electromagnetic spectrum

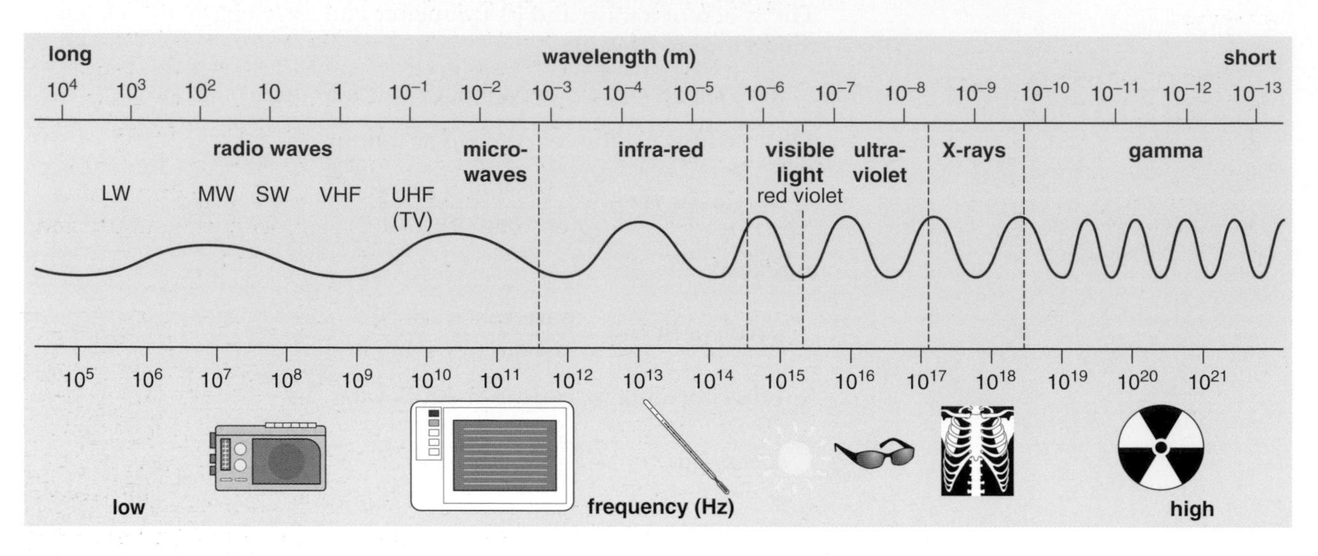

Figure 11.6 The electromagnetic spectrum

A prism will split white light up into a spectrum with the colours of the rainbow. (This is called **dispersion**.) Each colour of light has a different wavelength, and this spectrum of visible colours is part of a much larger spectrum of **electromagnetic waves** with an enormous range of wavelengths.

All electromagnetic waves:

- travel at the same speed, the speed of light: 300 000 000 m/s
- transfer energy from place to place
- raise the temperature of the material that absorbs them
- can be reflected and refracted.

Remember that wave speed, wavelength and frequency are related, but all electromagnetic waves have the same wave speed, so that in Figure 11.6 the waves with the shortest wavelength have the highest frequency.

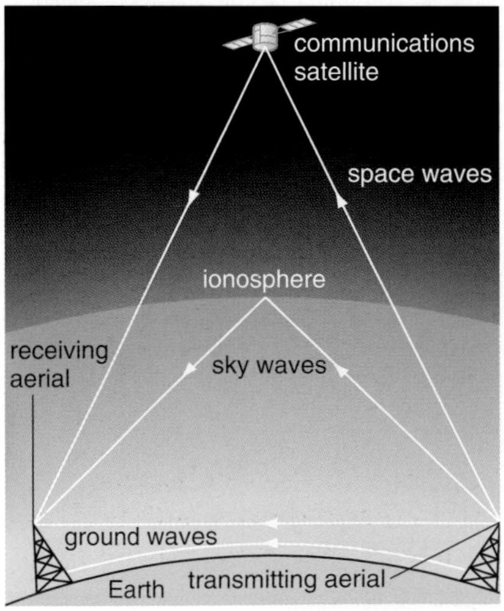

Investigating the waves in the electromagnetic spectrum

Radio and TV waves

Radio waves have the longest wavelengths. Long and medium radio waves are transmitted around the Earth by reflection from the upper atmosphere. These reflected radio waves are called sky waves.

The waves used by FM radio and television are shorter than 1 m. These ground waves cannot be reflected by the atmosphere. To receive a signal there has to be a large number of transmitters all in 'line of sight' of each other and your television aerial. Television signals can also be re-transmitted via satellites to distant parts of the world (there will be more about satellite communication in the long-distance communications section). Radio and TV waves are not dangerous because they carry very small amounts of energy.

Figure 11.7 Radio waves can travel by various routes

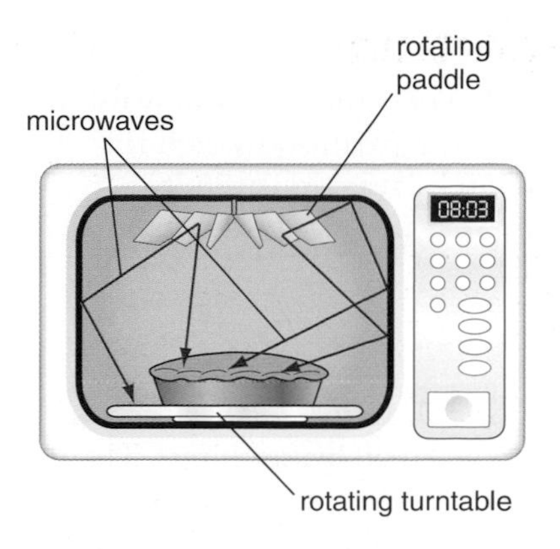

Figure 11.8 A microwave oven

Microwaves

Microwaves are used in microwave ovens (Figure 11.8). In a microwave cooker the waves come in from the top. They are reflected off the metal sides onto the food to be cooked. The glass door has a metal mesh in it. This stops the waves from escaping, which could be harmful. The frequency is chosen so that the microwaves penetrate the food and energy is transferred to the water molecules in it. As a result the food is cooked quickly and evenly from the inside. Normal cooking heats from the outside and it takes some time for the heat to travel to the centre.

Did you know?

The microwave oven was discovered as a result of radar research. In World War II scientists invented microwave transmitters to send out signals (radar) to detect enemy aircraft. After the war ended, radar research continued and an engineer just happened to walk past a radar transmitter when he noticed that the chocolate bar in his pocket started to melt.

Microwaves used in cooking are just short-wavelength (high-frequency) radio waves. So the radio and TV waves used to communicate with satellites are microwaves. Mobile phones use microwaves too, and just like TV transmitters, mobile phone signals need a good line of sight. You'll read more about this in the section on long-distance communication.

Figure 11.9 Infra-red sensory equipment is used to show where heat is lost in buildings and can help to locate people trapped in a smoke-filled house.

Infra-red waves

We know and feel infra-red radiation as heat radiation. Everything that is above absolute zero of temperature (–273 °C) emits infra-red radiation. Infra-red radiation in itself is not dangerous so long as you do not get too much of it. Stand in front of a bonfire too long and the radiated energy will certainly warm you up and could burn you.

Infra-red cameras detect heat (Figure 11.9). They are used by the fire service to find people in smoke-filled buildings and by police helicopters to locate suspects at night. Infra-red cameras show which houses are well insulated and which are not.

Visible light

The Sun is our main source of both light and heat. Its energy keeps us warm and is essential to life. Plants use visible light for photosynthesis to make their own food and oxygen. It is the only part of the spectrum we can detect with our eyes.

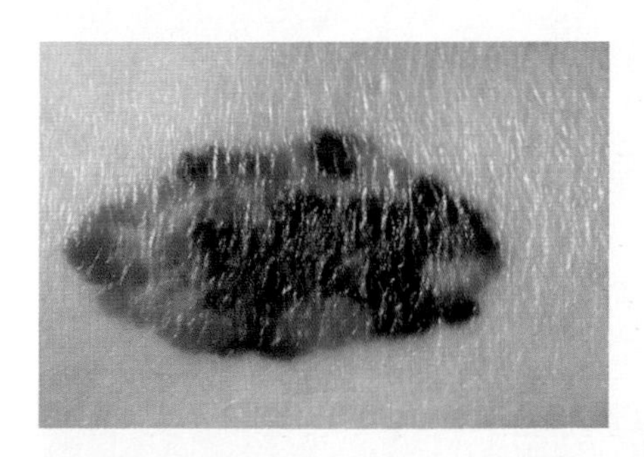

Figure 11.10 Spend too long in the sun and you have a bigger risk of skin cancer.

Ultraviolet radiation

At this end of the electromagnetic spectrum the waves become increasingly dangerous. Their wavelength gets shorter and shorter. As their frequencies increase, so does their energy and danger to life. Ultraviolet radiation damages your skin because the radiation has enough energy to **ionise** atoms in cells. A sun tan shows that your skin has already been damaged. Sometimes ionising radiation can cause cells to mutate. This can lead to cancer (Figure 11.10).

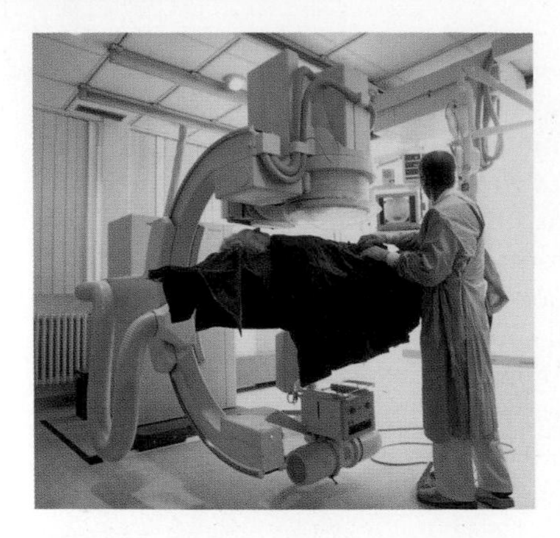

Figure 11.11 An angiography machine uses X-rays to diagnose heart conditions in patients.

NASA's pages on the electromagnetic spectrum: http://imagine.gsfc.nasa.gov/docs/science

Electromagnetic spectrum and the Kuiper Airborne Observatory (go to *Space* and then *The electromagnetic spectrum*): vathena.arc.nasa.gov

X-rays

Exposure to X-rays can cause cancer. However they are widely used in medicine where the benefits of the X-ray greatly outweigh the dangers from it (Figure 11.11). They can be used in carefully controlled conditions to cure cancers. Very powerful X-rays are used to detect flaws and fractures in metals.

Gamma rays

Gamma radiation comes from nuclei of radioactive materials such as uranium. Gamma rays are very dangerous to all living things. They can cause cancers or kill cells directly. Like X-rays they are used to detect flaws in metals. They can also be used to treat cancer, and to sterilise medical instruments (see the section on uses and dangers of radioactivity in Chapter 13).

Questions

5 Make a chart of the main features of the electromagnetic spectrum. List the main sections of the spectrum. For each section give:

a its range of wavelengths

b its uses and dangers.

Activity

1 In your group, each one of you chooses a section of the electromagnetic spectrum.

2 Use books or the internet to find more detail on how we use that part of the spectrum.

3 Share your findings within your group or class.

Reflection and refraction of light

Light can pass through glass. It can also be reflected from the surface and light rays can change their direction (Figure 11.12).

Figure 11.12 Reflection in a shop window

Practical work

Measuring refracted and reflected rays

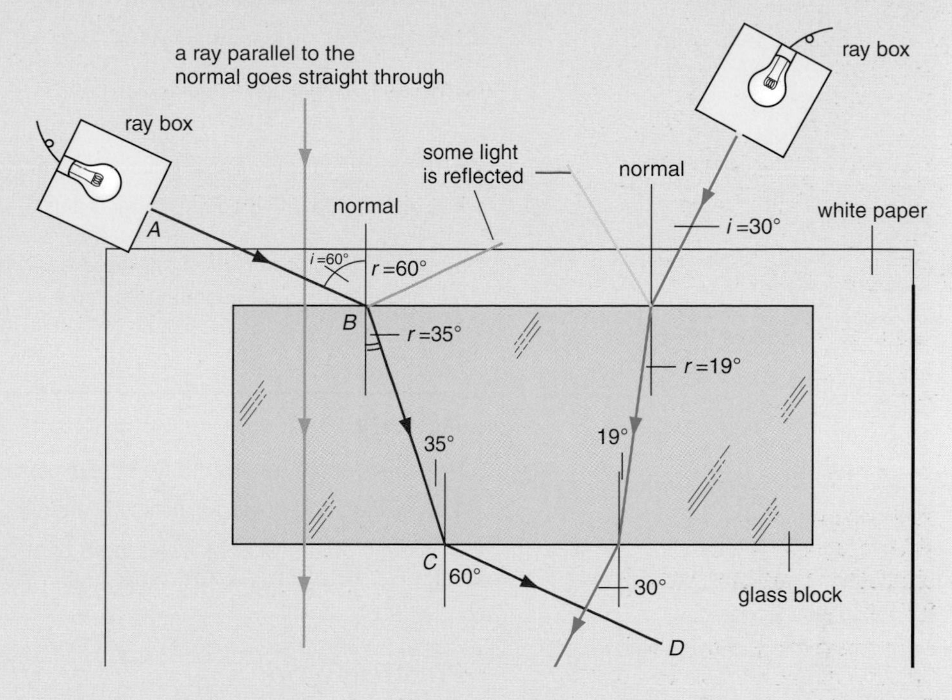

Figure 11.13 Refraction of light rays through a glass block

1. Place a glass block on a sheet of white paper.
2. Draw round it.
3. Use the ray box to shine a ray of light into the glass block.
4. Mark the ray going into the block (incident ray) and the ray coming out of the block (emergent ray).
5. Mark the position of the faint reflected ray.
6. Remove the block and join up the points to mark the path of the original ray.
7. Use a protractor to draw a normal line at the point where the rays enter and leave the glass block.

You can see that:

- There is a faint **reflected ray**.
- The emergent ray is parallel to the incident ray. It has been displaced sideways.
- The **refracted** ray is bent towards the normal as it passes from air to glass.
- The emergent ray (passing from glass to air) is refracted away from the normal.
- The angle the incident ray makes with the normal is called the **angle of incidence**.
- The angle the refracted ray makes with the normal is called the **angle of refraction**.
- The angle the faint reflected ray makes with the normal is called the **angle of reflection**.

8 Now replace the glass block on its mark.
9 Shine the ray at different angles of incidence and watch what happens to the refracted and reflected rays.
10 Shine the ray along the normal line.

- A ray travelling along the normal line is not refracted. It passes straight through.
- The angle of reflection is equal to the angle of incidence.
- Light can be reflected *and* refracted at the surface of the glass.

Summarise how you carried out the experiment and the conclusions you reached. Include your experimental results.

Activity

Use books or the internet to find the explanation of these observations.

1 A swimming pool looks shallower than it is.
2 A pencil in a beaker of water looks bent.

Practical work

Measuring total internal reflection

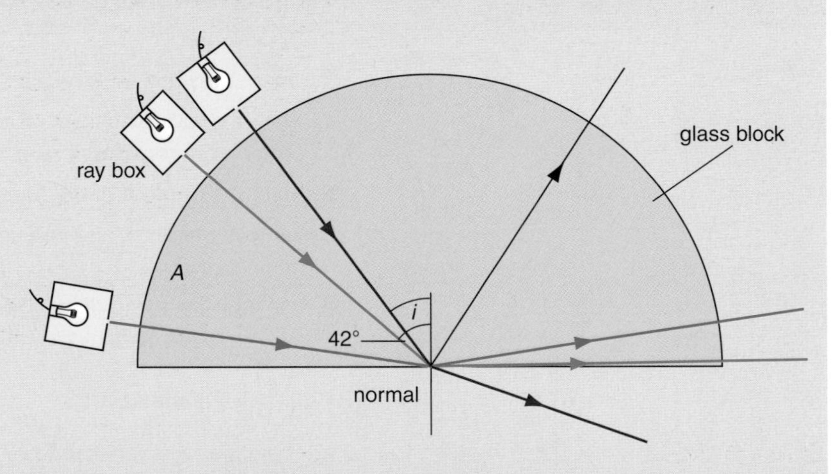

Figure 11.14 Refraction and reflection in a glass block

1 Place a semicircular glass block on a piece of white paper and draw round it.
2 Draw a normal at the midpoint of the straight side.
3 Shine a ray of light at a small angle to the normal that you have drawn.
4 Mark and measure the emergent ray.

You can see the following.

- The incident ray is always at right angles to the glass block and so is not refracted.
- The emergent ray is refracted away from the normal.
- There is a very faint reflected ray back into the glass block (shown in black).

5 Slowly increase the angle of incidence.

You will see the following.

- The angle of refraction gets bigger and the emergent ray gets closer to the straight side of the block.
- At a certain angle of incidence, the ray comes out along the edge of the block (blue).

6 When this happens, mark the position of the incident ray.
7 Increase the angle of incidence.

- The ray is always *reflected* back into the block for all angles greater than the one you have marked (green).
- This is **total internal reflection** because *all* (100%) of the light is reflected back into the glass.
- The angle you have marked is called the **critical angle**.
- The angle of incidence is equal to the angle of reflection.

The following are some important points to remember about total internal reflection.

- It occurs only when a ray of light travels from an optically dense medium (glass or water) to a less dense medium such as air.
- The angle of incidence must be greater than the critical angle.
- The critical angle for glass is 42°.
- The critical angle for water is 49°.

Summarise how you carried out the experiment and the conclusions you reached. Include your experimental results.

Question

6 Explain these terms:
  a critical angle
  b total internal reflection.

Activity

Use books or the internet to find out how total internal reflection is used in:

1 periscopes
2 car reflectors
3 binoculars.

In each case draw diagrams to explain how they work.

Optical fibres

Practical work

Examining an optical fibre

1 Examine an optical fibre as shown.

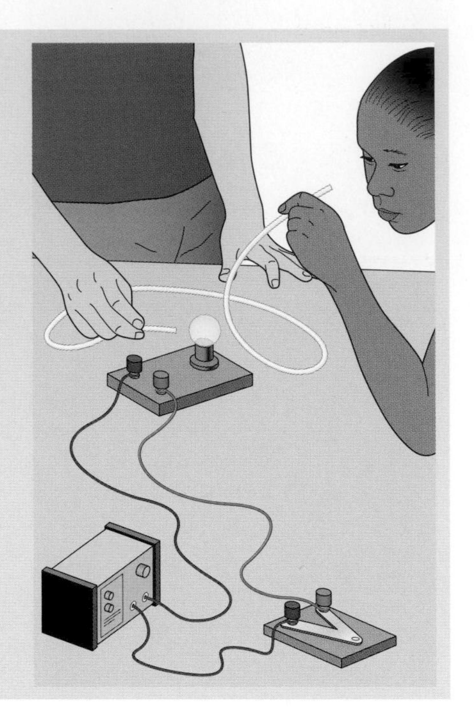

Figure 11.15 Light travelling through an optical fibre

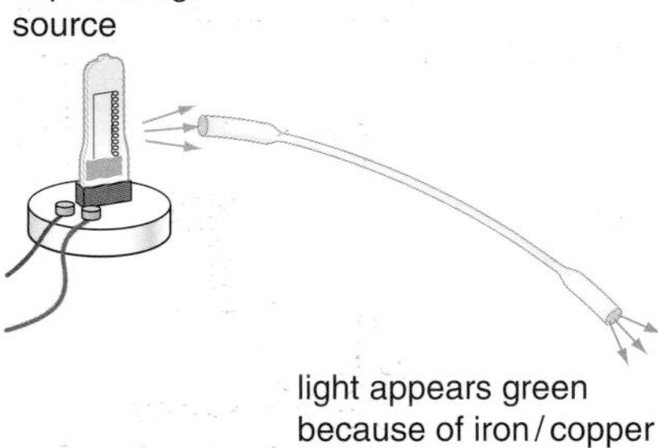

Figure 11.16 This optical fibre can be made from a glass rod.

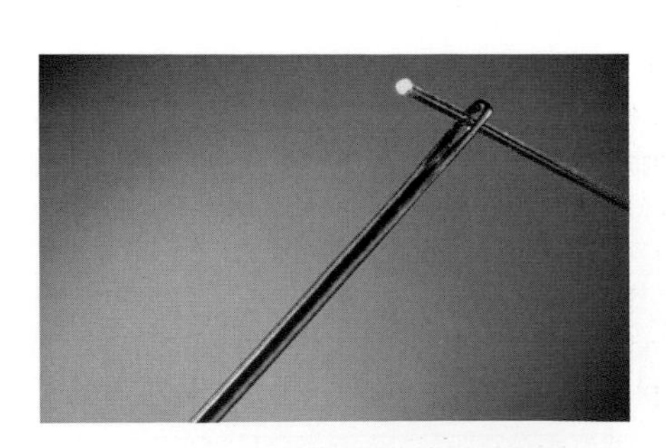

Figure 11.17 Optical fibres in the eye of a needle

We all know that light travels in a straight line, so how is it that the light comes out of the other end even when the fibre is bent? We need to find out how optical fibres are made.

A homemade optical fibre

Your teacher may make an optical fibre. Your teacher will:

- heat a glass rod in a Bunsen flame until it gets really soft
- quickly stretch the rod to pull it into a very thin fibre without breaking it
- when cool, shine a very strong light along the curved fibre.

If the fibre is touched with greasy fingers, you will see light escape at the greasy points.

Commercial optical fibres

The optical fibre is made from very pure glass. It starts off as a rod about 15 mm in diameter. It is heated until it melts then it is pulled out to make a very thin fibre about as thick as a human hair, and 1 km of this pure glass absorbs as much light as a single sheet of window glass.

The thin fibre is coated with another layer of very pure glass, but with a slightly lower refractive index (it is optically less dense). The whole fibre is then covered with a layer to protect it from bumps and scratches.

H

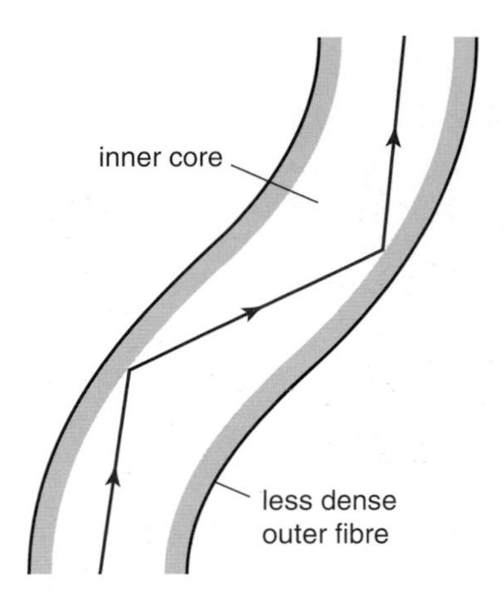

Figure 11.18 Total internal reflection allows light to pass along optical fibres.

How do optical fibres work?

Light enters one end of the optical fibre. It travels along it, but soon it strikes the outer layer of less dense glass. It strikes it at an angle greater than the critical angle. It is totally internally reflected. It travels towards the other side where it is totally reflected again. It continues along the whole length of fibre in this way until it reaches the other end (Figure 11.18).

Long-distance communications

Optical fibres

Optical fibres are far better at transferring information than copper wires. A single fibre can carry over 1.5 million telephone conversations, compared with 1000 conversations for copper wires. Most telephone calls, faxes, internet calls, etc. pass along optical-fibre lines. The fibres can carry ten television channels. An intercontinental optical cable carries many fibres so an enormous amount of information can be transferred. It is very cost-effective.

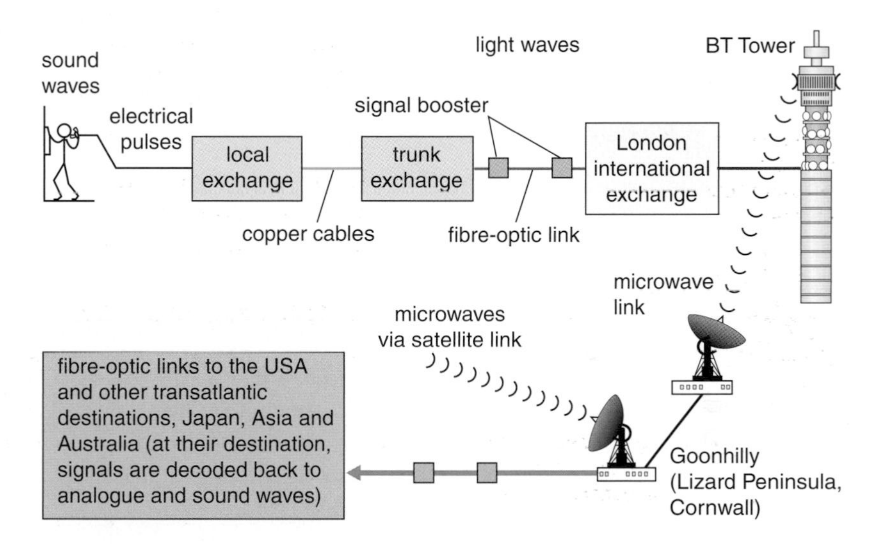

Figure 11.19 The path of an international telephone call as it leaves the country

For long-distance telephone calls, the electrical signals are converted to digital (on/off) pulses at the exchange. The digital signal is converted to light pulses by a **laser**. The infra-red laser flashes at high speed. Infra-red light is used because it passes through the glass better than visible light. Repeaters boost the signal at 30 km intervals along the fibre. At the far end, another decoder converts the digital signal from the laser into a changing voltage that is then converted into sound at the telephone earpiece.

Other advantages over copper wire are the following.

- Fibre-optic lines use less energy.
- They need fewer boosters.
- There is no crosstalk (interference) with adjoining cables.
- They are difficult to bug.
- Their weight is lower and so they are easier to install.

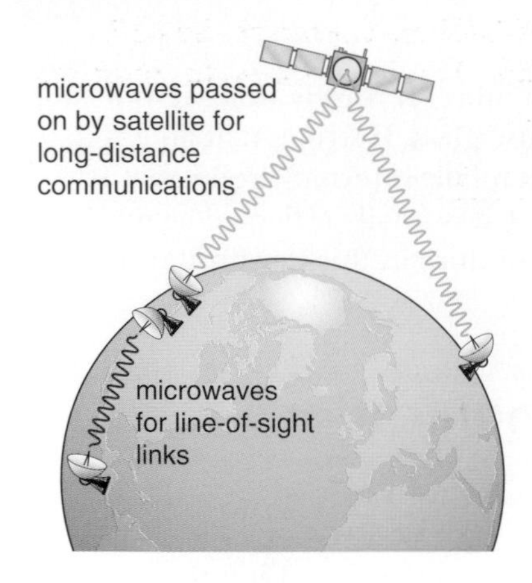

Figure 11.20 Microwave transmitters and receivers must be in a direct line with each other.

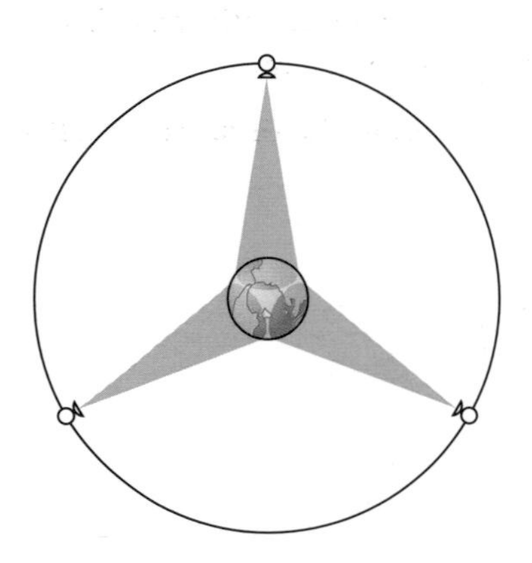

Figure 11.21 The Earth from above the North Pole; three geostationary satellites can send signals to most of the Earth.

Figure 11.22 Satellite dish at Goonhilly Down, Cornwall

Microwaves

Microwaves carry the signals for mobile phones, television and FM radio. Microwaves are wireless signals – you don't need a copper cable or optical fibre. One disadvantage of using microwaves is that there must be a clear path between the transmitter and your television aerial. To cover the largest area, television transmitters are tall and sited on hills.

The curvature of the Earth means that repeater stations have to relay the microwave signal to distant transmitters. Satellites must be used for long-distance communications around the world.

Theoretically, only three satellites are needed to transmit signals around the world. In practice, more are used. The satellites are placed in orbit at a height of 36 000 km. They orbit the Earth exactly in time with the Earth's rotation. This is called a **geosynchronous orbit.**

Here in the UK, TV, phone, fax and data signals are sent to satellites from one of three BT stations. The Goonhilly Down earth station is in Cornwall. It has over 60 satellite dishes. They vary in size from tiny 'receive only' antennas to satellite dishes over 30 m in diameter. The signal coming in is very much weaker that the one sent out.

Did you know?

The strength of the signal coming in from the earliest satellites was very weak. The energy was no more than the heat you would feel on Earth from a single-bar electric fire placed on the Moon.

Optical fibres or microwaves?

Both optical fibres (infra-red) and satellites (microwave) are used for international phone calls and TV broadcasts. It takes time for the signals to travel from an earth station up to one of the satellites and back again (Figure 11.23). Let's compare the time delay in sending a signal from A to B.

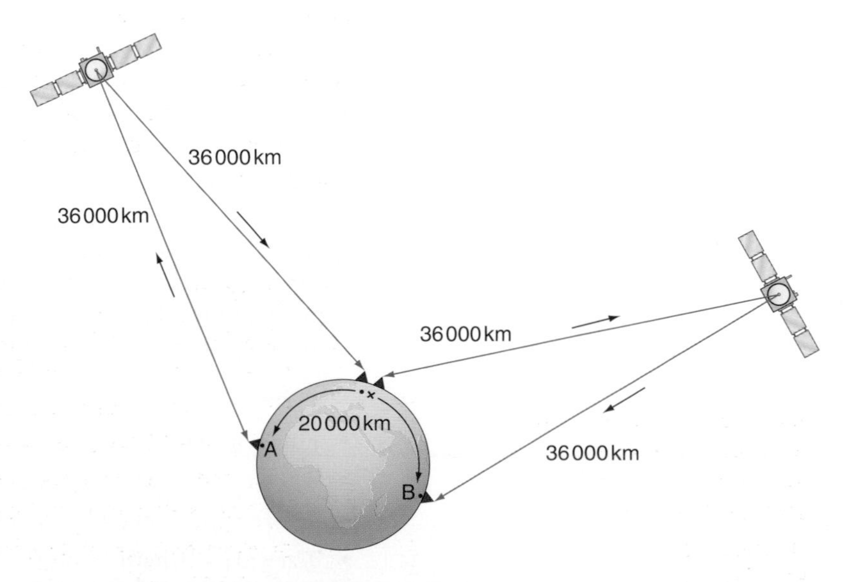

Figure 11.23 The satellite signal has much further to travel.

The satellites orbit at a height of 36 000 km. Therefore the path length is 4 × 36 000 km, or 144 000 km. This is for studio to studio via satellite. Use the following formula.

$$\text{speed (km/s)} = \frac{\text{distance travelled (km)}}{\text{time taken (s)}}$$

Formula first:

$$\text{time for travel} = \frac{\text{distance}}{\text{speed of light}}$$

Put the numbers in:

$$\text{time for travel} = \frac{144\,000\ \text{km}}{300\,000\ \text{km/s}}$$

$$= 0.5\ \text{s approximately}$$

An outside broadcast might increase the journey to 200 000 km, making the time for travel about 0.7 s. This time delay on news broadcast or telephone conversations will be quite noticeable. You may well have observed this effect on your television.

With optical fibres connecting the two studios, the distance travelled may be only 20 000 km.

Formula first:

$$\text{time delay} = \frac{\text{distance travelled}}{\text{speed of signal in glass}}$$

Put the numbers in:

$$\text{time delay} = \frac{20\,000\ \text{km}}{200\,000\ \text{km/s}}$$

$$= 0.1\ \text{s}$$

The time delay with optical fibres is only 0.1 s, and this is much less noticeable.

Will optical fibres take over?

Optical fibres can handle a tremendous number of voice and data calls. Because of their greater information capacity, no noticeable time delay and no need for repeater stations, there is a worldwide shift towards using optical fibres for long-distance applications. However microwaves and satellites will never be replaced by them. Microwave links often take over optical fibre traffic when the cable is being repaired.

History of telecommunications (go to *Galleries* and then *From buttons to bytes*): www.connected-earth.com

Optical fibres: en.wikipedia.org

Questions

7 Make a list of the advantages of using optical fibres rather than copper wires for communication.

8 Draw a diagram of an optical fibre. On your diagram draw the path of a single incident ray as it passes through the fibre.

9 Explain why repeater stations are needed for long-distance communication by microwaves.

10 Satellites must be used for long-distance microwave communication around the world. Draw a simple diagram to show how this is possible.

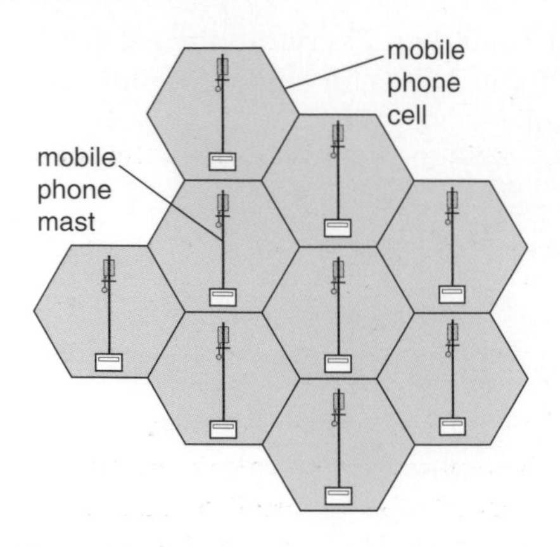

Figure 11.24 Each area is divided up into hexagonal cells with a base station. All base stations are connected to a control centre (MTSO).

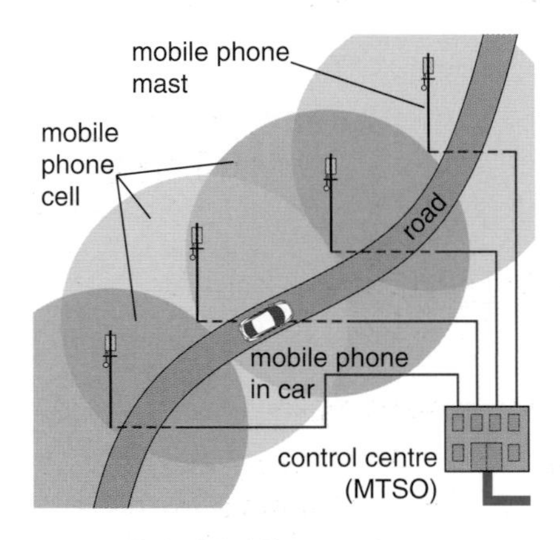

Figure 11.25 As you travel, the mobile phone signal is passed from cell to cell.

Mobile phones

There are millions of mobile phones (cell phones) in use all over the world. They are radios that can transmit and receive at the same time.

Mobile phone cells companies (carriers) divide your town or city into a number of cells (Figure 11.24). Each one covers about 25 km^2.

Each cell has a base station. It consists of a tower and a box or small building to hold the electronic equipment. The large number of cells means that mobiles can transmit at a much lower power, typically 0.6–1.0 W. Each carrier has a central office in each city called a mobile telephone switching office (MTSO).

This is what happens when you switch on and someone is trying to call you (Figure 11.25).

- Your phone listens for the base station code. If it cannot find one, you get a 'no service' message.
- Your phone and the base station code communicate and compare codes.
- Your phone sends a registration request. A sort of 'I am here if anyone calls' signal.
- The MTSO keeps track of your location on its database.
- When your friend calls, the MTSO looks in its database to find which cell you are in.
- The MTSO picks a frequency pair that your phone will use to take the call. Your phone and the tower switch to these frequencies.
- You are connected. You are talking by two-way radio.
- As you move towards the edge of your cell, the base station notices that your signal strength is getting weaker. The next base station notices your signal strength increasing. At some point your phone gets a signal from the MTSO to change frequencies. Your phone switches to the new cell.

Mobile phones and health

Some people are worried that mobile phones might affect their health. They think that because the transmitting aerial has to be held close to the head, it will increase the risk of brain cancer.

So far, all the research has shown that radio waves and microwaves cannot cause cancer. Radio waves are at the low-energy end of the spectrum. They are not ionising radiations like ultraviolet and X-rays, which can cause cancer. Microwaves can cause a heating effect in human tissue (just like in a microwave oven). There may be a long-term effect. This is one of a number of areas where science cannot always give a definite answer one way or the other. A long-term international study will assess the health of around 250 000 mobile phone users, but the results are not due until 2020.

Years ago, when people first started smoking they did not suffer any immediate effects. Now we know that smoking kills. Using mobile phones may affect us in the long term. The trouble is, if they are found to be dangerous, billions of us will be affected.

Children may be at risk because the soft bones in their skull may let more radio waves into their brains. The UK National Radiological Protection Board (NRPB) has recommended that mobile phones should not be given to children.

Always try to:

- keep your call as short as possible
- use an earphone/microphone cable to keep the transmitter away from your head
- limit calls inside buildings (where your phone needs to transmit at higher power), and use open spaces as much as possible.

features	model A	model B
Radio	no	no
Built-in speaker	yes	no
GPRS	yes	no
Sync ML*	yes	no
SAR Level (w/kg)	0.8	0.67
Lifeline pay monthly	2	1

Figure 11.26 SAR ratings for mobile phones

Safety tests

Mobile phones have to be tested for radiation. The 'specific absorption rate' is a measure of the amount of energy absorbed when the microwaves enter your head. To be licensed for sale, the phone must have a SAR value of less than 2 W/kg in the UK and 1.6 W/kg in the USA.

Sales brochures usually tell you all about the exciting features on their models. You should ask to see their SAR rating. They are often to be found in the small print at the back of the catalogue.

Satellites (go to *Electronics, Telecommunications* and then *How satellites* work): www.howstuffworks.com

Cell engineering (go to *Electronics* and then *Cell phone*): www.howstuffworks.com

Cell phone radiation (go to *Radiation*, *Radio wave surveys* and then *Mobile telephony and health*): www.hpa.org.uk

Terrestrial trunked radio (Tetra)

Tetra is the modern digital private mobile radio system for police, ambulance and fire services. Like mobile phones it needs a series of transmitter masts (Figure 11.27). Each handset has its own transmitter and receiving aerial. As with mobile phones, there are health concerns.

Planning requirements for communications masts

Planning requirements can differ in the various parts of the UK. All new ground-based masts will come under full planning control. Technical restrictions apply to the size, height and number of masts on a building. There are more stringent requirements in conservation and particularly scenic areas.

Public consultation requirements will be increased, especially for masts under 15 m high, which originally did not require permission. School governors must be consulted on proposals for masts near schools. The planning authority can also reject applications on amenity grounds.

There is a lot of concern about mobile phone and Tetra masts.

Figure 11.27 A terrestrial trunked radio (Tetra) phone transmitter mast

Activity

1. Research the arguments for and against the claims about the health risks of:
   - **a** mobile phone handsets
   - **b** mobile phone masts
   - **c** Tetra masts
   - **d** planning laws or lack of them concerning the position and building of the masts.
2. Nominate a spokesperson for each side.
3. Use a PowerPoint presentation to illustrate your case.
4. Select an impartial chairperson to ensure fair play.
5. Make your presentations to the class.

Questions

11 The electromagnetic spectrum is made up of seven regions of waves.

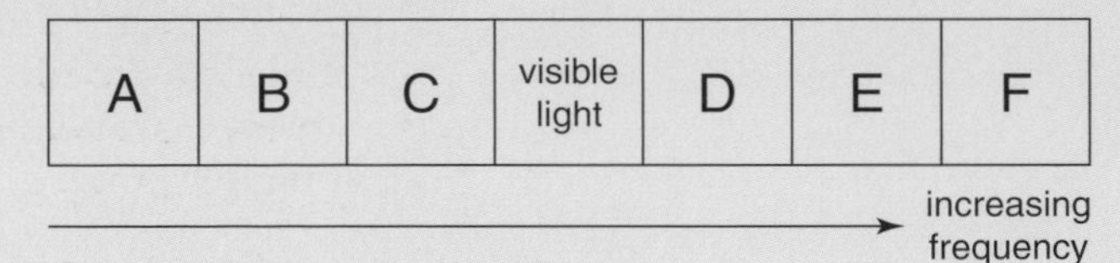

Region **A** contains waves of the **lowest** frequency and **F** contains waves of the **highest** frequency.

a Name region **E**.

b Name region **C**.

c Name the region most dangerous to humans.

d State **one** property that is the same for all electromagnetic waves.

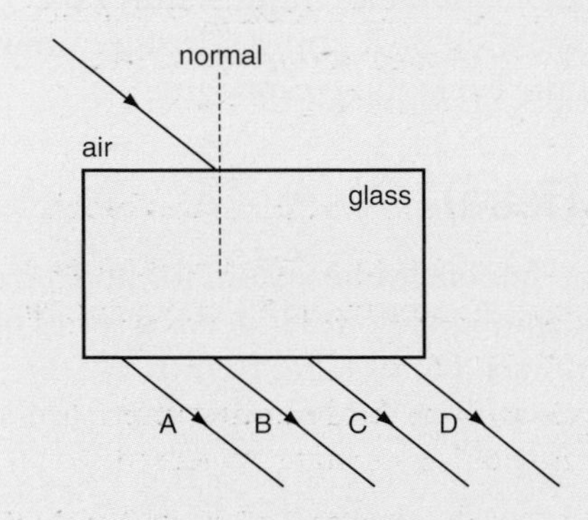

12 The diagram (left) shows a ray of light hitting the top surface of a glass block.

a Copy and complete the diagram showing **only** the correct ray both inside and outside the block.

b On your design mark with a cross (X) the angle of incidence at the top surface of the block.

c Give a reason why the ray of light is **not** totally reflected inside the block.

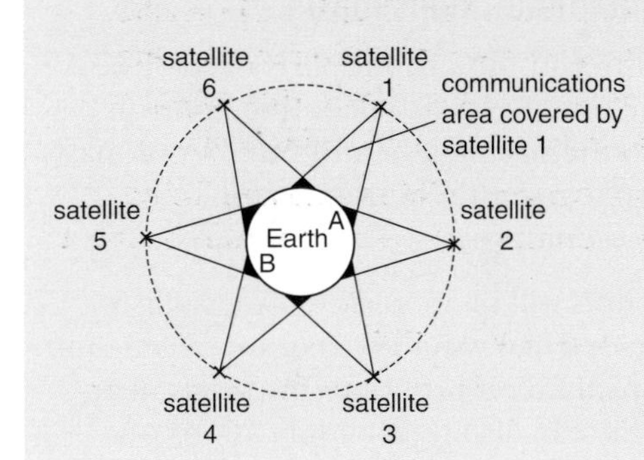

13 The diagram shows a ring of satellites around the Earth.

The satellites are placed in **geostationary orbit** 36 000 km above the equator. A signal of frequency 4.0×10^9 Hz carries telephone conversations, at a speed of 3.0×10^8 m/s, between Earth and the satellites.

a What is meant by a geostationary orbit?

b State **two** advantages of placing communication satellites in geostationary orbit.

c Explain the meaning of the phrase 'a signal of frequency 4.0×10^9 Hz'.

d Groundstation **A** cannot communicate directly with groundstation **B**. Explain how satellites **2**, **3** and **4** are involved in the communication between **A** and **B**.

Summary

1. Waves move energy from one place to another. When we make a wave in a slinky, each part of the slinky vibrates, moving to and fro as the wave passes through it.
2. All waves have a wavelength, frequency and amplitude.
3. Wave speed (m/s) = wavelength (m) × frequency (Hz).
4. Speed (m/s) = distance (m)/time (s).
5. All waves in the electromagnetic spectrum travel at the same speed, that is, the speed of light. The parts of the electromagnetic spectrum with the highest energy (shortest wavelength) are the most dangerous because they are ionising.
6. In order of decreasing wavelength and increasing frequency, the electromagnetic spectrum includes: radio, television, microwave, infra-red, visible light, ultraviolet, X-rays, gamma rays.
7. Rays of light can be reflected and refracted by a glass block.
8. A ray of light travelling from an optically dense medium to a less dense medium at an angle greater than the critical angle will undergo total internal reflection.
9. Laser light is sent along optical fibres for long-distance communication. The light is totally internally reflected.
10. Microwaves are also used in mobile phones and long-distance communications. The curvature of the Earth makes land-based repeater stations a necessity. Geostationary satellites give global communications coverage. To do this large, expensive transmitters and receivers have to be built.
11. There may be health risks from mobile phone and Tetra communications. Children and young people would be most at risk.
12. Phone masts are subject to planning laws.

Figure 12.3 The planets in the Solar System, and the Moon

What is it like on the other planets?

a **Mercury**: no atmosphere closest to Sun; no moons, no rings; rocky; surface temperature 350 °C day, –180 °C night

b **Venus**: thick dense atmosphere of sulphuric acid and carbon dioxide; greenhouse effect keeps heat in; temperature 450 °C; no moons; no life

c **Earth**: blue planet with swirls of cloud; supports life, only planet with water and oxygen; temperature 15 °C; one moon

d **Mars**: rocky; red dusty surface; planet may have had water in the past, but opinions differ; temperature –30 °C; no life found so far; two moons

e **Jupiter**: largest planet, all the other planets could fit inside it; surface all gas; atmosphere liquid hydrogen and helium; red spot is a giant storm three times bigger than Earth; temperature –150 °C; 28 moons discovered so far

f **Saturn**: gas giant like Jupiter; has rings, not solid but made of billions of tiny rocky particles frozen in ice; temperature −180 °C; has at least 30 moons

g **Uranus**: gas giant; pale green colour; rings, and 21 moons discovered so far; temperature −210 °C; its axis is tilted right over as if lying on its side as it orbits the Sun

h **Neptune**: very similar to Uranus; blue colour; at least eight moons; temperature −220 °C

i **Pluto**: dwarf planet; only 0.2 × size of Earth; rocky; most of the time its orbit is further out than Neptune, but from 1979 until 1999 its orbit was inside Neptune's; one moon; temperature −230 °C

Another planet?

It could be another planet, but it might be a rock or cigar-shaped piece of ice. Called 2003 EL61 and further out than Pluto, astronomers recently discovered an object about 1–1.5 times as big as Pluto. Estimates of the object's size are made by measuring its brightness. Astronomers are puzzled, because its brightness changes every 3 to 4 hours. A cigar-shaped object would reflect more sunlight when the longer side faces the Sun. If this object is long and thin, this opens the debate over how to define a planet.

Solar System:
http://pds.jpl.nasa.gov/planets/

Questions

1 Name the four inner planets.

2 Name the outer planets.

3 Which planets have rings and what are the rings made of?

4 Mercury is closer to the Sun than Venus. Why is Venus hotter than Mercury?

5 Describe the pattern of where the rocky planets are found and where the planets made of gas and liquid are found.

Asteroids

The **asteroid belt** is a collection of pieces of rock of various sizes that orbit the Sun between Mars and Jupiter. Some asteroids follow orbits that bring them close to Earth (Figure 12.4). Sometimes they bump into each other and could be pulled even closer by the Earth's gravity, and occasionally an asteroid could collide with the Earth. Some very large asteroids have created large craters on Earth. As more powerful telescopes are built, more are being discovered. Astronomers are looking out for those that might be on a collision course with Earth.

Figure 12.4 Ida, an asteroid about 56 km across. It has a tiny moon or natural satellite about 1.5 km long.

Did you know?

The recent discovery of an ancient impact crater off the north-west coast of Australia may explain the largest extinction of species the Earth has known. Around 250 million years ago 90% of marine life was wiped out as well as 70% of land species. Another asteroid impact in the Gulf of Mexico may have wiped out the dinosaurs about 65 million years ago.

Figure 12.5 The photographs, taken 6 days apart, that were used to discover Pluto. The arrows point to the position of Pluto on each of the two days. Pluto has moved against the background of the stars.

How are dim and distant asteroids and planets discovered?

Planets and asteroids are moving round the Sun. The name 'planet' originally meant 'wanderer', and planets were called this because of their movement relative to the star background (Figure 12.5).

The more distant the planet, the more slowly it moves against the background of stars. Taking photographs at intervals can detect movement, but it is a very slow process. Using computers to process the data has led to the detection of many more objects in our Solar System, and also of new planets orbiting other stars. Astronomers can also use computers to calculate the effect of the Earth's gravity on the trajectories (paths) of asteroids that could collide with the Earth.

Comets

Comets are made of a frozen mixture of ice and dust. They come from two places: the Oort Cloud, which is a spherical halo of comets surrounding the Solar System, and the Kuiper Belt, which is a ring of icy objects beyond the orbit of Neptune.

Comets have very elliptical orbits, which makes them travel far out into the Solar System and then swing back very close to the Sun. The solar radiation heats the ice and vaporises it. The dust particles are pushed away from the comet by **solar radiation pressure**. This means the comet's dust tail always points away from the Sun. When the comet is leaving the Solar System, its tail is in front of it. The tail can be many millions of kilometres long. It looks bright as it reflects light from the Sun.

H

Figure 12.6 The Hale–Bopp comet collided with Jupiter in 1997.

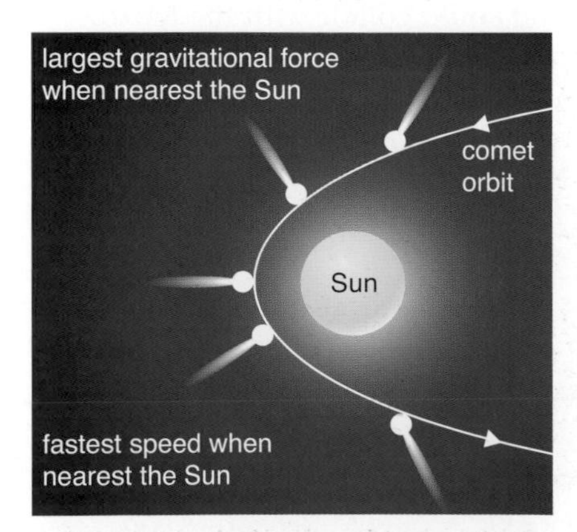

Figure 12.7 The path of a comet round the Sun is highly elliptical (oval-shaped).

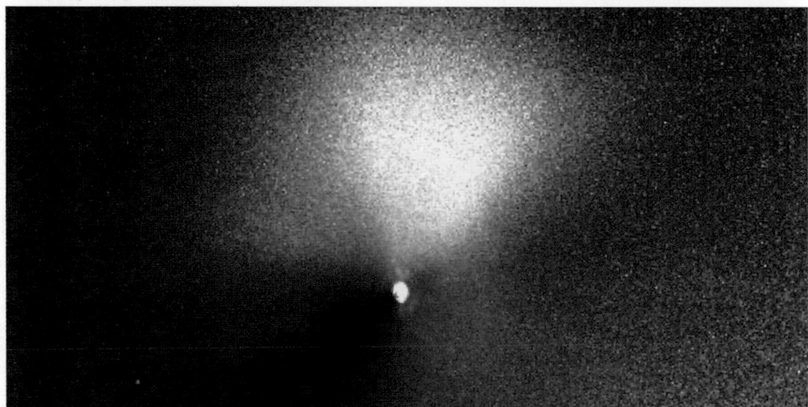

Figure 12.8 The US satellite Deep Impact, designed to find out what comets are made of, collided with the Tempel 1 comet. The first image was taken just before impact, and the second showed what happened afterwards.

Figure 12.9 The surface of Titan, moon of Saturn. Chunks of water ice, looking like rocks, are scattered over an orange surface. This surface might be a crust of mud or clay. Other parts of Titan's surface are probably frozen methane liquid.

Figure 12.10 The International Space Station is an artificial satellite in orbit around the Earth.

Figure 12.11 In theory, a shell fired at a high enough speed (C) would not fall to the ground but would orbit the Earth.

Moons

The Moon is the Earth's **natural satellite**. It orbits the Earth every 28 days. Samples of rock brought back by the American missions to the Moon show that the Moon was formed at about the same time as the Earth. Many of the other planets in the Solar System have moons.

Titan is Saturn's largest moon (Figure 12.9). The Cassini–Huygens spacecraft used the gravitational pull of the planets to accelerate it towards Titan. After a 7-year journey, it sent the Huygens probe to the surface of Titan. The probe carried cameras and scientific instruments.

Artificial satellites

Artificial satellites are launched by powerful rockets; they circle the Earth and are used for relaying television signals and telephone calls, and to make measurements of the Earth from their orbit (Figure 12.10). Global positioning satellites (GPS) help ships, planes and even walkers locate their position. Soon they will be used to charge motorists for using our roads.

Gravity

Gravity is the force of attraction between all pieces of matter. The force of gravity between everyday objects is very small. We notice it only when one of the objects has a very large mass. We feel the pull of the Earth because it has a large mass. The force of the Earth's gravity pulling on you is your weight.

Imagine that you have a large gun on top of a very high mountain. The faster you fire the shell, the further it travels before it falls back to Earth. Fire it even faster, and it would still fall towards the Earth but because the Earth is curved it would never actually hit the ground. It would stay at the same height above the ground and circle the Earth (see path C in Figure 12.11). The shell is now in **orbit**.

The Sun has a very large mass. It accounts for 98% of the mass of the entire Solar System. It is the force of attraction between the Sun and all the planets in the Solar System that keeps them in orbit round it. Artificial satellites are held in orbit by the Earth's gravity.

The gravitational force decreases with distance (Figure 12.12). This means the strength of the Sun's gravity is less for the planets further out. Comets move in elliptical orbits so sometimes they are very far from the Sun and sometimes very close. When they get closer to the Sun, the force of gravity increases and they speed up.

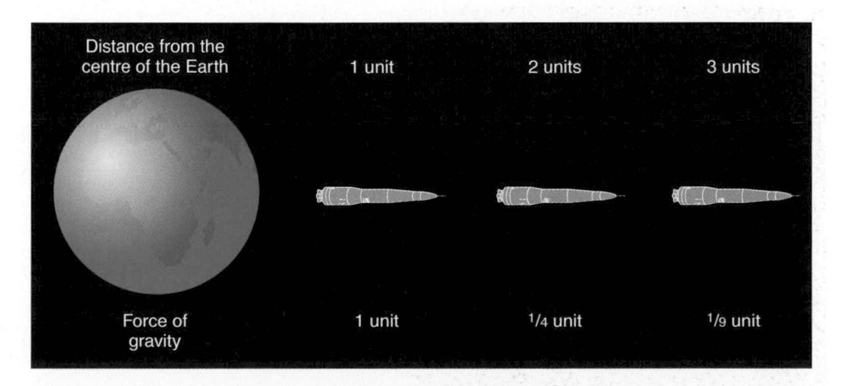

Figure 12.12 If the distance doubles, the force of gravity is reduced by a factor of four; if the distance trebles, the force is one-ninth of what it was before.

Questions

6 What keeps the planets in orbit round the Sun?

7 To keep in orbit at a height of 800 km above the surface of the Earth, a satellite's orbit speed must be 27 600 km/h.

a If the radius of the Earth is 6400 km, how far will the satellite travel when it makes one complete orbit? (The circumference of a circle is $2\pi r$, if the radius of the circle is r.)

b Calculate the time for one orbit at a height of 800 km above the surface.

8 The graph below shows how the orbital period of satellites depends on their orbital height above the ground.

a What is the orbital period of satellites that are at the heights:

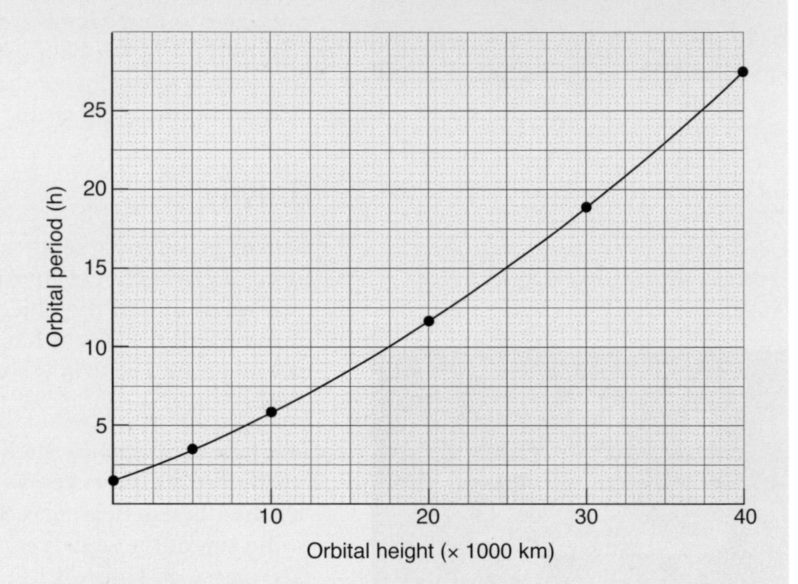

i 1000 km
ii 5000 km
iii 30 000 km?

b How does the orbital period depend on height?

c Is the orbital period proportional to the height above the ground? Explain your answer.

d What is the height of a satellite that orbits the Earth in a time of 24 hours?

Activity

Use books or the internet to find as many uses of satellites as you can.

How did the Solar System form?

The planets, moons and asteroids formed at about the same time as the Sun, about 4500 million years ago. There are a few theories on how the Solar System was formed. These theories are not based on experiment, because other planetary systems are so few and far away that we cannot see planets at different stages of formation. The formation of the Solar System is linked to the formation of the Sun. This is what astronomers think happened (Figure 12.13).

1 4500 million years ago a shock wave, in a spiral arm of our galaxy, triggered the collapse of a gas cloud. This developed into a doughnut shape, which flattened out

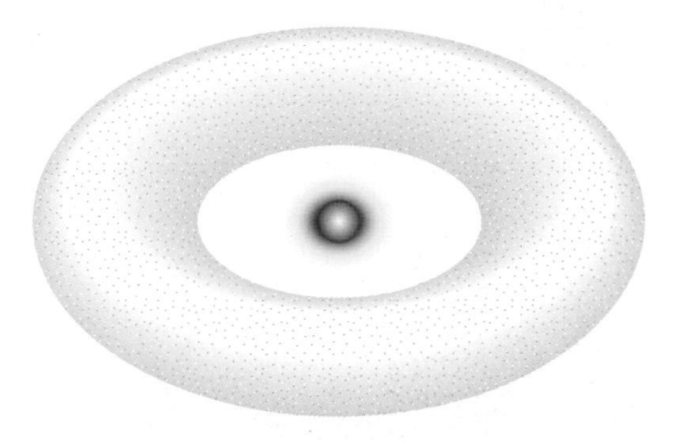

2 Enough hydrogen gathered in the centre for fusion to start in the Sun. Solid particles began to strike each other and stick together. This debris eventually formed planets and their moons.

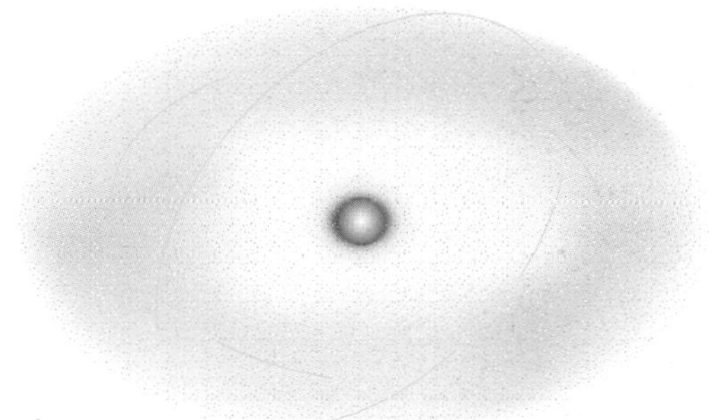

3 Eventually most of the gas and dust in the Solar System became attached to a planet. Left over dust and frozen gases formed the comets.

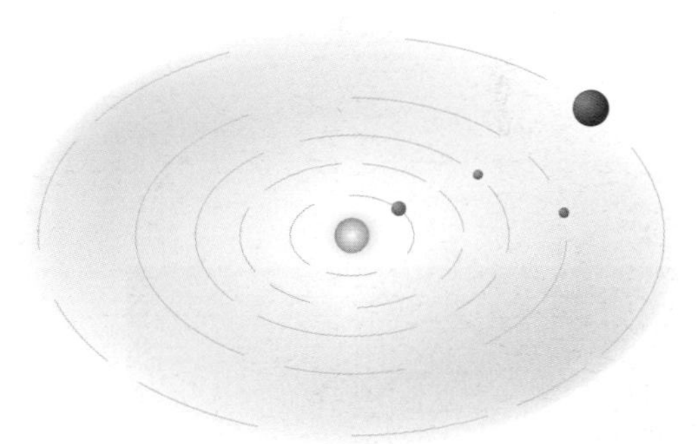

Figure 12.13 One theory on the formation of the Solar System

A star – the Sun – is born

Out in space there are vast clouds of gas called **nebulae**. The gas, which is mainly hydrogen and helium, may stay like this for millions of years. About 4500 million years ago, the gas and dust that was to become our Sun started to form. Something, maybe a shock wave from a supernova, caused the nebula to rotate and collapse in on itself.

The particles of gas and dust all had mass, and so were attracted to each other by gravity. This is called **gravitational collapse**. As the cloud got smaller, it got hotter and eventually the cloud became a star.

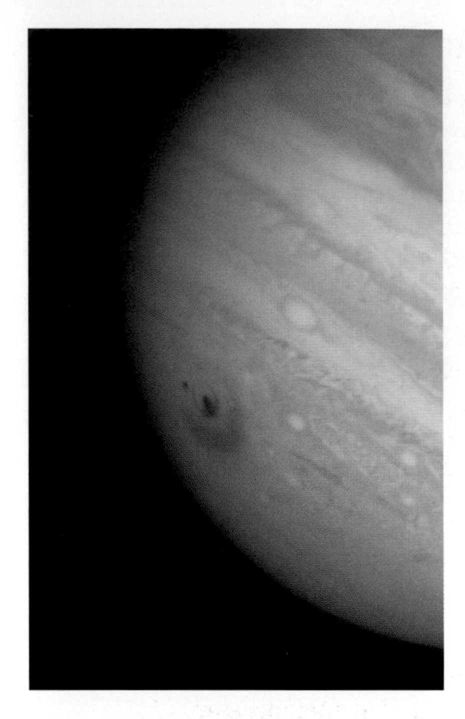

Figure 12.14 Jupiter, after being struck by the comet Shoemaker–Levy; the 'black eye' is 2500 km in diameter, and the outer ring is wider than the Earth.

Planets form

After the Sun formed there was still some remaining dust and gas in orbit in the shape of a disc. This material began to stick together to form a ring of debris around the Sun, and eventually planets.

Radiation from the Sun evaporated some gas material from the planets closest to the Sun. The inner rocky planets such as Earth were formed. In the outer Solar System the particles collided and stuck together as before. Being further away from the Sun and at a very much lower temperature, they kept their hydrogen and helium. These became the bigger gas giants, such as Jupiter and Saturn

Later, there was a collision when the proto-earth was hit by an object the size of Mars. It knocked out a big lump which became the Moon. The asteroids were probably formed when a rocky planet between Mars and Jupiter failed to form. Comets may be leftovers from the formation of the Solar System – the ice and gases that were too far from the Sun to be pulled by gravity into planets.

This theory explains the difference between the inner and outer planets, and why the Earth and Moon are similar and formed at the same time. But how did the Earth get its water? Well it could have come from icy comets. In 1994 the Shoemaker–Levy comet hit Jupiter (Figure 12.14). If a comet hit Earth, the ice in the comet would melt. Over time millions of comets could have hit the Earth.

Deep Impact and comet:
deepimpact.jpl.nasa.gov

Questions

9 How might the Earth have got its water?

10 What evidence is there to link the formation of the Solar System with that of the Sun?

11 Explain with the aid of a diagram how astronomers think the Solar System was formed.

Stars

How did we find out what stars are made of?

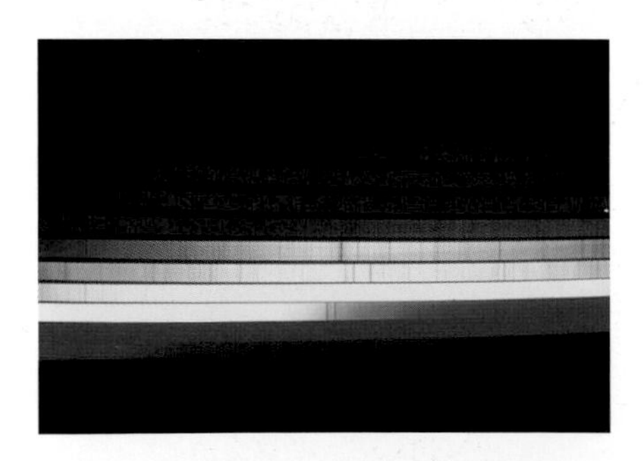

Figure 12.15 Spectrum of visible light from the Sun; the dark vertical lines indicate which elements are present in the Sun's outer layers.

A **spectrum** is produced when light is passed through a prism. Scientists fix special **spectrometers** to their telescopes to examine the light from the Sun and stars. When they look at the Sun's atmosphere they find it has a continuous spectrum crossed by some dark lines (Figure 12.15).

The dark lines correspond to bright lines of elements studied in the laboratory. About 140 years ago astronomers checked out all the lines and were able to say which elements were present in the Sun's atmosphere – mostly hydrogen. One set of lines was left. They did not match up with anything found on Earth at that time. This element was called 'helium' after the Greek word 'helios' meaning sun. Later on helium was found on Earth. It is the lighter-than-air gas used in party balloons.

The Sun and stars are composed mainly of the gases hydrogen and helium with small quantities of other elements, mainly carbon, nitrogen and oxygen.

Where do stars like our Sun get their energy from?

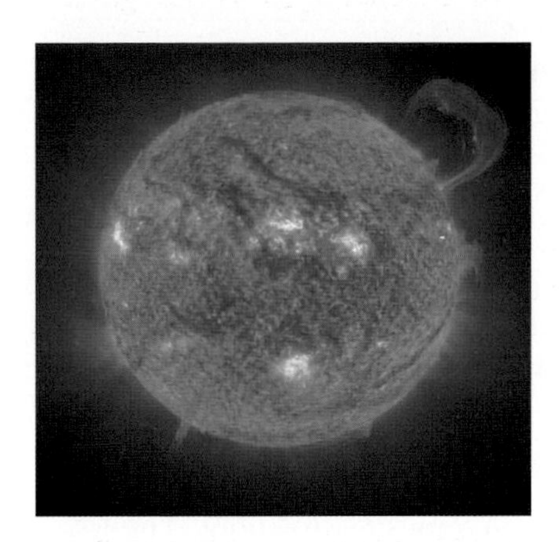

Figure 12.16 The Sun, showing a huge prominence curving in the Sun's magnetic field

Our Sun is a star just like billions of others in the Universe. At first scientists thought that it was burning like a fire. Did it burn coal – producing heat by a chemical reaction? A German scientist calculated that if the Sun's mass was all coal and if it had an unlimited supply of oxygen it would have burned for only a few thousand years. He went on to propose that the Sun's energy was fuelled by a limitless supply of solid particles from outer space falling into it.

The British scientist Lord Kelvin rejected this idea. He and Hermann von Helmholtz said that the mass of the Sun was not increasing. However the Sun was slowly contracting in size and this collapse would cause heating. They said the Sun's output energy could be maintained for at least several million years. The trouble was that geologists had found evidence that the Earth was much older than this – thousands of millions of years. How could the Earth have sustained life for millions of years before the Sun was apparently formed? This problem was not solved until the discovery of the sources of nuclear energy.

Did you know?

The Sun is losing mass at the rate of 4 million tonnes every second. It has been doing this for 4500 million years. It is only halfway through its life cycle.

We now know that the Sun and other stars are like a huge hydrogen bomb. The centre or core of the Sun is at a temperature of 14 million °C and at a very high pressure. At this temperature the hydrogen atoms join or fuse together to make helium. This process is called **nuclear fusion**. Each time it happens some mass is lost. It appears as a lot of energy. (Einstein's famous equation $E = mc^2$ applies here. Where E = energy measured in joules, m is the mass lost (kg) and c is the speed of light.)

Nuclear fusion produces energy: the star gives out light.

$$\text{hydrogen} \rightarrow \text{helium} + \text{energy}$$

The birth and middle age of a star

Figure 12.17 Stars are forming inside some of these gas clouds in the Eagle nebula.

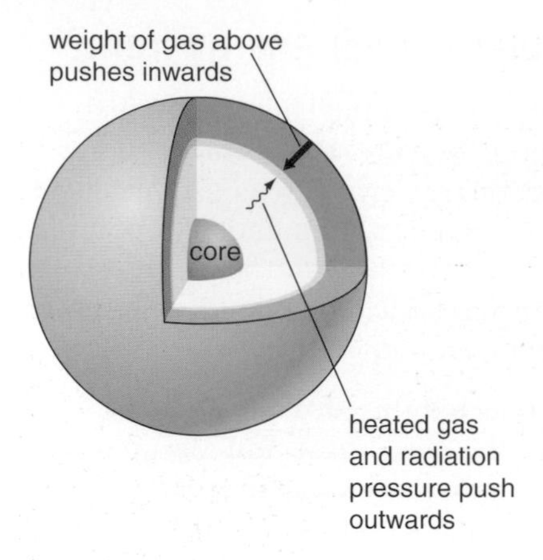

Figure 12.18 The outwards pressure of radiation and gas balance the weight of material pressing inwards.

Remember, from the section on the formation of the Solar System, that stars are formed when vast clouds of gas called nebulae collapse and warm up.

Gravitational energy is transferred to heat energy. It takes a long time but eventually the centre begins to glow dimly. At this stage it is called a **proto-star**. A long time later it reaches a high temperature. Nuclear reactions start to occur as the hydrogen fuses to become helium. It has now become a true star.

Stars have different colours, depending on how hot they are. Yellow dwarf stars like our Sun are the most common type with a surface temperature between 3000 °C and 6000 °C.

What happens to stars like the Sun in the end?

Inside a star there are forces acting in opposite directions. The gravitational force of all the mass of gas and dust crushes the star towards the centre. The other forces are caused by **gas pressure** and **radiation pressure**. Gas pressure is the outwards pressure caused by compressing the gas into a smaller volume. Radiation pressure is the outward force of the radiation produced by the nuclear reactions in the core. When the inwards and outwards forces balance the star is in a stable state – it doesn't shrink or expand (Figure 12.18). A yellow dwarf star (Figure 12.19) can be stable for a long time, around 10 000 million years. During this time, hydrogen is being converted into helium. Eventually the hydrogen gets used up. The nuclear reactions slow and stop, so the core collapses inwards as the gravitational force overcomes the radiation pressure.

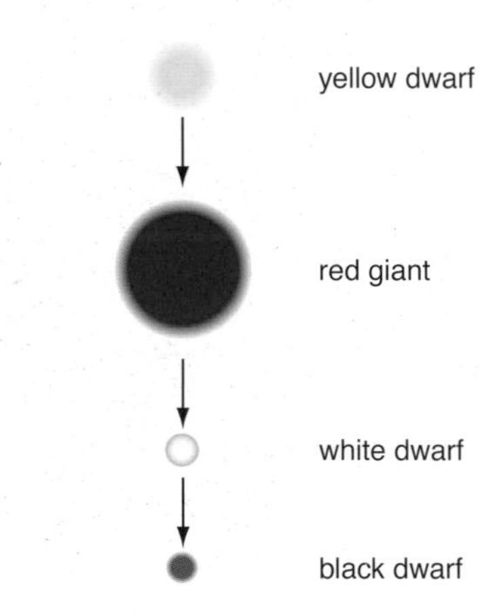

Figure 12.19 The life cycle of a yellow dwarf star

As the core shrinks the star heats up to a much higher temperature. This causes more nuclear fusion. Helium nuclei fuse to make heavier elements such as carbon and nitrogen.

Radiation pressure from the core is now greater. This expands the outer parts of the star, which is now classed as a **red giant**. This will happen to our Sun. It will expand outwards. Mercury, Venus and Earth will be burned up and become part of the Sun.

Later on the red giant loses its outer layers, blown away by radiation pressure to reveal the original core, which is now called a **white dwarf** star (Figure 12.20). After a long time a white dwarf cools down to become a red dwarf and finally a black dwarf.

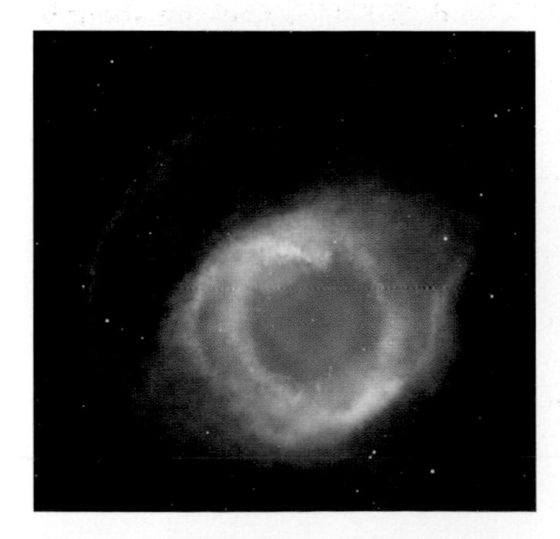

Figure 12.20 A white dwarf in the Helix nebula; dying Sun-like stars puff their outer layers into space.

Questions

12 How much energy is released when a mass of 1 kg is converted to energy?

13 The Sun is at a stable stage in its life. Explain, in terms of the forces acting on the Sun, what this means.

14 Explain how a star gets its energy.

Supergiants and supernovae

Figure 12.21 Multiple stages in the life cycle of blue stars. Upper right: a dark collapsing interstellar cloud. Centre left: young clusters of extremely hot massive (blue) stars destined to explode as supernovae. Upper left: this cluster with a blue ring around it is a blue supergiant nearing the end of its life.

Blue stars are much bigger and hotter than our Sun. Their surface temperature varies between 6000 °C and 25 000 °C. They are about 5 to 25 times larger and about 15 000 times brighter than the Sun. Because these big stars are brighter, they 'burn' faster and use up their hydrogen faster. The massive blue stars are stable for only about 100 million years. When they run out of fuel they expand to become a red **supergiant**. However their core is now much bigger and much hotter than a red giant – up to around 300 million °C. At these temperatures helium nuclei can combine to make heavier elements, which collect in the star's core. More and more energy is released by the nuclear fusion. The star gets hotter and brighter. (See Figures 12.21 and 12.22 on page 248.)

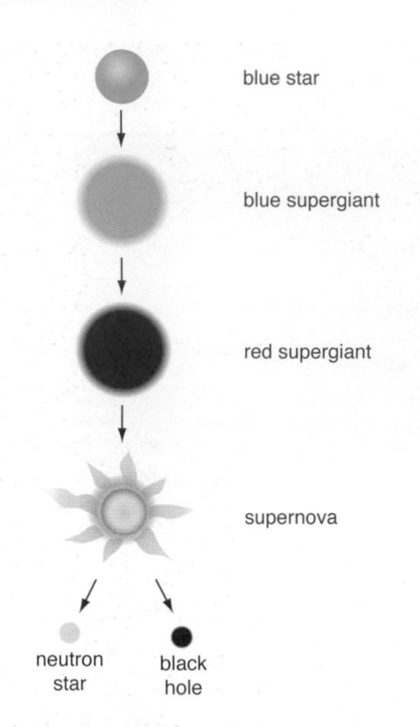

Figure 12.22 The life cycle of a large-mass blue star

Figure 12.23 Before (top) and after (bottom) the explosion of supernova 1987A

Eventually the nuclear reactions stop and the massive star begins to collapse. The temperature climbs as the core contracts and heats up. Very quickly everything collapses into the core at colossal speeds, causing the whole star to explode. The outer layers of the star are thrown into space producing a brilliant event known as a **supernova**. The exploding star is so bright that for a time its brightness can rival an entire galaxy of 100 000 million stars (Figure 12.23).

Did you know?

Nuclear reactors are releasing the energy that was locked in heavy elements by the death throes of massive stars. Uranium was created in a nearby star when it went supernova billions of years ago.

During the explosion, more heavy elements such as gold and uranium are formed when the energy of the explosion allows the nuclei of elements like iron and nickel to fuse together. The elements are thrown across space as dust ready to be made into new stars. The Sun, the Earth and all life on it have been made from elements formed in and released by these exploding stars. As more blue stars go supernova, the proportion of heavy elements in the Universe will increase as it gets older.

The next stage: neutron stars and black holes

Depending on its mass, the supernova may collapse into a **neutron star** or a **black hole**. A neutron star may only be about 25 km wide, but can have a mass similar to the mass of the Sun, about 10^{27} tonnes. Neutron stars do not shine but spin rapidly and give out a beam of radio emissions, like a lighthouse beam. As the neutron star rotates the emissions are detected as pulses by radio telescopes. They are called **pulsars**.

Black holes are formed from the supernovae of the biggest stars, some more than ten times larger than our Sun. If the core from a collapsed supernova is greater than three solar masses, it shrinks to form a black hole. A black hole with a mass which is six times greater than the Sun could have a diameter of only about 35 km. There is so much mass collapsed into such a small space that the gravitational force is huge. Nothing can escape from the pull of a black hole, not even light. At the black hole's centre the remains of the former star are crushed to a point of infinite density – called a singularity. You cannot see black holes, but you can see the effect they have on the stars and gas around them (Figure 12.24 opposite).

As the gas swirls into the black hole it heats up to many millions of degrees and emits X-rays. Satellite observations have found many places where we think black holes exist. Evidence from X-ray photographs leads many astronomers to think there is a super-massive black hole at the centre of every galaxy, including the Milky Way.

Questions

15 What did astronomers use to find out what the Sun and other stars are made of?

16 The Earth has a lot of heavy elements. Where did they come from?

17 Describe the life cycle of:
- **a** a yellow dwarf star
- **b** a blue giant star.

Activity

From the internet or reference books, write a paragraph on each of the following:

1. yellow dwarfs
2. supernovae
3. black holes.

Figure 12.24 An artist's impression of Cygnus X-1, which is a massive blue star. It is being pulled apart by what is thought to be a black hole, which is sucking matter out of the star.

Figure 12.25 The Milky Way galaxy

Figure 12.26 The Andromeda galaxy

The Universe

The light year: a new unit of distance

The Universe is so big we really do not know exactly how large it is. Measuring distances in kilometres results in impossibly large numbers. We use the **light year** instead. It is the distance that light travels in one year. Light travels at 300 000 km/s.

1 light year approximately = 10 000 000 000 000 km (10 million million kilometres = 10^{13} km).

Looking back into the past

Our nearest star (apart from the Sun) is Alpha Centauri. It is 4.31 light years away. Its light took 4.31 years to reach us. Whenever we look at it we see it as it was 4.31 years ago.

Galaxies

Look carefully at the sky on a clear dark moonless night and you will see a wide, faint band of stars. You will be looking edge-on at our galaxy, the Milky Way (Figure 12.25).

Our galaxy is enormous. It is 100 000 light years across and contains hundreds of billions of stars, with perhaps a billion planets and a million black holes.

How many galaxies in the Universe?

Our galaxy is not alone in the Universe. Our nearest neighbour is the Andromeda galaxy (Figure 12.26). It is 2.2 million light years away.

These are just two of the hundred thousand million galaxies out there. Some say there are more stars in the Universe than grains of sand on every beach in the world.

Questions

18 Astronomers use the term Milky Way to describe part of the Universe. What is the Milky Way?

19 Why can you say that when looking at the night sky you are looking into the past?

20 Astronomers find a red supergiant star. It is 4.9×10^{15} km from Earth. How long does it take light to reach Earth? (Speed of light = 300 000 km/s.)

It is difficult to answer the question 'How big is the Universe?', but according to the latest estimates, or as far as our best telescopes can make out, the outer edge is 12–15 billion light years away from us.

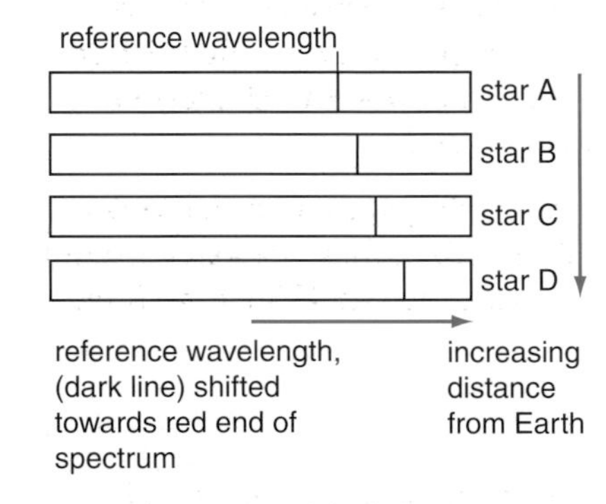

Figure 12.27 Spectra from distant stars show a 'red-shift'.

Looking at the light from distant stars

The spectrum of electromagnetic radiation from the Sun is crossed by a series of dark lines (as shown in Figure 12.15, on page 244). An American astronomer called Edwin Hubble used a very large telescope to look at light from distant galaxies. His telescope was fitted with a powerful spectrometer. He noticed that the distant stars had the same dark lines, but they were not in the same place. They were moved a little way towards the red end of the spectrum (Figure 12.27).

Hubble pointed his telescope in the opposite direction. He got the same effect. In fact all the stars in distant galaxies had the same set of dark lines shifted towards the red end of the spectrum.

Why are the lines shifted towards the long-wavelength red end?

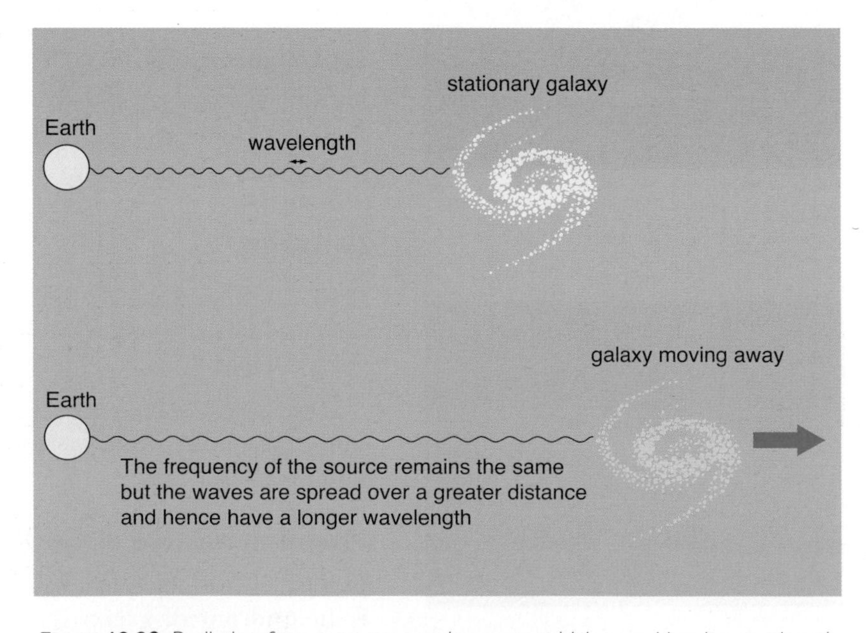

Figure 12.28 Radiation from a source moving away at high speed has its wavelength 'stretched' or 'red-shifted'.

Suppose the star and Earth are not moving either towards or away from each other. A dark line from the star corresponds to radiation of a certain frequency f and wavelength λ.

If the star is moving away from the Earth, the dark line will still correspond to radiation of that same frequency as it is emitted by the star, but each wave will have had to travel an extra distance. These waves have a longer wavelength. The red end of the spectrum has the longer wavelengths. So if the star is moving away from us, the dark line is shifted towards the red. This **red-shift** shows that the star is moving away from us. In whichever direction we look, the galaxies are moving away from us. The Universe is expanding in all directions. Think about a balloon, and imagine you put some marks on it to represent galaxies. As you blow up the balloon, the marks will move away from each other as the balloon expands.

60000
extrapolation
40000
20000
Speed (km/s)
Hydra
1000 2000 3000 4000
Distance (millions of light years)

Figure 12.29 Hubble plotted this graph and found that the speeds of galaxies are proportional to their distance away from us.

The more distant galaxies are moving away even faster

H

Hubble went on to examine stars from more distant galaxies. Figure 12.29 shows that the more distant the galaxy, the greater the red-shift. All the galaxies are moving apart (Figure 12.30). The further away, the faster they are moving. Hubble said that the Universe is expanding and has been for millions of years.

Some of the more distant galaxies are moving away at 90 000 km/s, about one-third the speed of light. Remember, these galaxies are not moving through space away from us. The space itself is expanding.

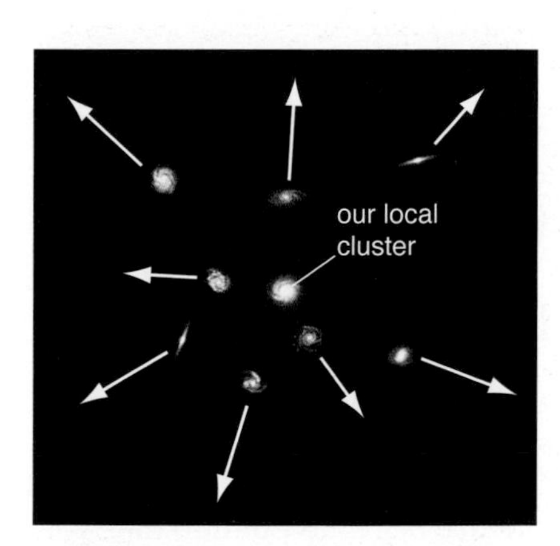

Figure 12.30 Galaxies are moving away from other galaxies.

Questions

21 Look at Table 12.2. Rhys has copied it from the board, but he had been talking and missed out some numbers. Use the graph in Figure 12.29 to work out the distance of the Ursa Major galaxy.

Table 12.2 Speeds and distances of galaxies

Constellation in which galaxy is found	Distance of galaxy (light years × 10^6)	Speed of galaxy (km/s)
Virgo	72	1 200
Perseus	400	
Ursa Major		15 000
Corona	1200	20 000
Boötes	2400	40 000
Hydra		60 000

1 light year = 10 million million (10^{13}) km. This is the distance that light travels in 1 year.

22 Use Table 12.2 and Figure 12.29 to work out the speed of the Perseus galaxy.

23 What is meant by 'red-shift'? What causes a red-shift?

24 Why are spectral lines from some stars 'red-shifted' more than others?

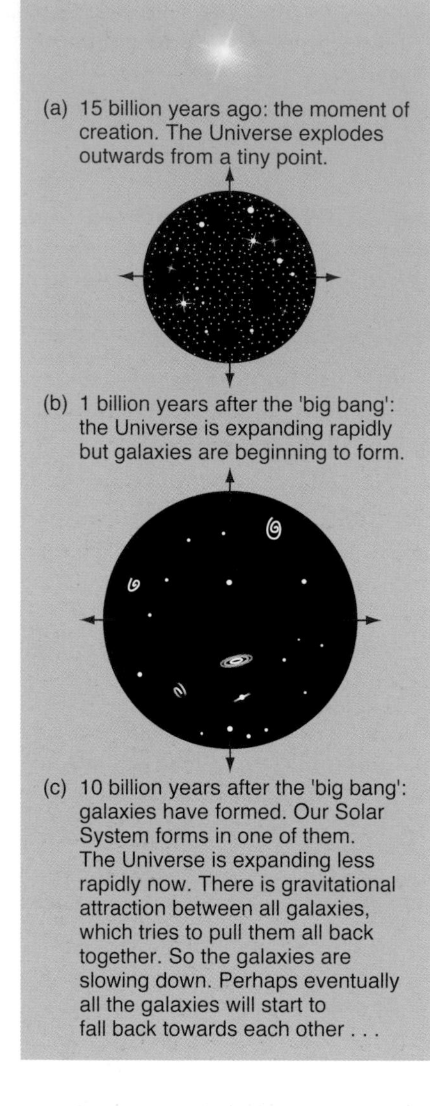

Figure 12.31 The formation of the Universe

How did it all start?

The red-shift tells us the Universe is expanding in all directions. If everything that is in it is moving apart, then it follows:

- At some time in the past the galaxies must have been much closer together.
- There must have been something that made them move apart in the first place.

If we could wind the clock back millions and millions of years, what might we find that caused the galaxies to fly apart?

It seems a reasonable guess that this was caused by an enormous explosion. It has been called the '**big bang**'. This was no ordinary explosion. We are talking about an explosion that threw out all the 100 000 million plus galaxies each containing 100 000 million stars, not forgetting all the planets, dust and gases that might be orbiting all these stars and, of course, the space itself. This is what is proposed by the 'big bang' theory. According to current measurements of the red-shift from distant galaxies, the 'big bang' happened about 12–15 thousand million years ago.

Just as when things are compressed they get hotter, when things expand they get cooler. The Universe has been cooling down ever since that explosion 15 000 million years ago to just 2.7 degrees above absolute zero. This produces a background radiation.

The results from these observations are thought to confirm the existence of the 'big bang'. They also give the age of the Universe as 13.7 billion years. This is in line with earlier estimates from the Hubble space telescope.

What is the Universe made of?

With trillions of stars and galaxies you would think that was enough. However, scientists now think that 23% of the Universe is made from another substance called **dark matter**.

Nobody knows what dark matter is. They cannot detect it either. Then a new discovery in 2003 was the mysterious **dark energy**, which accounts for 73% of the Universe. This bizarre force seems to be pushing the Universe apart at an accelerating rate. Gravitational pull should, of course, be making it slow down or contract.

The age paradox: will we have to change the theory again?

In the 1950s, Fred Hoyle and other astronomers thought that stars were being born, living and dying all the time. They called it the steady-state theory. Then all the evidence began to point to the 'big bang' theory.

Some scientists are concerned about the 'big bang' theory. They argue that the light from the most distant galaxies has taken thousands of million years to reach us. You would expect them to be composed of new stars. Observations show that there are older stars in

Questions

25 What is the 'big bang' theory? What does it predict about the size of the Universe?

26 How old is the Universe?

these distant galaxies. The oldest stars cannot be older than the world in which they live. They say there are problems with the idea of dark matter and the microwave background radiation in the Universe. These ideas are controversial, but who knows what the real story will be?

Animation tour of the 'big bang': www.superstringtheory.com

News on space discoveries: www.bbc.co.uk/science hubblesite.org

'Big bang' time machine: www.schoolscience.co.uk

Questions

27 You look at a star. It is exactly 100 light years away. Give the date and time, to the nearest hour, that the light left the star.

28 What is the name of our galaxy? How big is it?

29 Write a paragraph to say how the Universe began.

30 Just suppose the red-shift was caused by something else. Suppose we find that the galaxies are not moving away from us. How will this affect what we know about the Universe?

31 Stars are very massive. The force of gravity drawing together the matter from which the stars are made is very strong. The very high temperatures in the star create forces acting in the opposite direction to the gravity force.

a Name the matter (materials) from which stars are formed.

b Explain in terms of the forces acting on a star:

i what you understand by the 'stable period' of a star

ii why a star becomes a red giant.

c Describe the circumstances in which a neutron star is formed.

32 Read this passage carefully before answering the questions that follow. 'Scientists, looking at light from distant galaxies, have found that the light waves received are "stretched out". This causes the waves to shift towards the red end of the visible spectrum. This effect is called the **red-shift**. The size of the red-shift produced by a galaxy allows a scientist to calculate its speed. From the calculations of the speed of the galaxies, scientists now believe that 15 000 million years ago the Universe began as a single atom-sized concentration of matter and energy.'

a State how the wavelength of the light received is affected by the movement of the galaxy.

b The further away galaxies are, the greater is the red-shift. Explain what this suggests about the way the galaxies are moving.

c The Universe started with a 'big bang'. Give reasons why the red-shift supports this statement.

Summary

1 The Solar System contains planets, moons, satellites, asteroids and comets.

2 The planets orbit the Sun in nearly circular orbits. They all go round in the same direction.

3 Planets are held in their orbit by the gravitational attraction of the Sun.

4 The greater the distance from the Sun, the longer it takes to orbit the Sun.

5 Gravitational attraction keeps moons and satellites in orbit round a planet.

6 The Solar System was formed at the same time as the Sun, about 4500 million years ago. It was formed from a cloud of gas, ice and dust which collected together under the force of gravity.

7 The inner planets are rocky because most of their gas was evaporated by the heat of the Sun. The outer planets are mostly gas and liquid because they formed further away from the Sun.

8 Comets and asteroids are some of the leftovers from planet formation.

9 Comets' tails are bright because they reflect sunlight. The tail points away from the Sun because of radiation pressure from the Sun.

10 If the Sun produced light and heat by chemical reactions, this would conflict with what we know about the great age of the Solar System.

11 A spectrometer can be used to analyse the light from the stars. This tells us what the stars are made of.

12 The fusion of hydrogen into helium is the source of a star's energy.

13 A star is stable when the inward gravitational forces balance the outward radiation pressure.

14 Small stars like our Sun have a long stable stage in their lifetime. After this, a yellow dwarf star passes through red giant and white dwarf stages, eventually becoming a black dwarf.

15 Large mass stars or blue giants are much bigger and hotter. Their stable state is not as long as that of our Sun. They explode as a supernova. They collapse to form neutron stars; very big ones collapse to form black holes.

16 Elements heavier than helium have come from nuclear fusion reactions in stars and from supernovae.

17 The proportion of heavy elements in the Universe is gradually increasing.

18 We measure astronomical distances in light years. This is the distance light travels in one year. It is approximately 10 000 000 000 000 km.

19 Our galaxy contains hundreds of billions of stars.

20 There are about a hundred thousand million galaxies in the Universe.

21 Lines in the spectra of distant stars are moved to the red end of the electromagnetic spectrum. This tells us that the stars are moving away from us. The more distant the star the greater is the red-shift and the faster they are moving away.

22 The red-shift measurements provide evidence that the Universe started with an extremely large and very hot 'big bang'.

Chapter 13 Radioactivity

By the end of this chapter you should:

- know that radiation is produced by radioactive substances;
- know that this radiation comes from unstable atomic nuclei with an imbalance of protons and neutrons;
- understand the differences between alpha, beta and gamma radiation;
- understand the dangers of ionising radiation;
- understand that naturally occurring background radiation is all around us;
- understand how to protect yourself from ionising radiation, especially from the dangers of radon gas in the home;
- be able to discuss the risks and precautions needed for radioactive treatments in hospitals;
- know how long a radioactive material remains radioactive;
- know how we make use of radioactivity, for example in cancer treatment, carbon dating and tracers in industry;
- know about the problems associated with the disposal of radioactive materials.

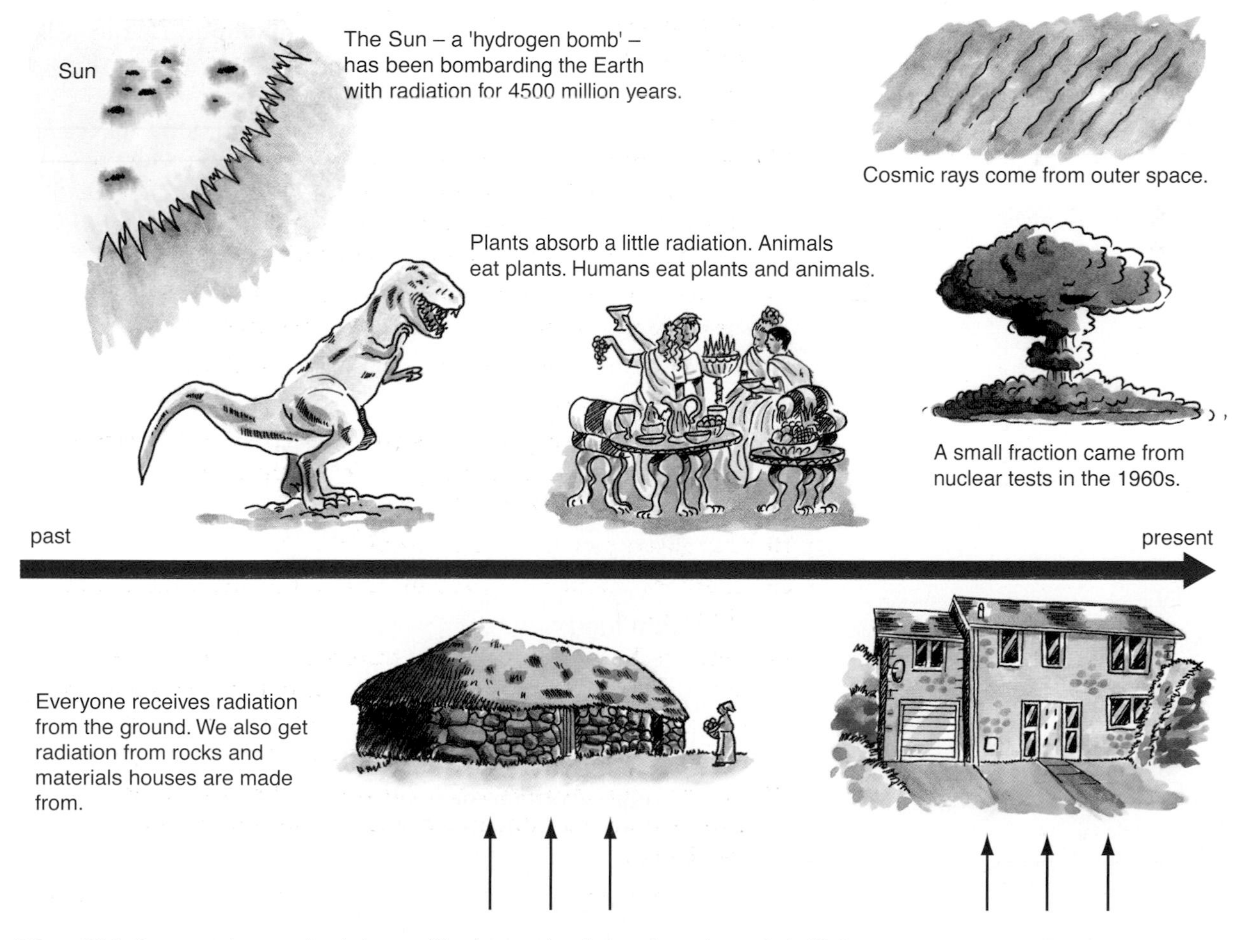

Figure 13.1 Everyone has received doses of background radiation throughout their lifetime.

Radioactive emissions

Radioactivity and radiation

We cannot see radiation, feel it, smell it, touch it or taste it. All radiation carries energy and can cause damage to cells in the body but some radiation is more dangerous than others. We know that infra-red radiation from a blazing fire can burn skin, but it is radiation such as gamma rays that can cause cancer.

Not all radiation comes from radioactive substances, but we know that the radiation from radioactive substances is harmful as it is **ionising radiation**.

Radioactivity comes from atoms that have unstable nuclei. The unstable **nucleus** will give off radiation until it becomes stable. Then the radiation will stop.

Radiation dose

Radiation affects your body. Radiation dose is measured in sieverts (symbol Sv). It is a big unit and a big dose of radiation. In practice, we use the millisievert (symbol mSv), which is one-thousandth of a sievert.

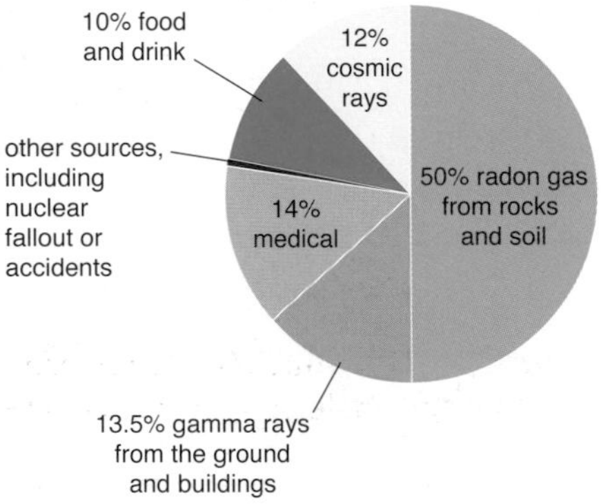

Figure 13.2 Sources of background radiation

Background radiation: it's been there all the time

From reading the papers and watching TV, you would think that all the radiation we get comes from nuclear power stations and bombs. This is not true: background radiation occurs naturally (Figure 13.1 on page 255). Most background radiation comes from the ground. The rocks and soil beneath us and the air we breathe are radioactive. Fast-moving particles from the Sun and outer space, called **cosmic rays**, are a part of background radiation. Our ancestors received over 85% of the amount of background radiation we receive today. Of the remaining 15%, we get 14% from medical and dental X-rays. Less than 1% comes from industry.

Questions

1 What is 'background radiation? Where does it come from?

2 What is your main source of radiation?

Did you know?

About 100 000 cosmic rays from outer space will pass through your body every hour. You get a bigger dose when you are in an aeroplane because there is less air between you and space to absorb them. A 1 h flight will increase your dose by about 0.005 mSv.

Activity

LoSalt is a salt-substitute product that includes some potassium chloride as a healthier alternative to salt. If this is very mildly radioactive, how healthy is this option? In your group investigate and discuss the merits of each case.

Radiation from food

We also get a little radiation from food. When plants grow they absorb carbon dioxide which includes a small amount of radioactive carbon. An isotope, potassium-40, accounts for most of the radiation in our food. Foods rich in potassium will contain more. Mussels, cockles and winkles contain more potassium. Eating lots and lots of them will increase your radiation dose.

What is it like inside an atom?

You should remember from Chemistry (Chapter 5) that the nucleus of an atom contains **protons** and **neutrons** surrounded by orbiting **electrons**. The nucleus is very, very small compared with the atom.

Suppose you could magnify the nucleus so that it was about the size of a 0.5 m (50 cm) television set in the middle of London. The atom is 100 000 times bigger than the nucleus, so the diameter of the atom would be about 50 000 m or 50 km. On this scale, an electron would be a very small object going round the M25 motorway.

Usually the atom is very stable. Radioactivity comes from unstable nuclei. Unstable or radioactive nuclei emit particles or radiation and then become more stable.

Radioactive decay of unstable nuclei

H

Remember that a useful way to show the composition of the nucleus is to use the **atomic number** or number of protons in the nucleus and the **mass number** or total number of protons and neutrons in the nucleus. This is shown here for the hydrogen nucleus in Figure 13.3.

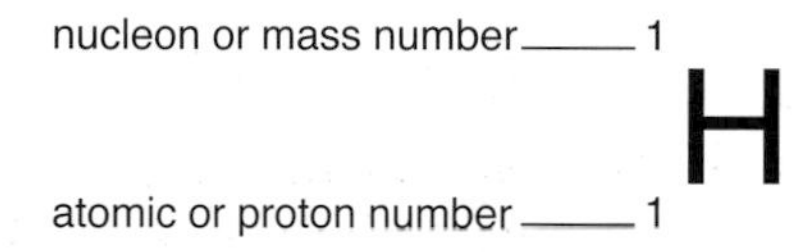

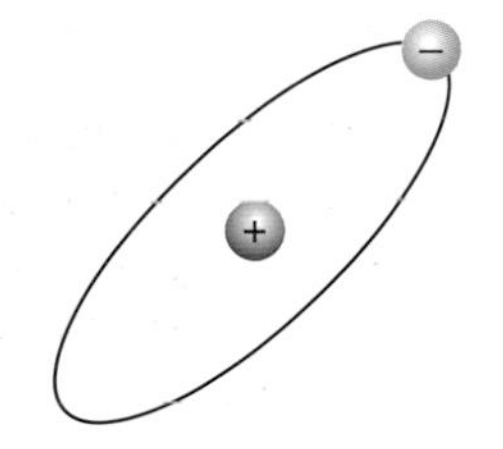

Figure 13.3 A hydrogen atom

Helium has two protons and two neutrons in its nucleus. There are two electrons in orbit round it. The helium nucleus is written as $^{4}_{2}He$ (Figure 13.4).

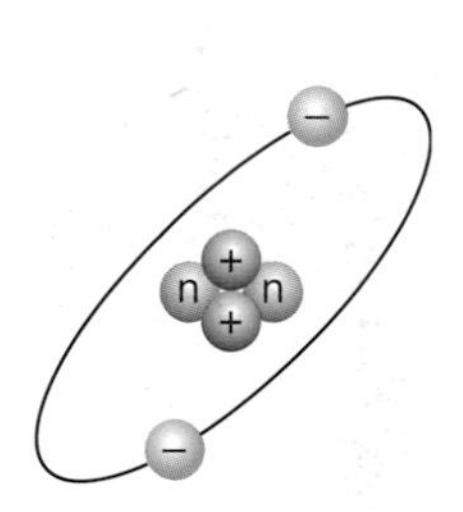

Figure 13.4 A helium atom

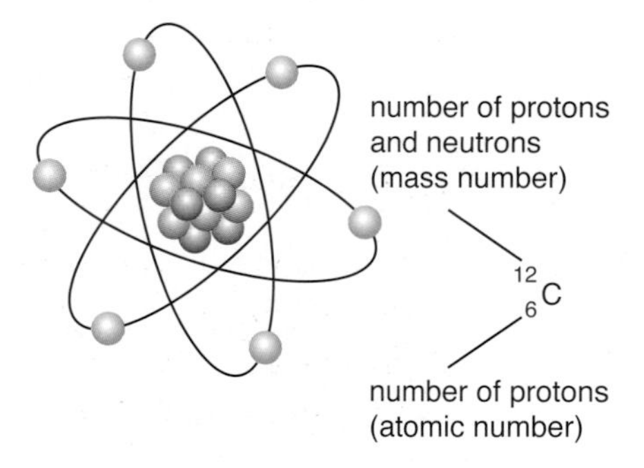

Figure 13.5 A carbon atom

Carbon is written as $^{12}_{6}C$. It has six protons and 12 – 6, that is 6 neutrons (Figure 13.5).

Some elements do not have equal numbers of protons and neutrons. Lithium is $^{7}_{3}Li$ and beryllium is $^{9}_{4}Be$. Gold is $^{197}_{79}Au$, and has many more neutrons than protons. It has 79 protons and 118 neutrons and is stable. It is not radioactive.

Isotopes

Different elements must have different atomic (proton) numbers. No two elements can have the same number of protons. But atoms of the same element can have different numbers of neutrons in their nucleus. They will have a different mass (**nucleon**) number.

Ordinary oxygen, ${}^{16}_{8}O$, has eight protons and eight neutrons. Some oxygen atoms have nine neutrons and are written as ${}^{17}_{8}O$. This is still oxygen because it has eight protons. It just has an extra neutron in its nucleus. ${}^{16}_{8}O$ and ${}^{17}_{8}O$ are **isotopes** of oxygen.

All elements have more than one isotope. Quite often isotopes with more than one extra neutron are unstable. Their nucleus breaks up into smaller parts. Very large atoms such as uranium, ${}^{238}_{92}U$, are unstable. When an unstable atom breaks apart it does so by giving off radiation. The radiation can be in the form of alpha (α), beta (β) or gamma (γ) radiation. This is called **radioactive decay**.

Questions

3 Explain the terms:
   a atomic number
   b mass number of an atom.

4 What is an isotope?

5 What happens when an unstable atom decays? What is the name of this process?

How radioactive is your house?

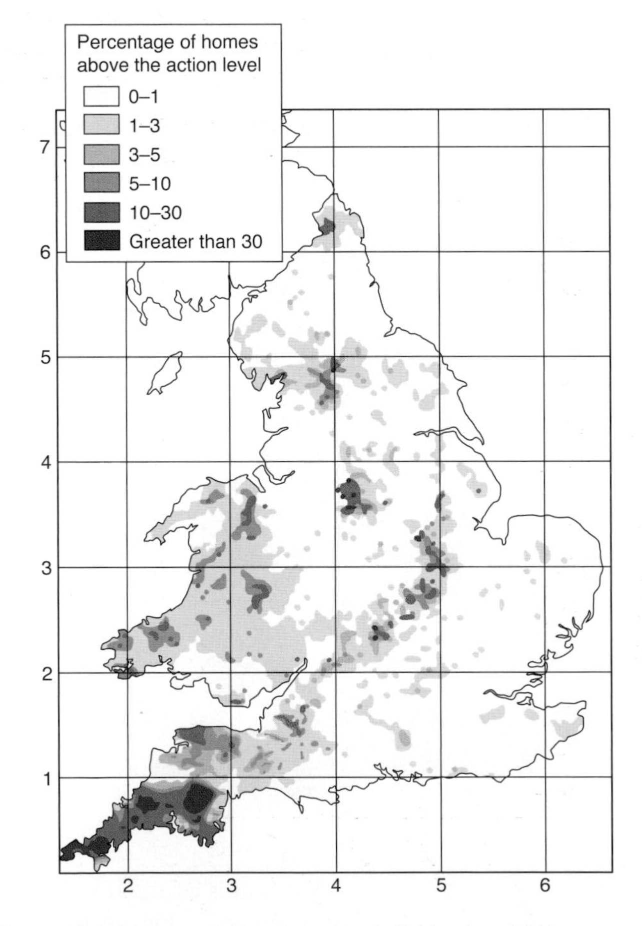

Figure 13.6 Radon emissions in homes in England and Wales

Some rocks are more radioactive than others, in particular rocks that are volcanic in origin. Granite rocks in Cornwall and parts of Scotland are the most radioactive. The map in Figure 13.6 shows that parts of England and Wales are more radioactive than others. The uranium in the rocks produces radon, a radioactive gas.

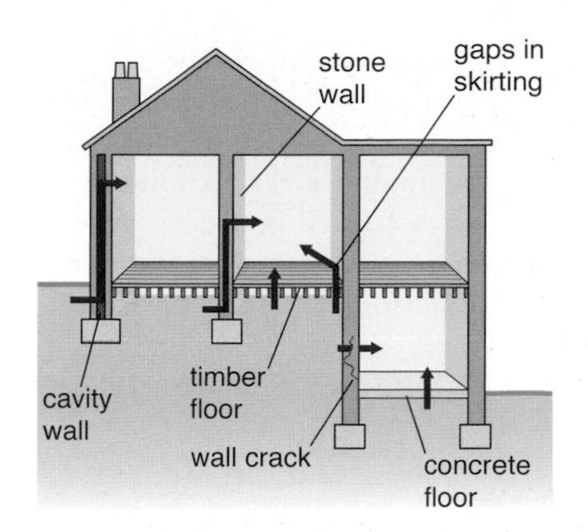

Figure 13.7 Radon can enter a house by many different routes.

Radon

Radon occurs naturally and is found everywhere. We get most of our background radiation from radon. The average dose in the UK is 2.5 mSv, but in Cornwall that rises to 7.8 mSv.

When uranium atoms decay, it becomes radium. When radium decays it becomes radon. This is a radioactive gas which we can inhale. Outdoors, radon blows away but the gas can get trapped in houses.

How does radon get into my house and is it a problem?

Radon gas can get into a house in a variety of ways (see Figure 13.7). In most houses, radon gas is not a problem. However, radon levels can vary from house to house. If the concentration is high, it can be a serious health risk. Radon can cause lung cancer.

The UK Government has advised that a level above 200 Bq/m^3 requires remedial action. One becquerel (symbol Bq) is equivalent to one disintegration (decay of one unstable atom) per second.

Figure 13.8 shows that the lifetime risk from exposure to average levels is very small. You are three times more likely to die as a result of an accident in your home. But, as Figure 13.9 shows, as the concentration of radon rises, so do the risks.

Figure 13.9 shows the risks for non-smokers. If you smoke 15 cigarettes a day, you increase your risk 10 times.

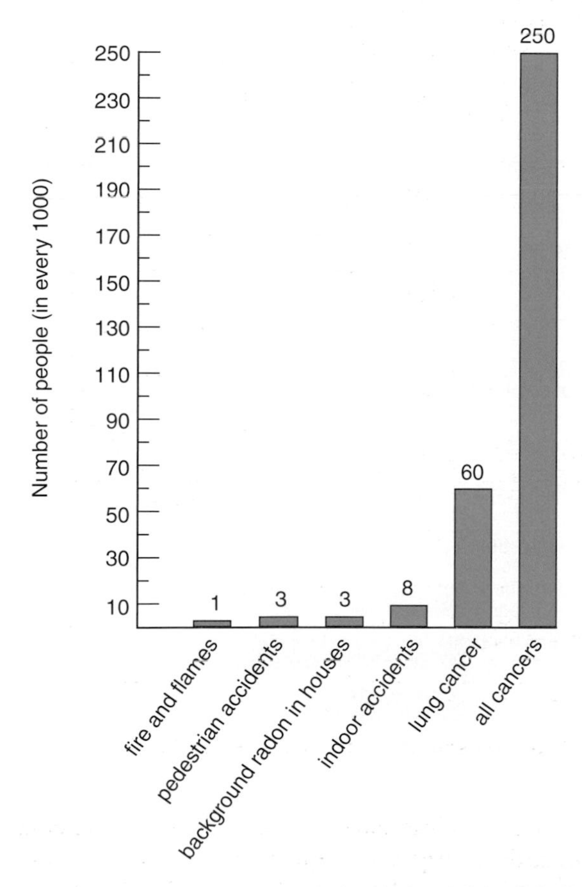

Figure 13.8 Lifetime risks of death from common causes (UK average for smokers and non-smokers)

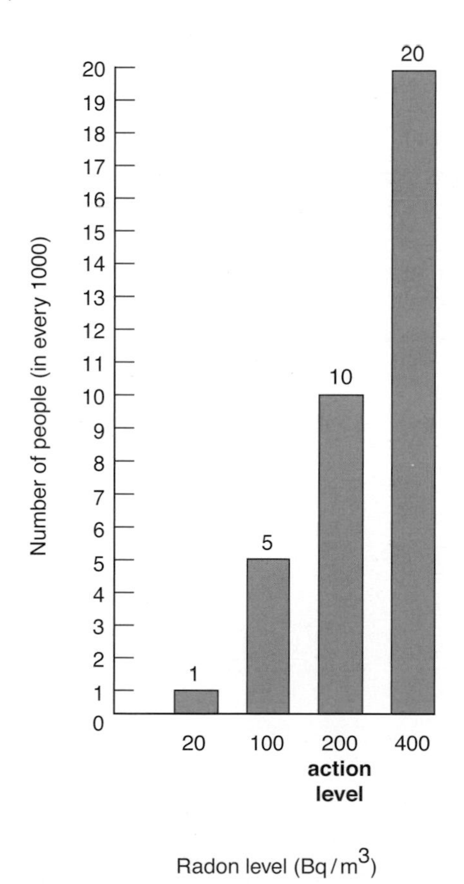

Figure 13.9 Lifetime risks of lung cancer from radon (for non-smokers)

Figure 13.10 The radon test pack

Activity

Megan's mum is worried about radon. She has seen the radon map of the UK and thinks she is in an 'at risk' part of Wales. Advise her on the steps she can take to find out about the risk and, if necessary, what she can do to make her house safe.

Maps of radon levels:
www.hpa.org.uk

Background radiation sources:
www.defra.gov.uk/environment/statistics/

Radon protection advice:
www.defra.gov.uk/environment/radioactivity/background/radon.htm

Search for other links for 'radon in homes', 'radon detectors', 'radon costs', etc.

What if I live in an 'at risk' area for radon?

The simplest way is to have your house tested for radon. The government can supply you with a detector pack (Figure 13.10). The pack contains two detectors. The detector is a piece of spectacle lens plastic in a plastic case about the size of a small doorknob. One is placed in the living room, the other in an occupied bedroom. After three months they are sent for analysis. The plastic records the radon levels. The gas leaves tracks in the plastic which are measured in the laboratory.

Questions

6 Why are large doses of radon dangerous?

7 How can you find out if your house is affected by radon gas?

8 Where does radon come from?

9 How does radon get into your house?

What happens if my house is above the action level for radon?

If your house is just above the action level, you can do the following:

- Improve the ventilation in the house.
- If you have a wooden floor, then increase the under-floor ventilation with extra airbricks.
- If the level is higher, then fit a fan-assisted airbrick.
- If you have a concrete floor then seal all the gaps so gas can't enter from the soil.

For more severe cases, a positive ventilation system can be installed. It has a ventilation fan which blows air into the house. A radon sump may also be installed under the floor.

For more up to date information, check out the Health Protection Agency and DEFRA websites.

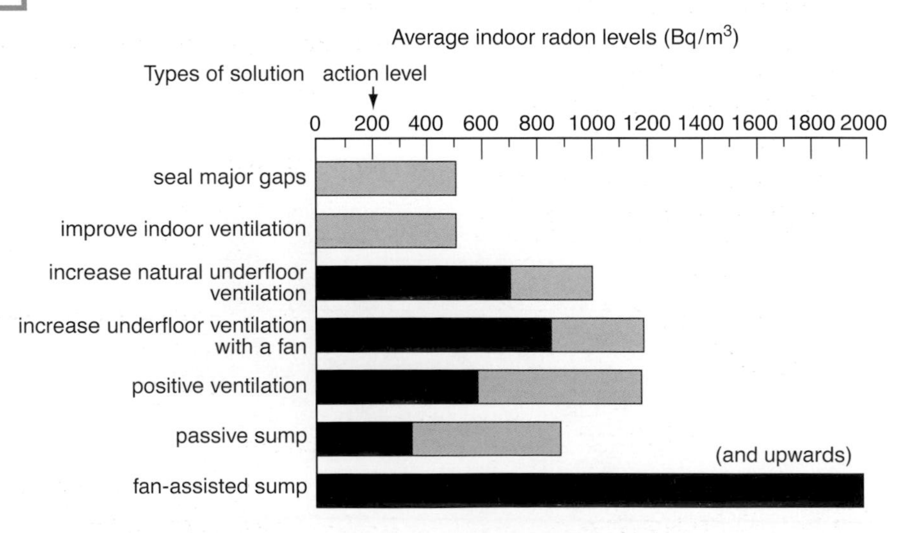

Figure 13.11 Effectiveness of radon-removal measures (black bar: high level of success; grey bar: some success)

Alpha, beta and gamma radiation

Three different types of radiation are emitted by radioactive substances: **alpha** (α), **beta** (β) and **gamma** (γ) (Figure 13.12). They all come from the nucleus of the atom. Gamma rays are electromagnetic radiation, but alpha and beta radiation are particles. It is called radiation because they are ejected in all directions from an unstable nucleus like the spokes (radii) from the hub of a wheel.

Alpha, beta and gamma radiation have different **penetration** – how far the radiation travels before it is stopped.

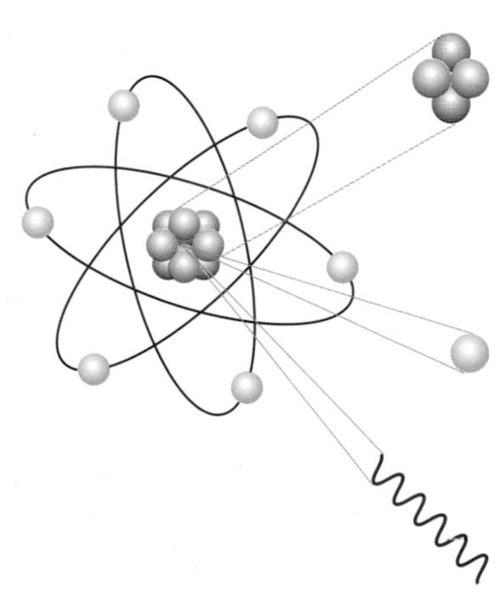

Alpha (α) radiation
These are particles, not rays, and travel at about 10% of the speed of light. An α particle is a helium nucleus (He^{2+}), a helium atom that has lost both electrons.

Beta (β) radiation
These are fast-moving electrons that come from the nucleus. They travel at about 50% of the speed of light.

Gamma (γ) radiation
This is an electromagnetic wave. It travels at the speed of light (3×10^8 m/s). It has very high energy.

Figure 13.12 Alpha, beta and gamma radiation

Did you know?

We use a Geiger counter to detect radiation from radioactive substances. It consists of a Geiger–Müller tube and a ratemeter that counts the particles detected. Most have a loudspeaker which makes a clicking sound to alert the user – the faster the clicks, the higher the radiation levels.

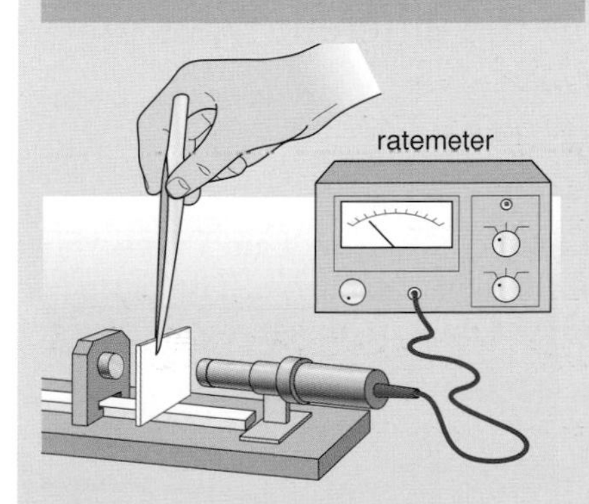

Figure 13.13 Testing to see which materials will stop radiation

Practical work

Investigating the penetration of alpha, beta and gamma radiation

This experiment must be demonstrated by a teacher.

Set up the apparatus in Figure 13.13 and measure the background radiation count.

Alpha radiation

1 The source (americium-241) emits α radiation only.
2 With tweezers, place it close to the Geiger–Müller tube, and measure the count rate.
3 With tweezers, carefully place a piece of card between the α radiation source and the tube.
4 Measure the count rate.

Beta radiation

5 The source (strontium-90) emits β radiation only.
6 Repeat the experiment. Try to stop the radiation with card.
7 Then try to stop the radiation with sheets of aluminium of increasing thickness.

Gamma radiation

8 Get a source (cobalt-60) that emits γ radiation only.

9 Repeat the experiment. Try to stop the radiation with card and then aluminium.

10 Then try to stop the radiation with sheets of lead.

If your school has a radium source, you can use this experiment to show that radium emits alpha, beta and gamma radiation.

These experiments show the relative penetrating ability of the three types of radiation (Figure 13.14).

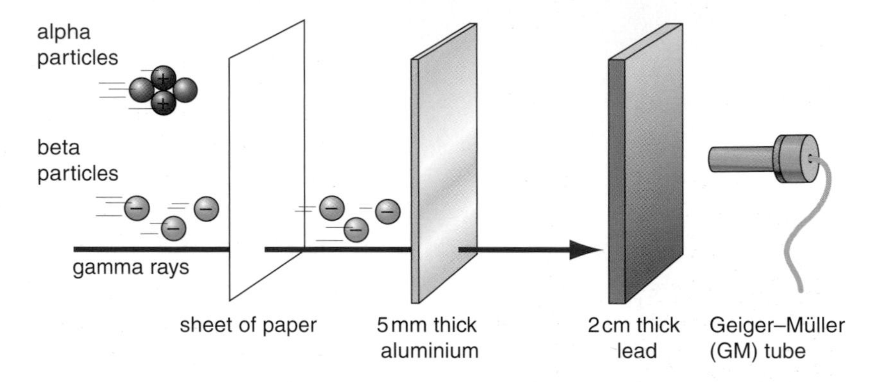

Figure 13.14 The penetration of alpha (α), beta (β) and gamma (γ) radiation

Questions

10 Why must you measure the background count rate when making measurements with radioactive sources?

11 Copy and complete the following table.

Radiation	Particle	Wave	Charge	Stopped by
Alpha				
Beta				
Gamma				

The dangers of exposure to the different types of radiation

Alpha (α) particles

Alpha particles are absorbed by a few centimetres of air. Unless you are very close to the source, most will never reach you. The particles are stopped by paper or skin. However they are dangerous if they get inside you.

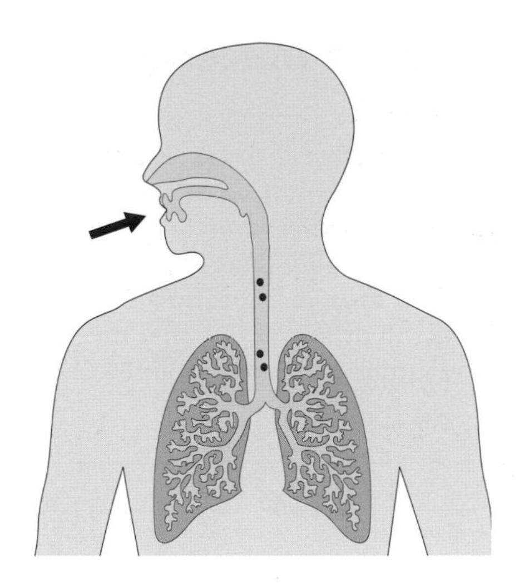

Figure 13.15 Radon gas can enter the lungs.

This is why radon gas is dangerous. You can breathe radioactive radon gas into your lungs (Figure 13.15). As it decays, it gives off alpha particles right inside the lung. Alpha particles are heavy (remember they are large helium nuclei) and can cause ionisation of atoms in human cells. Alpha radiation can cause 20 times as much damage as either beta or gamma radiation.

Beta (β) and gamma (γ) radiation

Beta radiation can penetrate thin sheets of aluminium. Gamma radiation will even pass through lead. Beta and gamma radiation are both harmful. They can cause radiation burns. The skin becomes red and sores develop. There are delayed effects that develop later, such as cataracts in the eyes and cancer.

Protection from radiation

The radioactive sources used in school are placed in a lead-lined box (castle) and in a locked and clearly labelled store. This applies to all radioactive material. For industrial use, with stronger sources, the lead boxes are much larger and thicker. Your teacher also uses long tongs or tweezers to keep at a distance from the source.

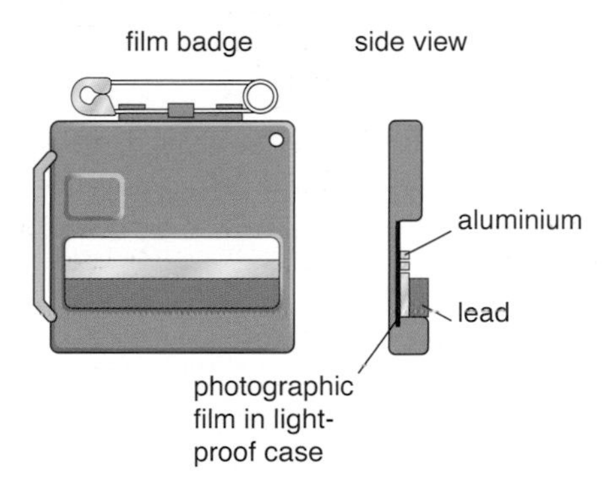

Figure 13.16 Dosemeter film badge

A lot of radioactive material is used in hospitals and industry. The workers wear a film badge (dosemeter) (Figure 13.16) so they can monitor the dose received. Anyone receiving more than the permitted dose cannot work in a radiation area.

Even patients receiving radiation treatment for cancer have to be protected. Radiation is used to kill cancer cells. It also kills normal cells. There is always a risk with any treatment.

The radiation beam is accurately focused on the cancer. The beam is then moved round the patient so that only the cancer, at the centre of the beam, gets the full dose. The other parts get a much smaller dose which does not cause another cancer. The radiographers must work behind lead glass screens to reduce their own dose.

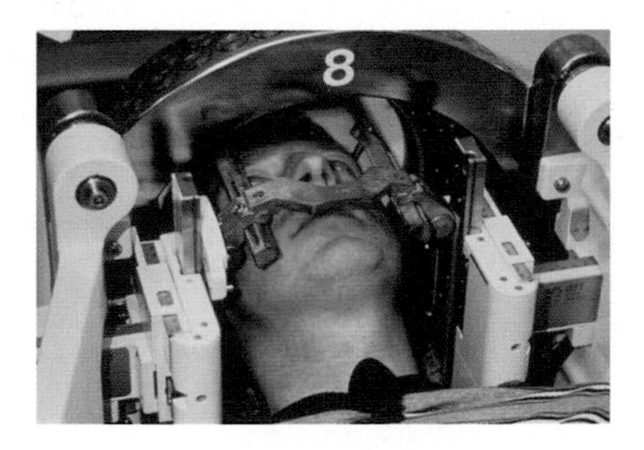

Figure 13.17 Gamma rays can be used to destroy cancer cells; the patient has cancer of the lymph nodes.

Questions

12 Describe an experiment to find the penetrating ability of of each type of radiation.

13 Explain why radioactive sources are kept in 'lead castles'.

14 Radioactive sources must be handled safely. How should this be done?

Activity

Should doctors give radiotherapy to a sick patient who has had an operation for cancer? They know that it will make the patient feel ill; it will prolong their life; but may cause additional complications. Set up a discussion group arguing the case for and against the decision.

The half-life of radioactive materials

What is a half-life?

When a radioactive material decays it gives off radiation. The process is completely random. We cannot predict when an individual atom is going to decay. There are billions of them in a sample of radioactive material. However, even in a very small sample, the odds are that some atoms will decay. We say that the **half-life** is the time taken for half the atoms to decay (Table 13.2).

Table 13.2 The half-lives of some useful radioactive materials

Element	Half-life	Uses
Americium-241	460 years	In smoke / fire alarms
Carbon-14	5 730 years	Occurs naturally in the atmosphere, absorbed by plants and animals; used for carbon dating
Technetium-99	6 hours	Medical tracer
Plutonium-239	24 000 years	Artificial, used in nuclear power industry

It is a bit like throwing a lot of dice at the same time. Some will turn up a six, but we cannot say when a particular dice will be a six. What we can be sure of is that the more dice there are, the more sixes we will get.

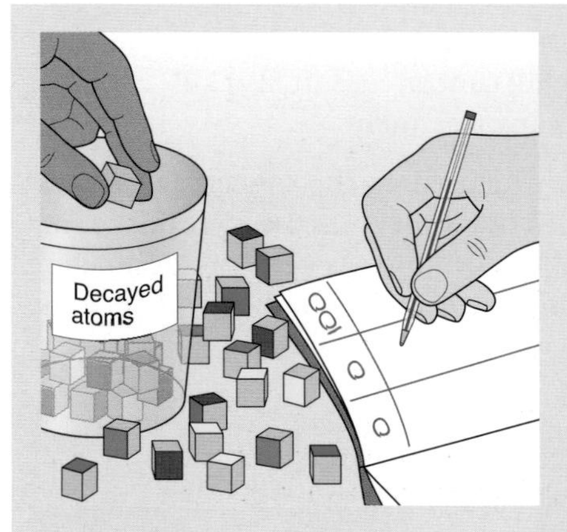

Figure 13.18 To find the 'half-life' of 100 dice

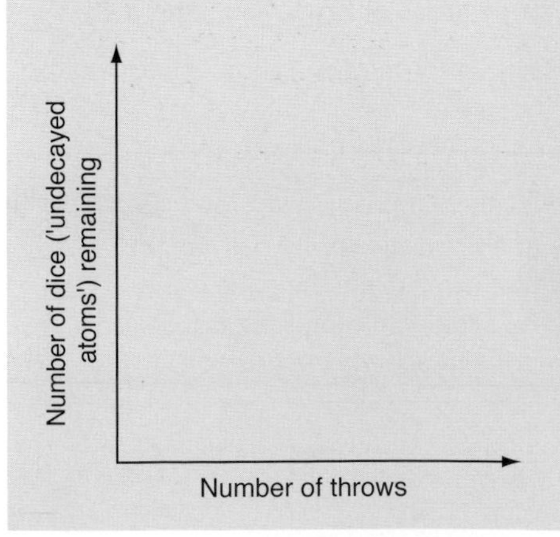

Practical work

The half-life of dice

We can use 100 dice or wooden cubes with one side marked, to simulate decay over a half-life.

1 Copy the table below.

Number of throws	Number of dice (atoms) decayed	Number of dice (atoms) remaining
0	0	100
1		

2 Put all the dice in the beaker. Carefully shake them onto the table.

3 Count those that have landed marked side up. These represent atoms that have 'decayed'. Remove them and record the results (Figure 13.18).

4 Put the remaining dice back in the beaker. Shake them out, count the 'decayed' ones and remove them as before.

5 Continue until there are no 'undecayed' dice left.

6 Draw a graph with axes as shown on the left.

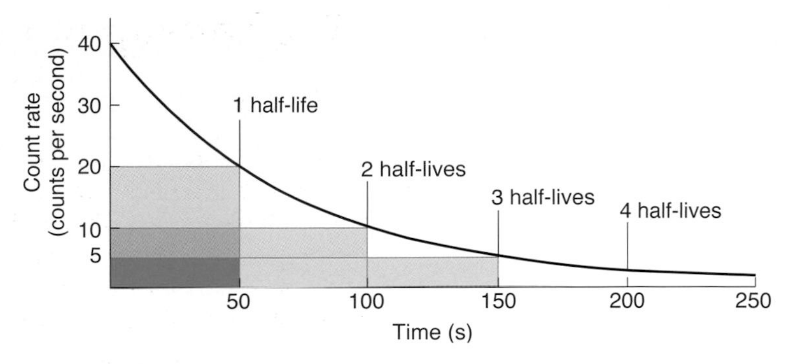

Figure 13.19 Decay of a radioactive material

The half-life is the time for half of the atoms (or dice) to decay. After one half-life, the **activity** (the number of nuclei that decay in one second) will have decayed to half its original value. After the second half-life, the activity will have decayed to half this value or one-quarter of its original value. After the third half-life it will decay to one-eighth of its original value.

The activity of radioactive substances is measured as **counts per second** (or counts per minute) on a Geiger counter. On the graph in Figure 13.19, the initial count rate was 40 counts/s. It takes 50 s for the count rate to fall to 20 counts/s, that is, one-half of its original value. It takes a further 50 s to halve again to 10 counts/s. The half-life is 50 s.

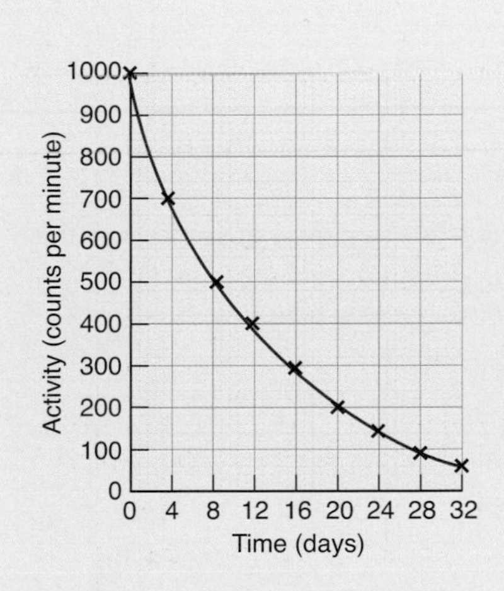

A radioactive decay curve for iodine-131

Questions

15 In Figure 13.19, what is the count rate after:
  a three half-lives
  b four half-lives?

16 The activity of a radioisotope is 3200 Bq. It has a half-life of 6 h.
  a What is its activity after 6 h?
  b What is its activity after 12 h?
  c What is its activity after 24 h?

17 In your dice experiment in the practical work above, how many throws of the dice were needed for half of them to 'decay'?

18 **a** What is the half-life of iodine-131 (see graph)?
  b What is the count rate after two half-lives?

19 The table on the next page shows the results of an experiment to measure the half-life of a radioactive material. The count rate was recorded every minute. The background count rate was 5 counts/s.
  a Complete the table.
  b Plot a graph of count rate (*y*-axis) against time (*x*-axis).
  c Work out the half-life of the source.

Count rate (counts/s)	Corrected count rate (counts/s)	Time (min)
105		0
64		1
39		2
25		3
17		4
12		5

20 a Three radioactive substances have to be stored safely. Details of the substances are given below.

Substance	Half-life (years)	Type of radiation given out
A	5000	Alpha (α)
B	4	Beta (β)
C	156	Gamma (γ), alpha (α)

Which of the following containers would you use for **each** substance:

i aluminium

ii thin plastic

iii lead-lined?

iv Give a reason for your answer to part iii.

b Copy and complete the table below for substance **B**.

Date	Mass of original substance left (kg)
1 March 1992	8
1 March 1996	
1 March 2004	

c A Geiger counter was used to measure the activity (in counts per minute) from a radioactive sample over a period of years. Over this period, the background radiation was regularly measured at 4 counts/min. The table of results is shown below.

Time (years)	0	1	2	3	4	5	6
Recorded activity (counts/min)	124	80	52	34	23	16	12
Activity due to sample alone	120						8

i Copy and complete the table giving the activity of the sample alone.

ii Explain what is meant by background radiation.

iii On graph paper, plot the values for the activity for the sample alone against time, and join them with a smooth curve.

iv Find the half-life of the substance from your graph.

Uses and dangers of radioactivity

Using radioactivity

Radioactive materials have many uses in our lives. Many of the things we use either contain radioactive materials or have been made by processes involving them. They have saved many lives both in their medical use and in essential safety devices such as smoke alarms.

Power generation

Radioactive materials are used in nuclear power stations. Nuclear power plants generate power by **nuclear fission** (the splitting) of an unstable atom. The atom releases neutrons which cause other atoms to undergo fission. This produces a lot of heat in a **chain reaction** (Figure 13.20).

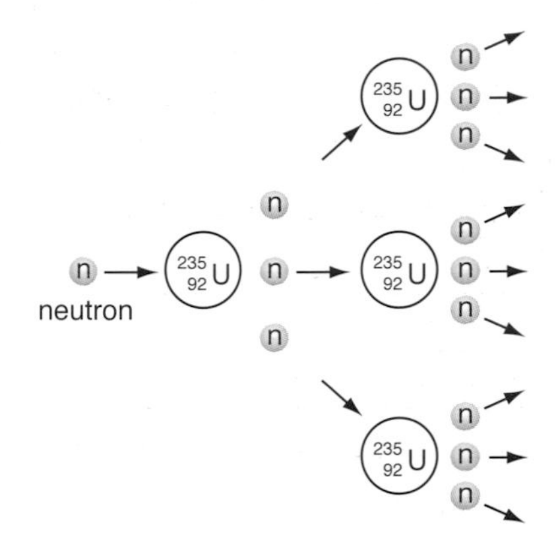

Figure 13.20 A chain reaction in uranium-235

The heat produced is used to generate steam which is used to generate electricity as in a fossil fuel power station.

Radioisotope thermoelectric generators

A radioisotope thermoelectric generator (RTG) is a simple device. It uses the heat from radioactive decay. The heat released is converted into electricity using a thermopile. It is a completely different process of heat generation from that used in nuclear power plants.

RTGs are used in unmanned spacecraft and in remote mid-ocean navigation buoys. They were once used to power heart pacemakers. Unfortunately there is a problem if the wearer dies and the pacemaker is not removed before cremation. Pacemakers now use much safer long-lasting batteries.

Radioisotope thermoelectric generators (go to *Nuclear facts* and then *Nuclear power in space*): www.nuc.umr.edu

Medical uses

You've already seen that gamma rays can be used to destroy cancer cells.

Radioactive tracers are used a lot in medicine. They can check that the internal organs of the body are working properly. The patient swallows or is injected with a radioactive substance. As it decays, the radiation is detected outside the body. You can see where the tracer becomes concentrated – or where it doesn't go (Figure 13.21). The radioactive isotope must have a very short half-life. Any radioactivity must decay quickly so that there is little chance of it damaging healthy cells.

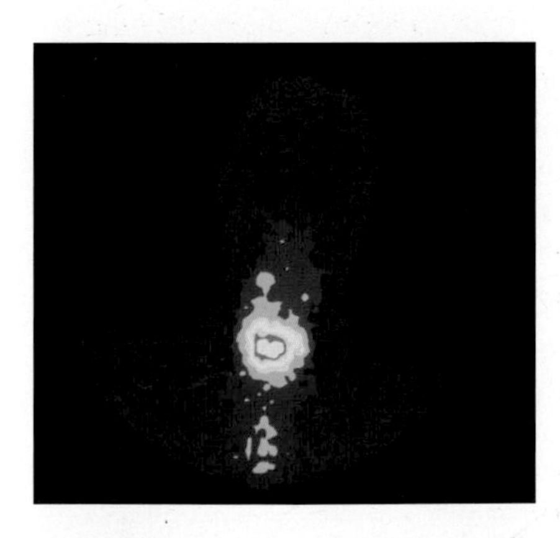

Figure 13.21 Scan of a person injected with iodine-131; this has shown that the patient has an enlarged thyroid gland.

Using a radioactive tracer can:

- detect problems with the heart
- detect bone cancers
- detect problems with kidneys
- find underactive thyroid glands in babies
- be used in research to find causes and cures for diseases like cancer, AIDS and Alzheimer's disease.

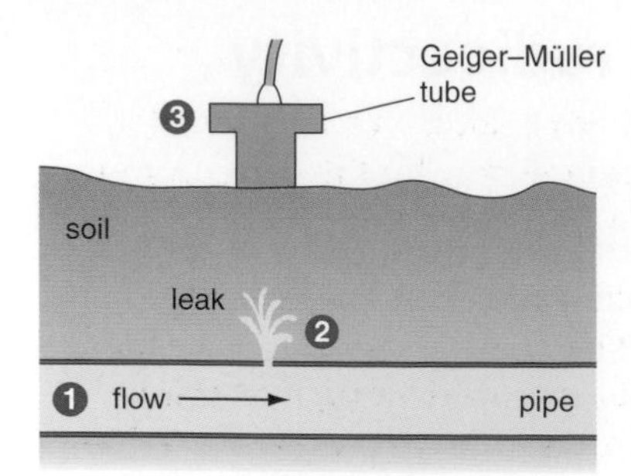

❶ Small amount of radioactive isotope (tracer) emitting penetrating γ rays is fed into pipe. Gamma rays must be used as they penetrate soil.

❷ Radioactive isotope leaks into soil.

❸ Geiger–Müller tube is used to detect radiation and position of leak.

Figure 13.22 Using a radioactive tracer to detect a leak in an underground pipeline

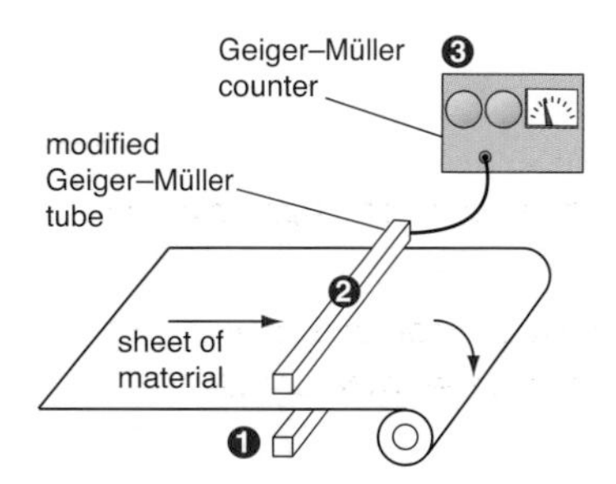

❶ Radioactive source emitting less penetrating beta radiation.

❷ Long, modified Geiger–Müller tube detects radiation penetrating the sheet of material.

❸ Geiger–Müller counter measures radiation level (the thicker the sheet, the lower the reading). Information is fed back to adjust the thickness of the material if necessary.

Figure 13.23 Thickness control using radioactive materials

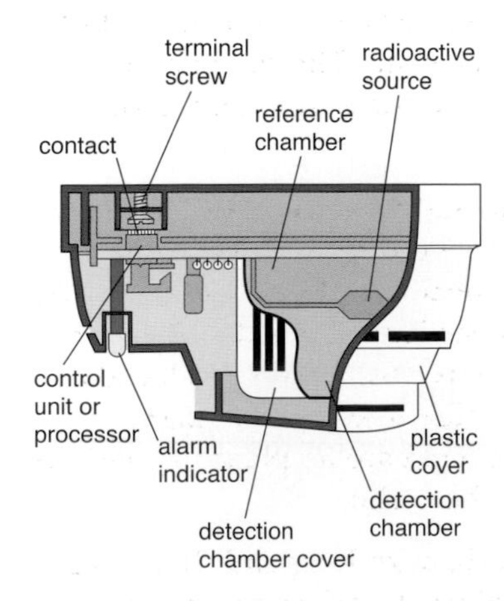

Figure 13.24 A smoke alarm

Other uses of radioactive tracers

Scientists use radioactive tracers to find out how plants absorb fertiliser, the best time to apply it and the right amount to use. The farmer uses the correct amount of fertiliser, saves money and cuts down on pollution. Other research helps scientists understand how plants get diseases and how to cure the diseases.

A leak in an underground oil or gas pipe can cause a lot of pollution. If a leak is suspected, a radioactive tracer is put in the pipeline. The operator, using a Geiger counter, walks or rides over the pipeline (Figure 13.22).

Thickness control

Sheet materials such as aluminium foil, paper, polythene and other plastic materials are made at high speed. The Geiger counter in the machine measures the amount of radiation passing through the material (Figure 13.23). The thicker the sheet, the less radiation will pass through to the detector. This information is used to keep the sheets at the correct thickness. Beta sources are the most useful because of their penetrating ability. Gamma rays are not stopped at all by such thin sheets of these materials, and alpha particles are stopped even by a very thin sheet.

Checking welds

Bridges, boilers, ships, submarines, pipelines and oil refineries are just some of many structures where sheets of thick steel are welded together. In many cases, a poor weld would be disastrous. Gamma ray sources and detectors are used to check the quality of the weld.

Sterilising to kill harmful bacteria

Medical instruments are placed in a package and sterilised with gamma radiation. Any harmful bacteria on the instruments and the inside of the package are killed.

Other products which are sterilised this way include: baby powder, cosmetics and contact lens solution. No chemicals are added to the products. They can be sterilised in their packaging. It makes them very safe to use. They do not become radioactive.

Food

Irradiation is an alternative to processing and sterilising some foods. When strawberries, onions, potatoes and spices are exposed to gamma radiation, harmful bacteria are killed. No chemicals are added and the food keeps its nutritional value. The food does not become radioactive.

Smoke alarms

Every house should have at least two smoke alarms (Figure 13.24). They have saved many lives.

The smoke alarm contains a tiny amount of americium-241. The source emits alpha particles which ionise the air in the alarm. This means that a small current flows. If smoke gets in the alarm it stops the current, and this switches on the loudspeaker circuit.

Radioactive dating

Rocks contain uranium-238. It decays to lead, with a half-life of 4500 million years. Scientists measure how much of the uranium has changed to lead. They then work out the age of the rock.

Radiocarbon dating is used on materials that have been alive. All plants and animals absorb carbon when they are alive. One carbon isotope, carbon-14, is radioactive; it is absorbed along with the more common carbon-12 isotope. When living things die they stop taking in carbon-14, and as it decays, so less and less carbon-14 remains in the sample. Scientists measure how much carbon-14 is left, compared with the amount of carbon-14 in living things today. Carbon-14 has a half-life of 5700 years. It can be used to date wood, paper, cloth and dead bodies. The age can be estimated within the range 1000–50 000 years.

Questions

21 List five uses of radioactive materials.

22 How are radioactive tracers used in:
  a medicine
  b plant research?

23 Explain how a smoke alarm works.

24 What technique do scientists use to find out the age of an Egyptian mummy?

25 Aled's dad has found out that his smoke alarm contains americium-241. It emits alpha (α) radiation. Should he be worried about the radiation? Explain away his fears.

Activity

Irradiated food: good for killing off the bacteria or is it just another way to let food suppliers get away with lower hygiene standards? Make out a case for both views. What is the opinion of the class?

The disposal of radioactive waste

You cannot get rid of radioactive materials by burning or making them react with other chemicals. Only time can reduce the radioactivity. Some materials have half-lives of thousands and even millions of years. The disposal of radioactive waste products is a problem. Most radioactive waste comes from nuclear power stations, but hospitals and industry also use radioactive substances and so produce low-level and intermediate-level radioactive waste.

Nirex is a company owned by the UK nuclear industry. It was set up to dispose of intermediate- and low-level waste. The radioactive waste has to be kept away from people and the environment for a very long time.

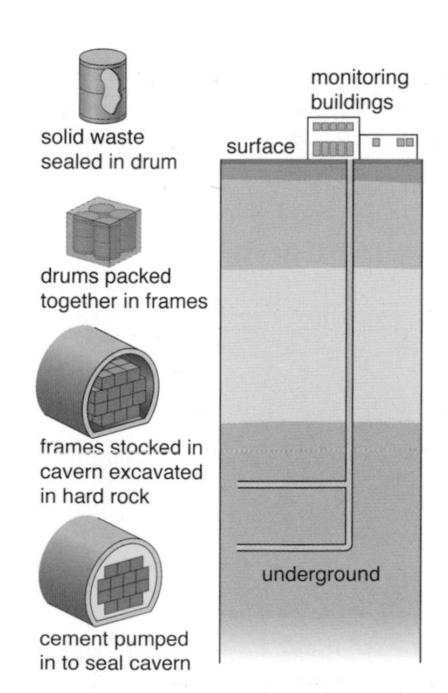

Figure 13.25 Radioactive waste could be stored underground like this

There is a danger that when radioactive waste is stored the radioactive substance might leak out and contaminate the ground and water supplies. Naturally no one wants to live near a nuclear waste dump. At present most of the low-level nuclear waste is disposed of at Drigg in Cumbria. High-level waste is currently stored where it is produced or where it is treated. There were plans to bury intermediate waste underground (Figure 13.25), but as yet there are no plans to build the underground store at Sellafield or anywhere else.

High-level waste contains a lot of material with long and short half-lives. It comes from the reprocessing of used nuclear fuel from

Waste disposal site at Sellafield, Cumbria (go to *About* and then *Background information on radioactive waste*):
www.nirex.co.uk

Dealing with waste (go to *Physics*, 14–16):
www.schoolscience.co.uk

A world of atoms (go to *Physics*, 14–16):
www.schoolscience.co.uk

power stations. It is still very radioactive. The radiation makes it very hot. It must first be cooled in large tanks. It is then sealed into a type of glass inside a steel container. It must be kept in an air-cooled store for 50 years or more. Some people are very worried that we are leaving our problems for future generations to deal with.

Activity

There are rumours that a radioactive waste disposal site for low-level waste only is to be built near you. It will be discussed at the next council meeting. In your group, role-play the discussion between those in favour and those against.

Questions

26 How is low-level waste contained?

27 How must high-level waste be treated before it can be stored?

28 Workers in nuclear power stations wear special badges. These measure the nuclear radiation that a worker may be exposed to. The badge consists of a photographic film behind different covers as shown in the figure (left).

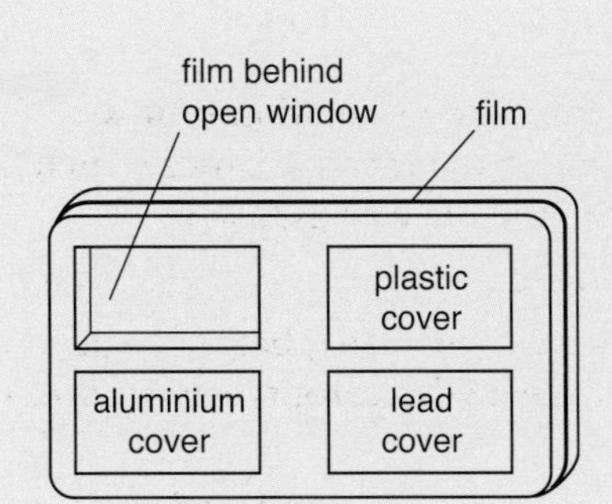

- **a** State what type of radiation, alpha, beta or gamma will be stopped by the plastic.
- **b** What type of radiation will pass through the plastic but will be stopped by the aluminium?
- **c** What type of radiation could pass through all three materials?

29 Technetium-99 is a gamma (γ) emitter with a half-life of 6 h. It is used to study blood flow around a patient's body. A small dose of technetium-99, with an activity of 120 counts/min, is injected into a patient's bloodstream.

- **a** Estimate the activity of the technetium-99 in the patient's blood 12 h after the injection.
- **b** The patient returns for another examination a week later. Explain why another dose of technetium-99 has to be given.
- **c** State two ways in which gamma (γ) radiation is different from other radioactive emissions.

30 Radioactive sodium is a beta (β) emitter and has a half-life of 15 h.

- **a** Explain what you understand by the term beta emitter.
- **b** The background radiation in a laboratory is 20 counts/min. Owing to a leakage of radioactive sodium (half-life 15 h), the count rate rose to nine times the background level.
  - **i** Calculate the count rate in the laboratory after the leakage.
  - **ii** Calculate the count rate from the leaked sodium alone.
  - **iii** On graph paper draw, as accurately as possible, the decay curve for the activity of the leaked sodium. (Use the axes shown on page 271 and a suitable scale.)
- **c** A count rate detector, placed in the laboratory, reads the activity in the laboratory. Explain why the detector gives a reading of 60 counts per minute after 30 hours.
- **d** If the safety level is 1.5 times the background reading, show on your graph the time when it would be safe to enter the laboratory.

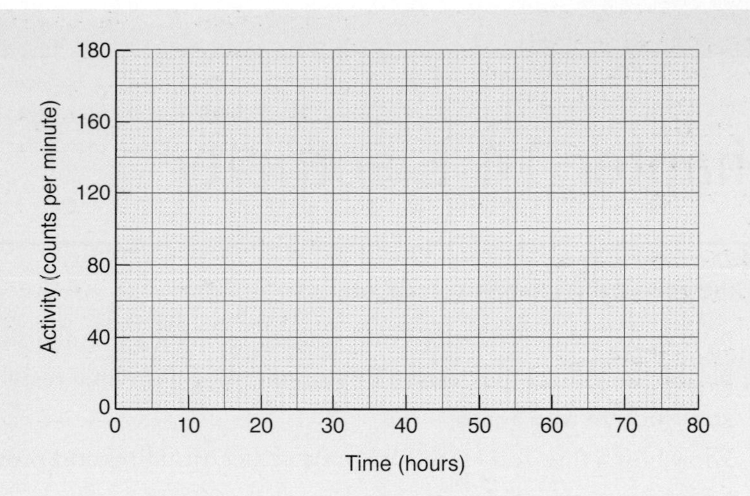

Summary

1. We are exposed to background radiation. Most of it is natural; the rest usually comes from medical sources such as X-rays. Only 1% comes from the nuclear industry.
2. Radiation comes from unstable nuclei.
3. The three types of radiation are alpha (α), beta (β) and gamma (γ) radiation. They are all dangerous, some more so than others.
4. Our biggest source of background radiation is radon. It is a gas which comes from the ground.
5. Radon detectors are available and simple measures can remove the gas from houses.
6. Alpha particles are helium nuclei. They are stopped by paper or a few centimetres of air. They are dangerous when ingested.
7. Beta particles are fast-moving electrons from the nucleus. They are stopped by a few millimetres of aluminium.
8. Gamma rays are high-energy electromagnetic radiation. They are very penetrating. Some will even get through lead.
9. All radioactive substances take time to decay.
10. The half-life is the time it takes for half the atoms to decay.
11. The half-lives of radioactive materials can vary from very small fractions of a second to millions of years.
12. There are many uses of radioactivity because of its penetration ability, effect on living things, or decrease in activity with time.
13. The disposal of radioactive waste presents a problem for us and for future generations.
14. Intermediate-level waste from materials with a long half-life will have to be stored for many tens of years or more.
15. High-level waste has to be cooled before being made into a glass-like substance. It has to be stored in safe conditions for a very long time.

Chapter 14 Electricity

By the end of this chapter you should:

- be able to make calculations of resistance from measurements of current and voltage;
- be able to control the current in a circuit using a variable resistor;
- know how to wire a plug;
- know how a fuse and earth wire prevent electrical fires and protect people from electric shocks;
- be able to choose the correct fuse for an appliance;
- understand the difference between fuses and circuit breakers;
- discuss why different countries have different voltage electricity supplies.

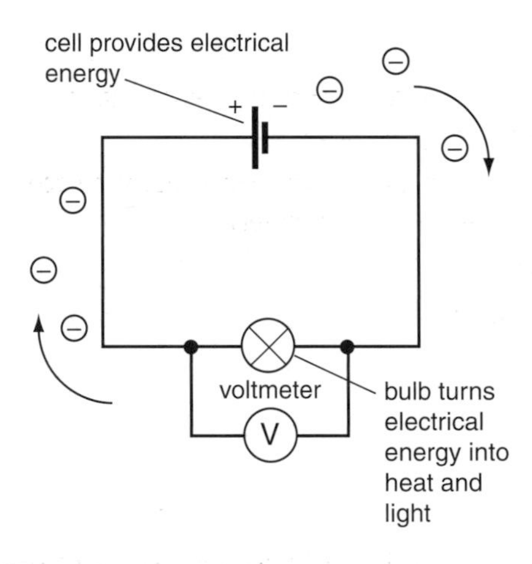

Figure 14.1 A voltmeter measures voltage.

Figure 14.2 A digital voltmeter (left) and a digital ammeter (right).

Simple electrical circuits

Measuring voltage

A flow of electric charge (electrons) is called an electric **current**. Energy is needed to make the electrons flow round the circuit. The chemicals in the cell have chemical energy. In a closed circuit, this energy is transferred to the electrons.

Electrical potential energy is measured with a **voltmeter** (Figures 14.1 and 14.2). We say that there is a **voltage** across the cell. The unit is the **volt**.

The greater the voltage of a cell or battery the more energy it is able to give to the lamps or other components in the circuit. Car batteries usually give 12 V. The cell in a wristwatch might be only 1 V.

Voltmeters can be either digital or analogue. The voltmeter must be connected the right way: the + side of the voltmeter to the + terminal of the battery.

Measuring electric current

We measure electric current with an **ammeter** (Figure 14.3). The ammeter must always be connected in series. The unit of electric current is the ampere (**amp** for short).

There are two types of ammeter: digital and analogue. The + side of the ammeter must always be connected to the + terminal of the cell or battery.

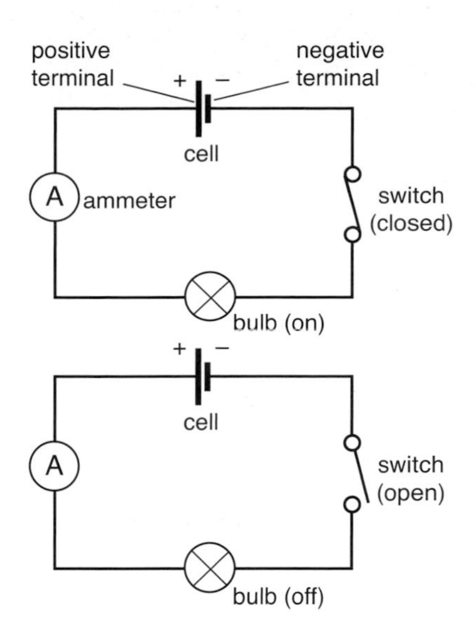

Figure 14.3 An ammeter measures electrical current.

cell/battery		ammeter	
indicator lamp		voltmeter	
filament lamp		microphone	
switch		bell	
resistor		buzzer	
variable resistor		loudspeaker	
diode		motor	

Figure 14.4 Circuit symbols

Prefixes

These are used to describe very large or very small currents and voltages.

$$1\,\text{kV (kilovolt)} = 1000\,\text{V}$$

$$1\,\text{mV (millivolt)} = \frac{1}{1000}\,\text{V}$$

$$1\,\mu\text{A (microamp)} = \frac{1}{1\,000\,000}\,\text{A}$$

Practical work

Measuring current and voltage in a circuit

1 Set up a circuit containing a battery and a lamp.
2 Use an ammeter to measure the current in the circuit.
3 Use a voltmeter to measure the voltage in the circuit.

Questions

1 Draw a circuit diagram of a 12 V battery connected in series with a lamp, a fuse and an ammeter (see Figure 14.4).
2 Draw a circuit diagram of a 12 V battery in series with a motor, switch and fuse. A voltmeter is connected across the battery.
3 How many milliamps are there in 1 A?
4 How many volts are there in 10 kV?
5 What is a microvolt?

Resistance

Suppose that you put a lamp in series with a cell and note its brightness. Now put another lamp in series with the cell. The lamps are not as bright as with one lamp in the circuit. The very thin wire in the filament of the lamp reduces the current. We say the thin wire has a **resistance**. Putting two lamps in series increases the resistance even more.

Practical work

1 Connect a lamp in series with a switch and an ammeter.
2 Record the reading on the ammeter.
3 Add a second and then a third lamp to the circuit.
4 Note the reading on the ammeter each time.

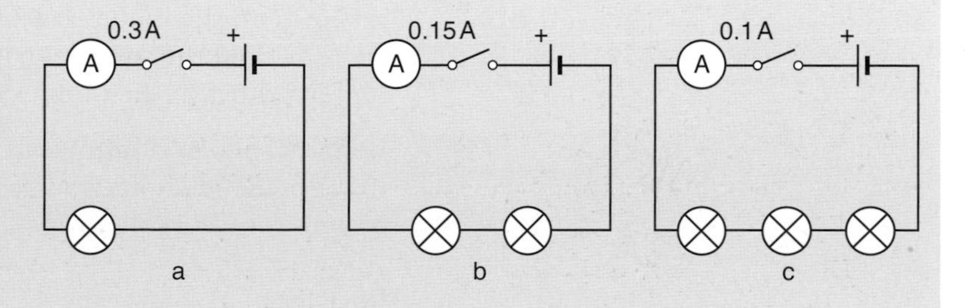

Figure 14.5 Increasing the resistance decreases the current.

By joining the lamps in series we have increased the length of the thin resistance wire in the circuit.

- What were the readings on your ammeter?
- What happens to the current in a circuit when the resistance is increased?

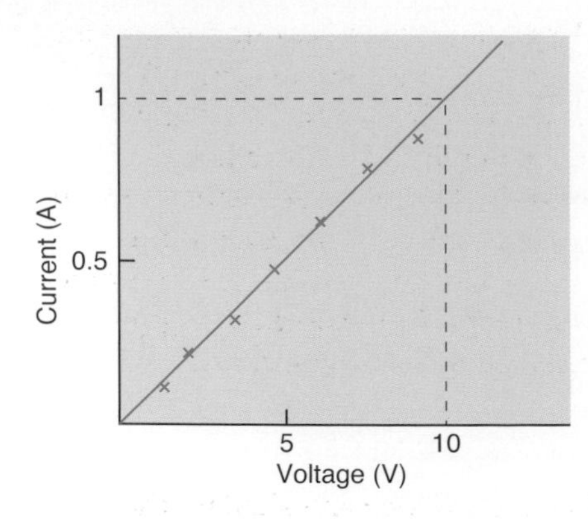

Figure 14.6 Current–voltage graph for ohmic resistor

Ohm's law and resistance

In the 1820s, a German physicist called Georg Ohm investigated the resistance of various materials. The unit of resistance, the **ohm** (symbol Ω), was named after him. He discovered that the current in a metal wire is proportional to the voltage across it (providing the temperature remains constant). This result is known as Ohm's law.

Proportional means that if you double the voltage, the current is doubled. The graph in Figure 14.6 is a straight line through the origin.

Measuring resistance

The circuit in Figure 14.7 shows how an ammeter and voltmeter can be used to measure resistance. To find the resistance, use the following formula.

$$\text{resistance }(\Omega) = \frac{\text{voltage (V)}}{\text{current (A)}}$$

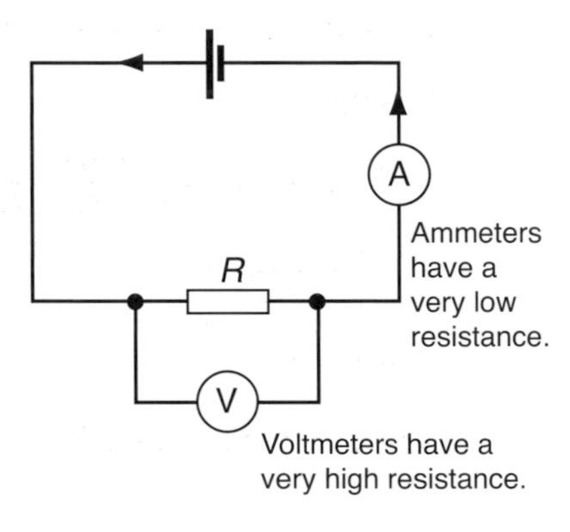

Figure 14.7 Measuring resistance with an ammeter and voltmeter

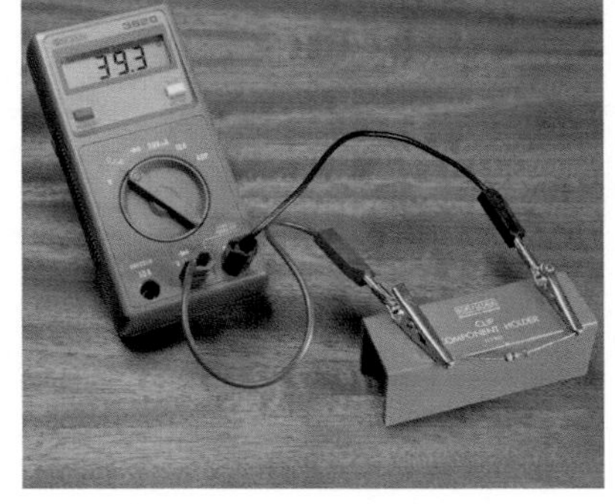

Figure 14.8 A multimeter can be set to read voltage, current or resistance.

Using an ohmmeter (multimeter)

A more convenient way to measure resistance is to use a multimeter (Figure 14.8). When it measures resistance, it uses its own internal battery. The meter has a scale marked in ohms. When a low resistance is connected there is a large current and a low resistance is shown on the meter scale. If a high resistance is connected the current from the internal battery is very small and the meter gives a high reading on the ohms scale.

Before using an ohmmeter, briefly touch the leads together to check that the meter shows zero resistance, indicating that the internal battery is in working order.

Resistors

Resistors are useful components. They can be used to reduce the current in a circuit. Wire resistors are sometimes made with nichrome wire. Nichrome is an alloy of the metals nickel and chromium.

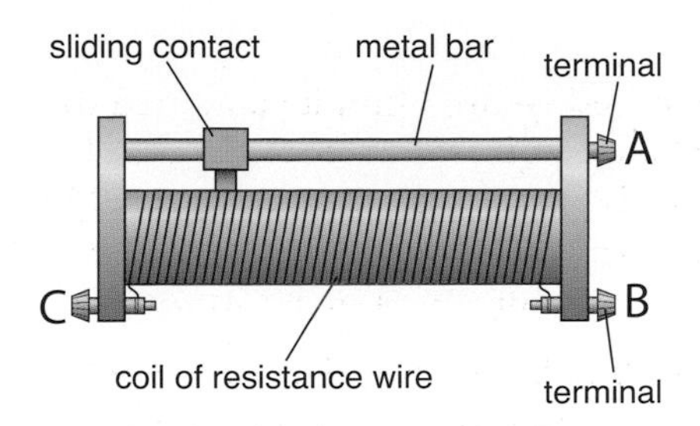

Figure 14.9 A variable resistor

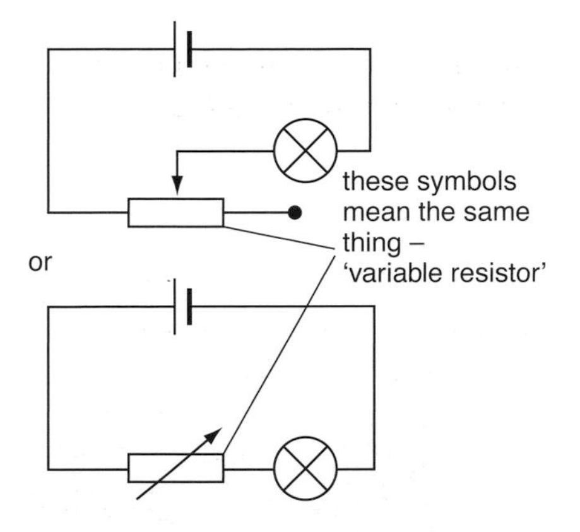

Figure 14.10 These circuits show the symbols for a variable resistor.

A rheostat is a variable resistor. A resistance wire is wound round an insulating core. The ends are connected to terminals C and B. Connecting to terminals B and C will provide a fixed resistance. Terminal A is connected to a thick metal bar that has a low resistance. The sliding contact joins the bar to the resistance wire. If now terminals A and C are connected to the circuit, then if the slider were halfway along the bar only half the resistance wire would be in the circuit. With the slider in the position shown in Figure 14.9, there is only about one-quarter of the total resistance.

Calculating resistance

If we measure the voltage and the current, we can calculate the resistance from the formula given earlier:

$$\text{resistance } (\Omega) = \frac{\text{voltage (V)}}{\text{current (A)}}$$

Example

A 12 V car battery is connected to a car headlamp. The current is 6 A. What is the resistance of the headlamp?

Formula first:

$$\text{resistance } (\Omega) = \frac{\text{voltage (V)}}{\text{current (A)}}$$

Put the numbers in:

$$\text{resistance} = \frac{12}{6}\,\Omega$$

The resistance, $R = 2\,\Omega$

Questions

6 A 12 V car battery is connected to a car sidelight. A current of 1 A flows. What is the resistance of the sidelight?

7 Gwyn's Gran has an electric scooter. It has a 24 V motor and draws a current of 5 A. What is the resistance of the motor?

Practical work

Investigating how current changes with voltage for a resistor at constant temperature

1. Connect the circuit as shown in Figure 14.11.
2. Adjust the variable resistor so that there is a very small current in the circuit.
3. Record the current and voltage in a table.
4. Change the variable resistor so that the voltage across the resistance wire increases a little.
5. Record the current and voltage.
6. Repeat a few more times. Carefully check that the wire does not get warm.
7. Draw a current–voltage graph. It should look like Figure 14.6.

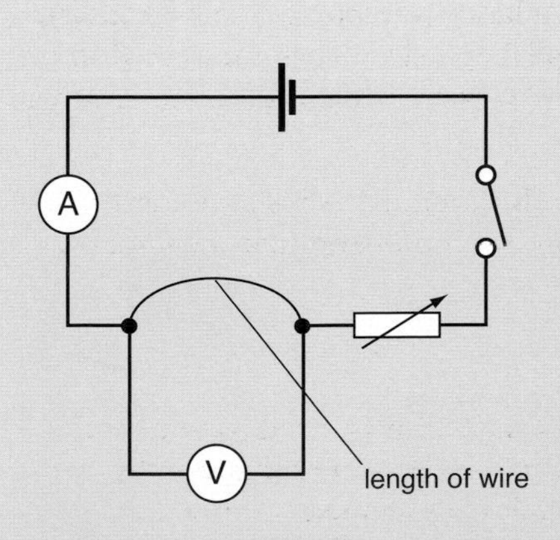

Figure 14.11 Apparatus to show how current changes with voltage

Question

8 Rhys measured the voltage across the ends of a conductor as the current through it was changed. The table shows the measurements from his ammeter and voltmeter.

Voltage (V)	1.0	2.0	3.0	4.0	5.0	6.0
Current (A)	0.5	1.0	2.0	1.9	2.5	3.1

Use his results to plot a graph of voltage (y-axis) against current (x-axis). Draw the line of best fit.

a What law may be deduced from the graph?

b Use the graph to find the resistance of the conductor.

Resistors get hot

When a current flows through a wire, the wire will get hot. For the same current, the greater the resistance the hotter it will get. The thin wire in a lamp has a high resistance The heat produced by the current is enough to make it glow white-hot. It gives out light as well as a lot of heat.

Practical work

Drawing a current–voltage graph for a filament lamp

1. Connect a circuit as shown in Figure 14.12.
2. Adjust the variable resistor so that there is a very small current in the circuit. The lamp should not light up.
3. Record the current and voltage.
4. Use the variable resistor to increase the current.
5. Repeat until the lamp glows brightly.
6. Each time record the voltage and current in a table.
7. Draw a current–voltage graph. It should look like Figure 14.13.

You might think that, since the wire in the lamp is made from metal, the graph should be a straight line, but it is most definitely a curve. This is because Ohm's law applies only when the temperature is constant. In the filament lamp the wire gets hotter as the current increases. The resistance of the wire increases with temperature. A filament lamp does not obey Ohm's law. It is a non-ohmic conductor.

Drawing a current–voltage graph for a diode

1. Repeat the experiment with a diode (Figure 14.14).

This gives a completely different type of graph (Figure 14.15). If the voltage is applied in the reverse direction, very little current flows. A current flows in the forward direction only when the voltage is greater than 0.6 V.

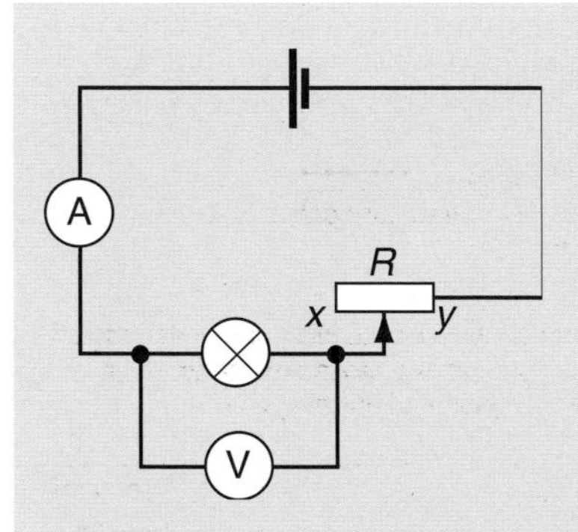

Figure 14.12 Filament lamp circuit

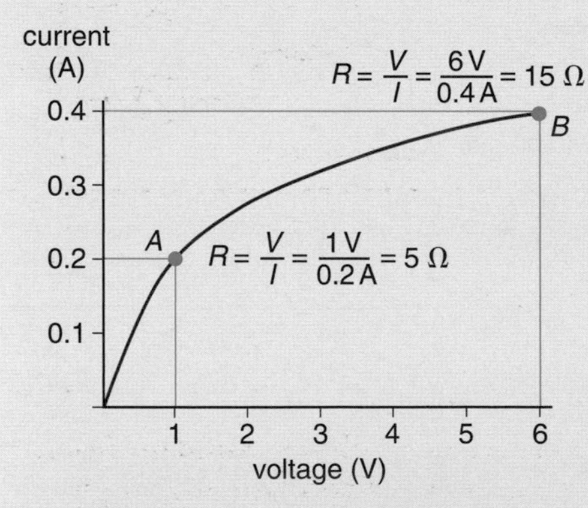

Figure 14.13 Current–voltage graph for a light bulb

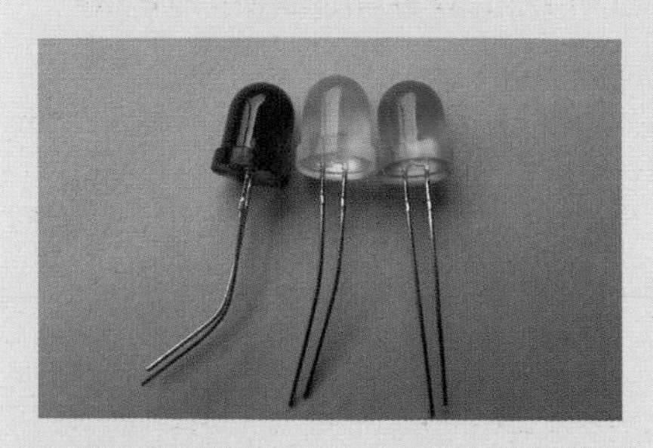

Figure 14.14 Red, yellow and green LEDs are now being used in traffic lights.

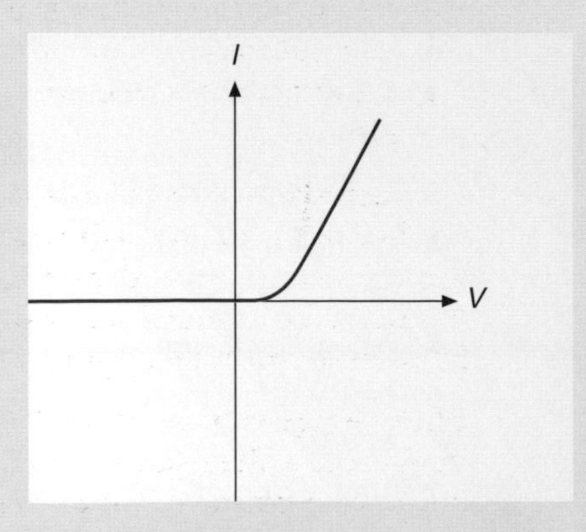

Figure 14.15 Current–voltage graph for a diode

Safety features in mains circuits

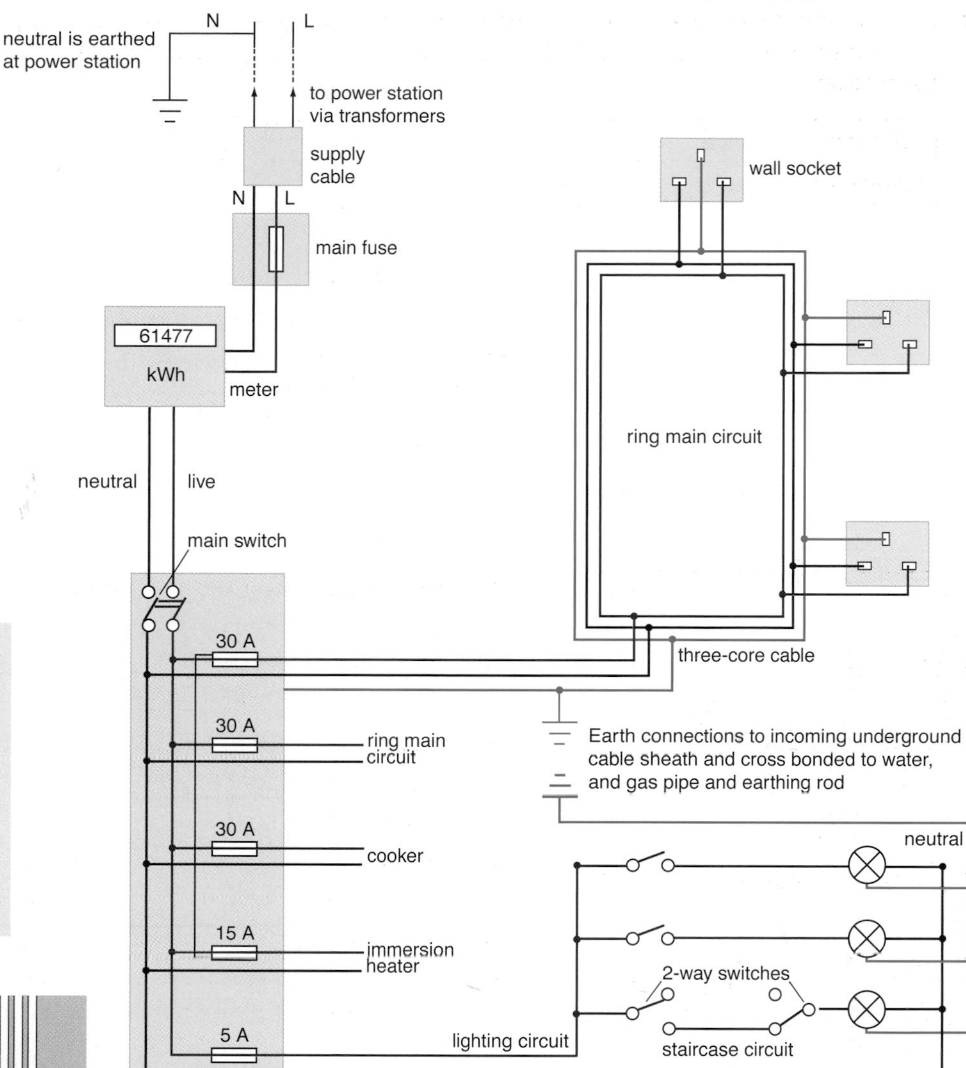

Figure 14.16 The electricity supply for a house

Questions

9 What happens to a resistor when a large current flows through it?

10 Explain the meaning of the term 'a non-ohmic resistor'.

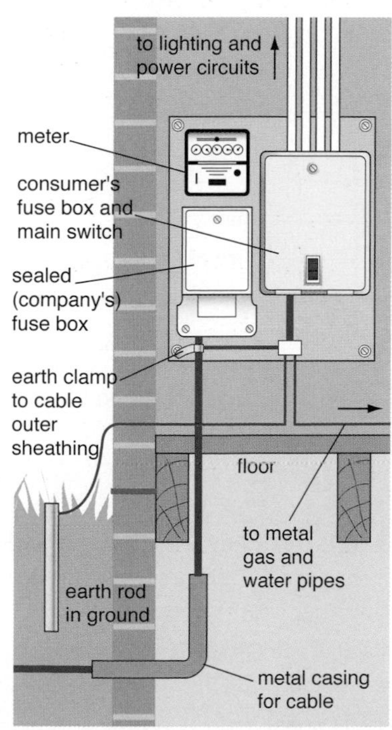

Figure 14.17 Protective multiple earth connections are there for your safety.

The mains supply to our homes

The mains supply in our homes passes through the company fuses, meter and a consumer unit (Figure 14.16). There are two supply leads. One, called the **neutral** (symbol N), is connected to earth at the local substation. The other, the **live** (symbol L) carries alternating current. It changes from positive to negative 50 times per second.

For safety, your house is also fitted with an **earth wire**. It is earthed by connecting it to an earth connection on the supply cable and/or to a special copper rod driven into the ground. It will also be connected to a metal water or gas pipe which enters your house from underground (Figure 14.17).

The average voltage of the UK electricity supply is 230 V. This is a dangerous voltage and could kill you, and so the wires must be covered in plastic or other insulating material. An **insulator** has a very high resistance.

Figure 14.18 Double insulation symbol

Some appliances do not need an earth connection. They have a double layer of insulation to protect you. A double-insulated appliance will have the symbol shown in Figure 14.18 on it.

Inside your house, separate circuits distribute the electricity. Two or more ring main circuits will supply all the wall sockets. Each circuit is protected by its own (30 A) fuse or circuit breaker. Ring main circuits carry an earth wire for safety. Lighting circuits carrying a smaller load have a thinner wire and a 5 A fuse. New installations will also carry an earth wire.

When replacing the fuse:

- always turn off the supply at the main switch
- always replace the fuse wire with genuine fuse wire of the correct rating
- if in doubt, consult a qualified electrician.

Flexible wires to appliances

The wires are colour-coded. The live wire is covered with brown plastic insulation, the neutral with blue. The earth wire has green and yellow stripes. If the outer insulation is damaged, the wires must be shortened or replaced. Never use a lead with the inner coloured wires showing. If you have to wire a plug:

- make sure that the wires are in the right place; looking at the rear of the plug with the cover off, it's bRown, Right, with bLue on the Left
- do not strip the insulation back too far
- make sure the screws are tight and all the wire trapped in the hole
- make sure the outer insulation is under the flex grip and secure
- choose the right fuse
- if in doubt, consult a qualified electrician.

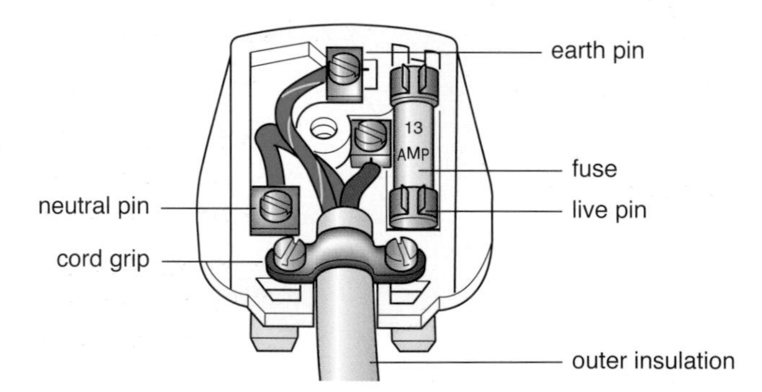

Figure 14.19 A correctly wired plug

The **fuse** contains a very thin wire. If too large a current flows, it will heat up and melt or 'fuse'. The outer glass cartridge protects it from mechanical damage.

Safety and the earth wire

If the appliance has a metal case there is a risk that it could become 'live' if the live wire touches it. Should this happen, anyone touching the case would get an electric shock which could

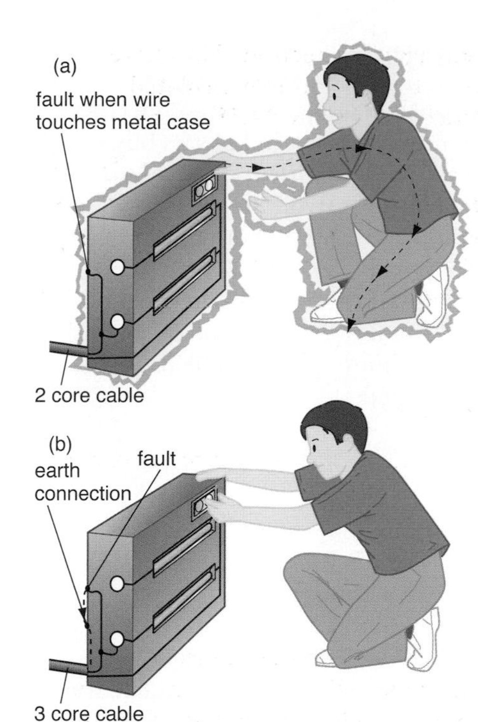

Figure 14.20 Earthing a metal case

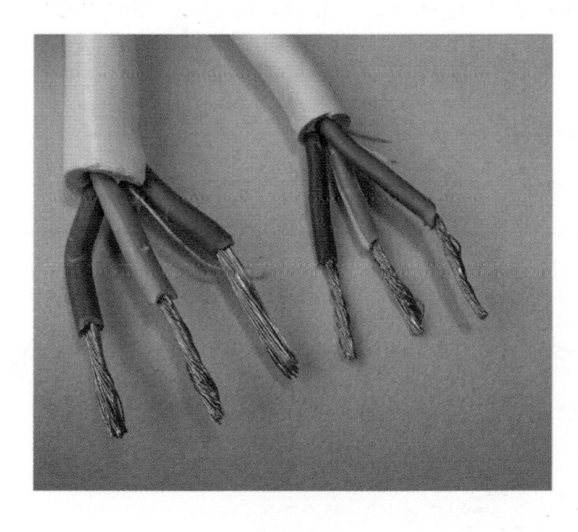

Figure 14.21 Core wires of various diameters and current ratings

kill. To prevent this, the metal case is connected to the earth wire in the plug (Figure 14.20). Now if the live wire touches the case, the current flows directly to earth. The large current 'blows' the fuse, cutting off the supply.

Fuses protect the wires

Appliances are supplied fitted with a flexible lead (wire) of the right diameter (thickness) (Figure 14.21). If the appliance develops a fault and too high a current passes through the wire, it could catch fire. But a fuse of the correct size will blow and prevent this happening. The same is true of the fuses in the main distribution box. Always replace the fuse with one of the same value.

Remember, the fuse is there to protect the wires from overheating. It will not protect the appliance from developing a fault. A fuse will *not* protect you from getting a shock if you touch the live wire.

Which fuse to use?

Domestic appliance fuses usually have one of these values: 3 A, 5 A or 13 A. Suppose that the appliance is rated at 1 kW. Which fuse should you use?

Formula first:

Remember from Chapter 10 that:

power = voltage × current

So:

$$\text{current (A)} = \frac{\text{power of appliance (W)}}{\text{voltage of appliance (V)}}$$

Put the numbers in:

$$\text{current (A)} = \frac{1000\ \text{W}}{230\ \text{V}}$$

The answer is:

current = 4.35 A

This is the current taken by the appliance when it is used normally. Therefore a 5 A fuse should be used to protect the wire to the appliance, which should be flexible.

Safety note: a damaged wire should always be replaced with the same or a higher rating. Never use a thinner wire, or insulation tape. If in doubt consult an electrician. Mistakes can kill.

Questions

11 Why are electrical wires covered in plastic?

12 A 13 A plug is fitted with a fuse. What is the job of the fuse?

13 What are the colours of the live, neutral and earth wires?

14 A fan heater is rated at 1500 W. What size of fuse should be fitted in the plug?

15 What is a 'ring main' circuit?

Figure 14.22 Always follow the safety instructions for extension leads.

Extension leads wound on drums

Long extension leads are wound onto drums for convenience. As any appliance could be plugged into it, you should always buy an extension lead with 13 A wire. It will be safe with appliances up to the maximum of 3 kW. You should always unroll the entire length of the wire when using the extension lead with large motors or heaters. This is because the wire will heat up when a current flows. When the wire is tightly wound inside the drum, it cannot cool down. Overheating will make it melt and catch fire. Good quality extension leads will have a label stating the maximum power to be used, both when they are coiled and when uncoiled.

Circuit breakers

Miniature circuit breakers (MCB) are used in modern consumer units (Figure 14.23). They replace fuses. The circuit breakers contain an electromagnet. When the current exceeds the rated value, the strength of the electromagnet is enough to separate the contacts. The circuit is switched off. Circuit breakers are better than fuses because they can be reset and act faster than fuses. Circuit breakers, like fuses, do not provide protection should you touch a live component.

Figure 14.24 A residual current device

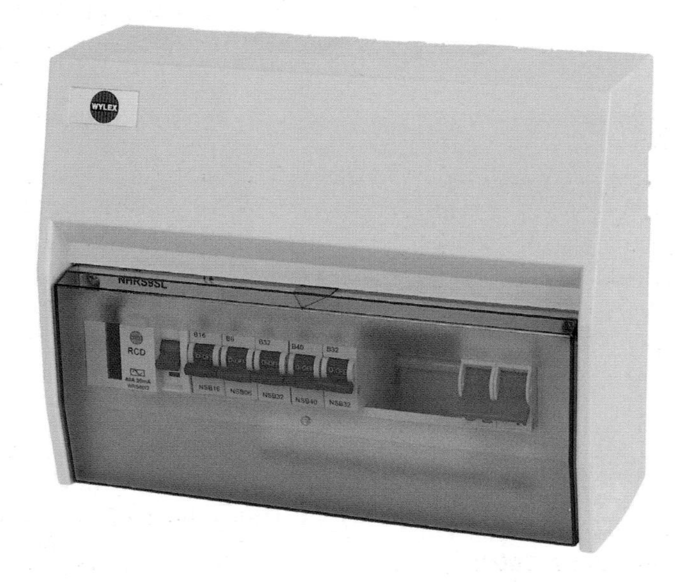

Figure 14.23 A typical consumer unit with resettable circuit breakers

Questions

16 Aled's dad has a small 20 m lawn, an electric lawnmower and a 50 m extension lead. Name a piece of electrical equipment he should buy. Outline how he should use all the equipment safely.

17 Name two devices used to protect wires and cables from overheating.

Residual current devices (RCDs)

These devices are designed to help protect you against electric shock (Figure 14.24). This type of circuit breaker measures the current in the live and neutral wires. If everything is working correctly, then these two are equal.

Suppose that you accidentally touch the live wire. Some current will flow through your body to earth. There will be a smaller current in the neutral wire compared with the live. The RCD will detect this difference in current and immediately throw the switch to break the circuit. The RCD is very sensitive. It can detect a current difference of 30 mA ($\frac{30}{1000}$ A). It will switch off in a few thousandths of a second.

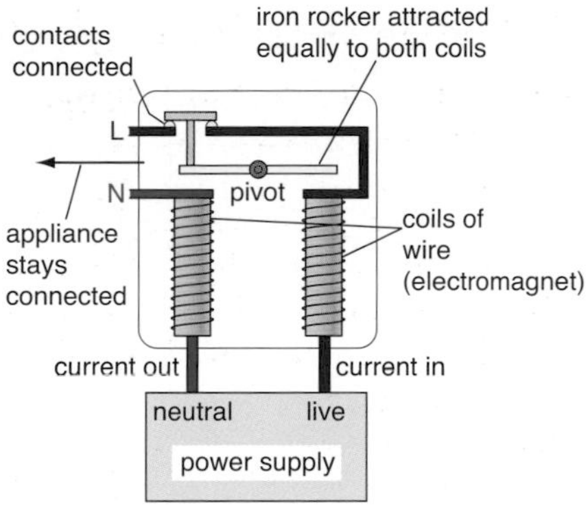

contacts connected
iron rocker attracted equally to both coils
L
N
pivot
appliance stays connected
coils of wire (electromagnet)
current out
current in
neutral
live
power supply

Figure 14.25 Residual current device when there is no leak to earth

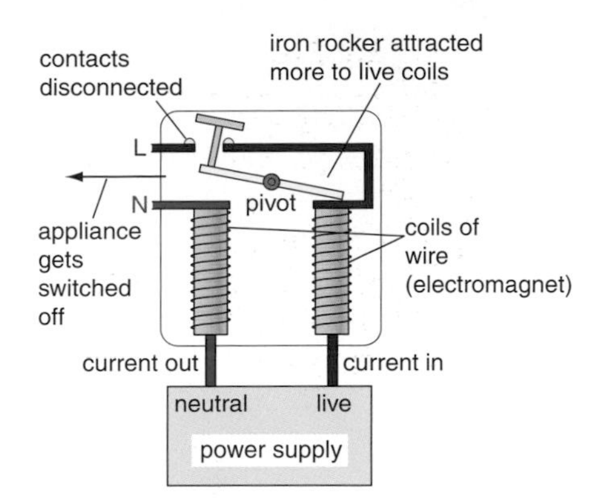

Figure 14.26 When there is a leak to earth, there is less current in the neutral wire, so the iron rocker is more strongly attracted to the live coil; the movement of the rocker disconnects the circuit.

You will get a shock, but the device cuts off quickly; the shock will not kill you. Always use the test button to check that the RCD is working correctly. Always use an RCD when working with power tools. You should always use an RCD device when using any electrical appliance outdoors.

Activity

If a 230 V electricity supply is dangerous, should we change to the safer 120 V as used in some other countries? Discuss the consequences of such a change and the arguments for and against it.

Why can we not use a lower voltage?

The simple answer is that the UK and most of Europe has always used a mains supply of 230 V. To change to a lower supply voltage now, with so many millions of appliances already in use, would be far too expensive. We would need a lot of transformers to make existing appliances work on a lower voltage. New appliances would need thicker flexible wires – remember a lower voltage means higher current to get the same power.

Many building site contractors in the UK use power tools that work at 110 V. It is a much safer voltage for use in damp conditions. A special transformer is used to convert from the existing 230 V. A lower voltage decreases the risk of an accident being fatal.

Not all countries supply their domestic electricity at 230 V. The US has opted for the lower and some say safer 120 V. Because of this, some makers of electrical equipment fit a voltage changeover switch so that appliances can be used in different countries. Usually the switch has two settings: 230 V or 120 V. Using the 120 V setting in the UK will destroy the appliance.

Circuit breakers (go to *Electronics* and then *Circuit breakers*): www.howstuffworks.com

Electricity (go to *Site map* and then *Copper and electricity*): www.schoolscience.co.uk

Questions

18 Circuit breakers have been developed to protect the users of electrical equipment.

a State **two** advantages of circuit breakers.

b Miniature circuit breakers (MCB) perform the same function as fuses. They are located in the live wire. Explain how miniature circuit breakers protect electric circuits and equipment.

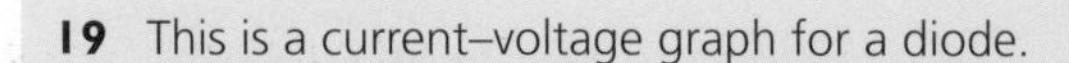

19 This is a current–voltage graph for a diode.

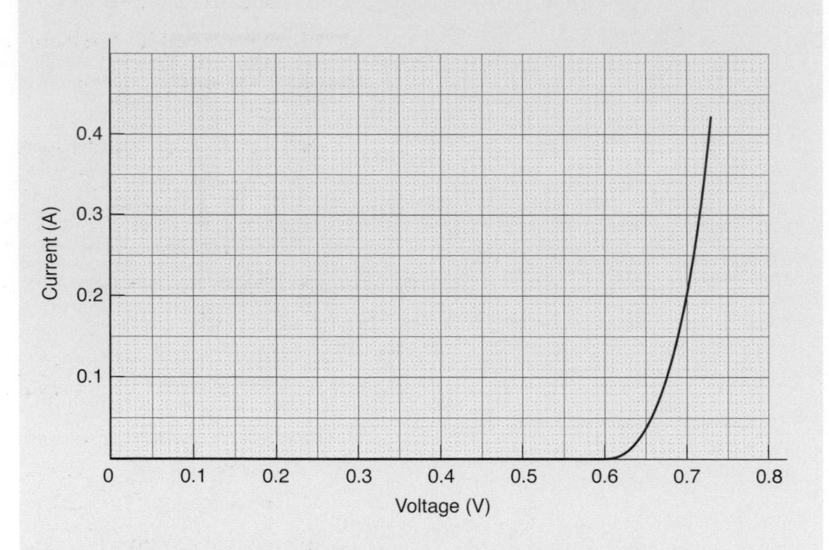

a What is the current at 0.3 V?

b What is the voltage when the current is 0.2 A?

c Write down the equation connecting **resistance**, **voltage** and **current**.

d Calculate the resistance of the diode when the current is 0.2 A.

Summary

1. An electric current is a flow of charge.
2. Electric current is measured using an ammeter connected in series with the circuit.
3. The amp is the unit of electric current.
4. The voltage across a component is measured in volts using a voltmeter. The voltmeter must be connected in parallel with the component.
5. A resistance reduces the flow of current in an electric circuit.
6. The unit of resistance is the ohm (symbol Ω).
7. Resistance can be calculated using the following formula.

$$\text{resistance } (\Omega) = \frac{\text{voltage (V)}}{\text{current (A)}}$$

8. The longer a wire, the greater is its resistance.
9. The thinner a wire, the greater is its resistance.
10. If the current–voltage graph of a resistor is a straight line through the origin, then the resistor obeys Ohm's law.
11. Some resistors do not obey Ohm's law. The current–voltage graph is not a straight line.
12. The cables supplying electricity round our homes have three wires: live, neutral and earth.
13. The earth wire is connected to an earth point on the supply cable and is cross-bonded (connected) to a metal water or gas pipe or to a metal stake.

14 The flexible wires to appliances are colour coded: brown = live, blue = neutral, green and yellow = earth.

15 Always buy extension leads fitted with 13 A wire.

16 Always fully unwind the extension lead from its containing drum.

17 A fuse is a very thin wire which melts when an excess current flows through it. It is used to protect cables and wires from overheating. When used with an earth wire, a fuse will prevent the body of a metal appliance from becoming live.

18 The correct value of a fuse is calculated from the following formula.

$$\text{current (A)} = \frac{\text{power (W)}}{\text{voltage (V)}}$$

19 Modern consumer units contain circuit breakers, which switch off when an excess current flows. They react faster than fuses. They can be reset.

20 A residual current device (RCD) is a fast-acting circuit breaker. It measures the current in the live and neutral wires. If they differ, perhaps because a person has touched a live wire, the current is immediately switched off.

21 An RCD should always be used when using appliances outdoors.

22 The 230 V mains electricity is dangerous. Some countries have adopted a lower 120 V system.

Chapter 15 Forces and motion

By the end of this chapter you should:

- know how we measure and display motion;
- understand how forces affect the motion of objects;
- understand why moving objects reach a steady speed;
- know where forces come from;
- understand how objects gain or lose energy;
- know how we keep ourselves safe in and around cars.

Distance, speed and acceleration

Speed

You are late for school. If you travel at a fast speed the journey takes a shorter time. We can calculate speed using the following formula.

$$\text{average speed} = \frac{\text{distance travelled}}{\text{time taken}}$$

If distance is measured in metres (m) and time in seconds (s) then speed is measured in metres per second (m/s).

Example

You travel 400 m in 80 s on your bicycle. What is your speed in metres per second?

Formula first:

$$\text{average speed (m/s)} = \frac{\text{distance travelled (m)}}{\text{time taken (s)}}$$

Put the numbers in:

$$\text{average speed (m/s)} = \frac{400\ \text{(m)}}{80\ \text{(s)}}$$

$$= 5\ \text{m/s}$$

If distance is measured in kilometres and time is measured in hours, speeds are usually written in km/h. If a car takes 1 h to travel 80 km, its speed is 80 km/h. What is its speed in m/s?

$$\text{average speed (m/s)} = \frac{\text{distance travelled (m)}}{\text{time taken (s)}}$$

$$= \frac{80 \times 1000}{1 \times 3600}$$

$$= 22.2\ \text{m/s}$$

Figure 15.1 Model car

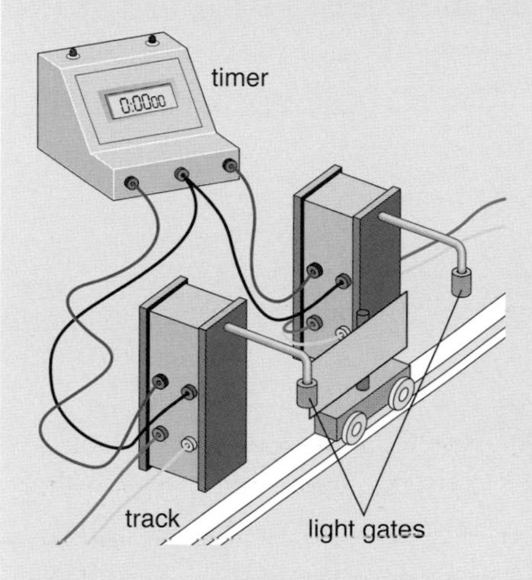

Figure 15.2 Timing the cars electronically

Practical work

Finding the speed of a model car

Method 1

You will need a model car, a metre rule and a stopwatch.

1. Mark out a test track, say 2 m or 5 m long, the longer the better.
2. Use the watch to time your car over the measured track.
3. Calculate its average speed using the following formula.

$$\text{average speed} = \frac{\text{distance travelled}}{\text{time taken}}$$

4. Record your results and conclusions.

Method 2

If your school has a timer and timing gates.

1. Set up a pair of light gates a measured distance apart, and connect them to an electronic timer.
2. The timer starts when the card breaks the infra-red beam in the first gate. When the card breaks the beam in the second gate, the timer stops and displays the time interval.
3. Calculate the car's average speed between the two light gates.

Question

- Copy and complete this table.

Distance (m)	600	250		1
Time (s)	20		150	100
Speed (m/s)		5	15	

Did you know?

Pendine Sands west of the Towy estuary were once used as a test track for world land speed records. Now the cars are so fast that a much longer stretch of perfectly flat surface is needed.

Acceleration

At the start of a bike ride your speed is zero. You press hard on the pedals and your speed increases. You go faster and faster. This change in speed is called **acceleration**. It has been caused by the unbalanced force from your legs, through the pedals and chain to the wheels.

When you put the brakes on you slow down. This reduction in speed is called **deceleration**. It is the opposite of acceleration. Sometimes it is called a negative (–) acceleration.

Calculating acceleration

A racing car can accelerate at a greater rate than a family car. It can reach any speed in a shorter time than the family car. This is why car manufacturers compare the time taken to reach 60 miles per hour. We can find acceleration using the following formula.

$$\text{acceleration} = \frac{\text{change in speed}}{\text{time taken to change speed}}$$

We usually measure speed in m/s and time in s, so the unit of acceleration is the unit of the change in speed (metres per second) divided by the unit of the time taken for the speed to change (seconds). Therefore the unit of acceleration is metres per second squared (symbol m/s^2).

An acceleration of $2\,m/s^2$ means that there is an increase in the speed of the car of 2 m/s every second. If the car starts from rest (0 m/s), after 1 s its speed is 2 m/s. After another second the speed is 4 m/s, and after the third second it is 6 m/s. Then it is 8 m/s, 10 m/s, etc.

Example

Suppose that your car starts from 0 m/s. After 2 s, it is travelling at 6 m/s. What is its acceleration?

Formula first:

$$\text{acceleration } (m/s^2) = \frac{\text{change in speed (m/s)}}{\text{time taken to change speed (s)}}$$

change in speed = final speed – initial speed

Put the numbers in:

$$\text{change in speed} = 6\,m/s - 0\,m/s = 6\,m/s$$

time taken for change = 2 s

The answer is:

$$\text{acceleration } (m/s^2) = \frac{6\,m/s}{2\,s} = 3\,m/s^2$$

Calculating deceleration

A car is travelling at 20 m/s. The driver puts the brakes on, and the car comes to a standstill in 2 s. What was the acceleration?

Formula first:

$$\text{acceleration } (m/s^2) = \frac{\text{change in speed (m/s)}}{\text{time taken to change speed (s)}}$$

change in speed = final speed – initial speed

Put the numbers in:

$$\text{change in speed} = 0\,m/s - 20\,m/s = -20\,m/s$$

time taken for change = 2 s

$$\text{acceleration } (m/s^2) = \frac{-20\,m/s}{2\,s} = -10\,m/s^2$$

The minus sign (–) shows that this is a deceleration.

Questions

1 What happens to a car when it accelerates?

2 What happens to a bike when it decelerates?

3 What do you need to apply to the bike pedals to make it accelerate?

4 What are the units of acceleration?

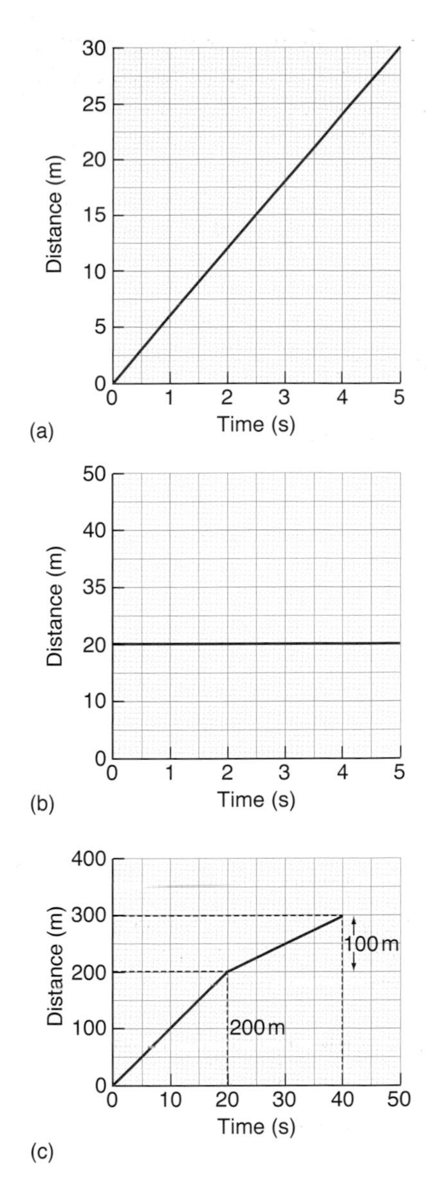

Figure 15.3 Distance–time for (a) steady speed, (b) zero speed and (c) a change in speed

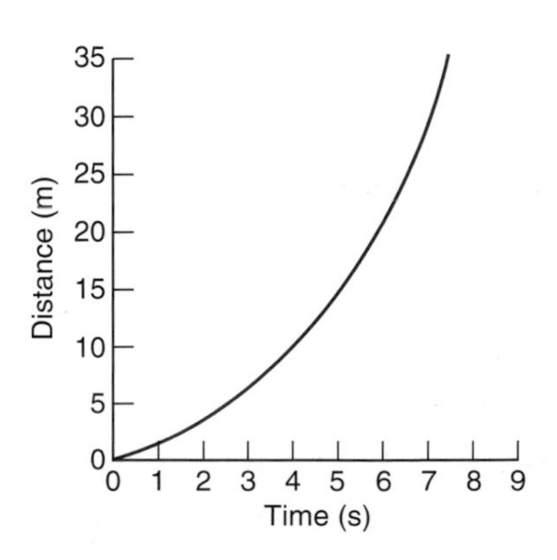

Figure 15.4 Distance–time graph for increasing speed

5 Beth is a keen cyclist. She reaches a speed of 15 m/s in 15 s on her bicycle. Her dad has creaky knees, but still likes to ride his bike. It takes him 15 s longer to reach a speed of 15 m/s.
  a What is Beth's acceleration?
  b How long does it take her dad to reach 15 m/s?
  c What is his acceleration?

6 A car travels at 15 m/s. It accelerates at 1.5 m/s^2. What is its speed after 1 s, 2 s, 5 s and 10 s?

Distance–time graphs

Look at the graph in Figure 15.3a. It shows that the object takes 5 s to travel a distance of 30 m.

$$\text{average speed (m/s)} = \frac{30\,\text{m}}{5\,\text{s}}$$

$$= 6\,\text{m/s}$$

This is the value of the **gradient**, or slope, of the graph. The gradient of a distance–time graph gives the speed of a body.

In the graph in Figure 15.3b, the distance travelled is always the same. The object has not moved; it is stationary. When a distance–time graph is horizontal, the object is not moving. The gradient is zero so the speed is zero.

In the graph in Figure 15.3c, the slope of the graph changes. For the first 20 s,

$$\text{average speed} = \frac{200\,\text{m}}{20\,\text{s}}$$

$$= 10\,\text{m/s}$$

In the next 20 s, the distance travelled is from 200 m to 300 m, that is, only 100 m. The average speed drops to:

$$\text{average speed} = \frac{100\,\text{m}}{20\,\text{s}}$$

$$= 5\,\text{m/s}$$

On a distance–time graph, the steeper the slope, the greater is the speed.

Changing speed

The graph in Figure 15.4 is not a straight line. It is a curve that gets steeper over time. Remember that the steeper the gradient, the greater is the speed. This graph shows that the speed is increasing all the time. The moving object is accelerating.

Questions

7 Beth is on the bus on her way to school. Here is a distance–time graph of her journey.

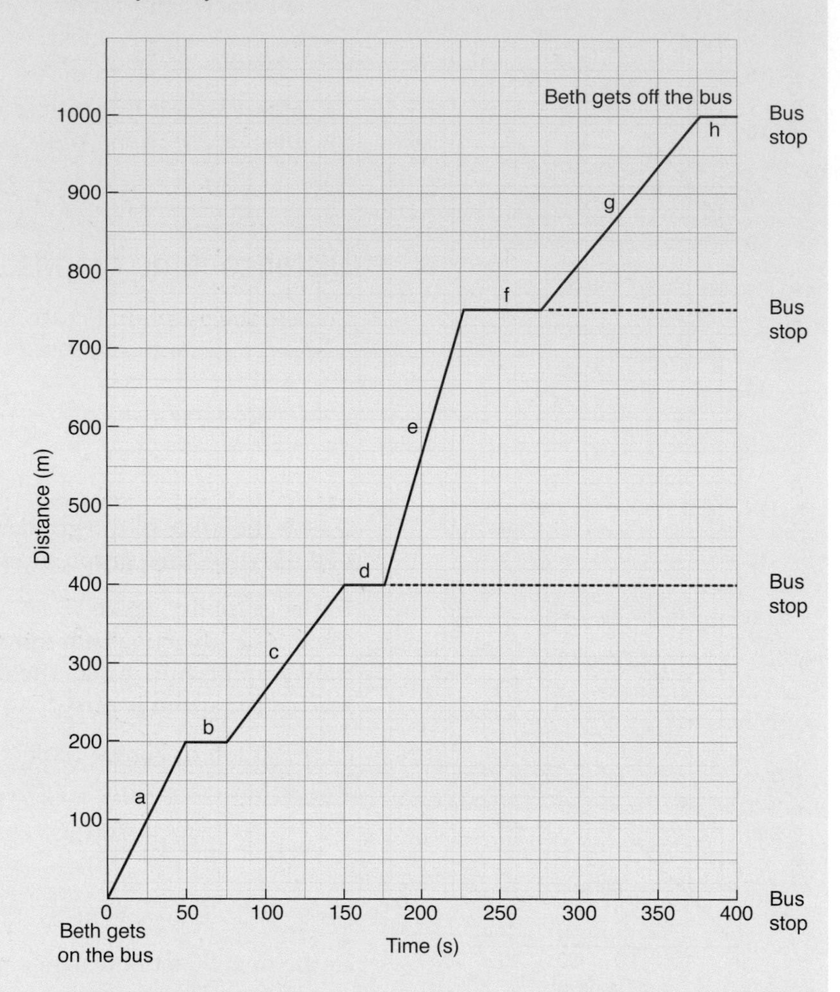

a How far did Beth travel on the bus?

b In which section of the journey was the bus travelling at its:

i slowest speed

ii fastest speed?

c What was the speed of the bus in section (a) and section (b)?

d Suggest what might have happened at section (b).

e How far was it to the first bus stop?

f How long was the bus stationary at section (f)?

g What was the average speed of the bus over the whole journey?

8 a The table below shows the distance travelled by a cyclist over time. Use the information in the table below to draw a graph of distance travelled, *y*-axis, against time, *x*-axis.

b Use the graph to find:

i for how long the cyclist was not moving

ii the distance travelled between 80 s and 140 s.

Distance (m)	0	100	200	300	400	400	400	500	600	700	800
Time (s)	0.0	7.5	15.0	22.5	30.0	40.0	60.0	80.0	100.0	120.0	140.0

9 A cyclist travels along a straight road. The letters on the diagram show the position of the cyclist every 10 s along the road. The numbers show the distance in metres.

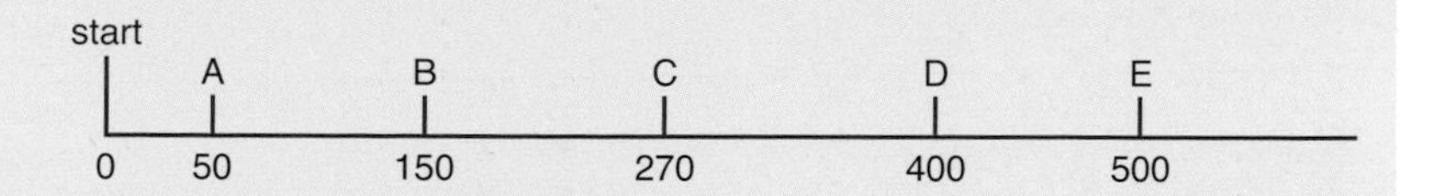

a Write down the time taken by the cyclist to travel between **C** and **E**.
b Write down, in words, the equation connecting distance, time and speed.
c Use your equation to calculate the cyclist's average speed between **C** and **E**.

Speed–time graphs

Safety is all important when driving a heavy goods vehicle (HGV). Drivers must take regular rest breaks and must not exceed permitted speeds. A **tachometer** is fitted inside the cab. The vehicle's speed and rest breaks are recorded on a tachograph. The graph on the tachograph in Figure 15.5 is an example of a speed–time graph.

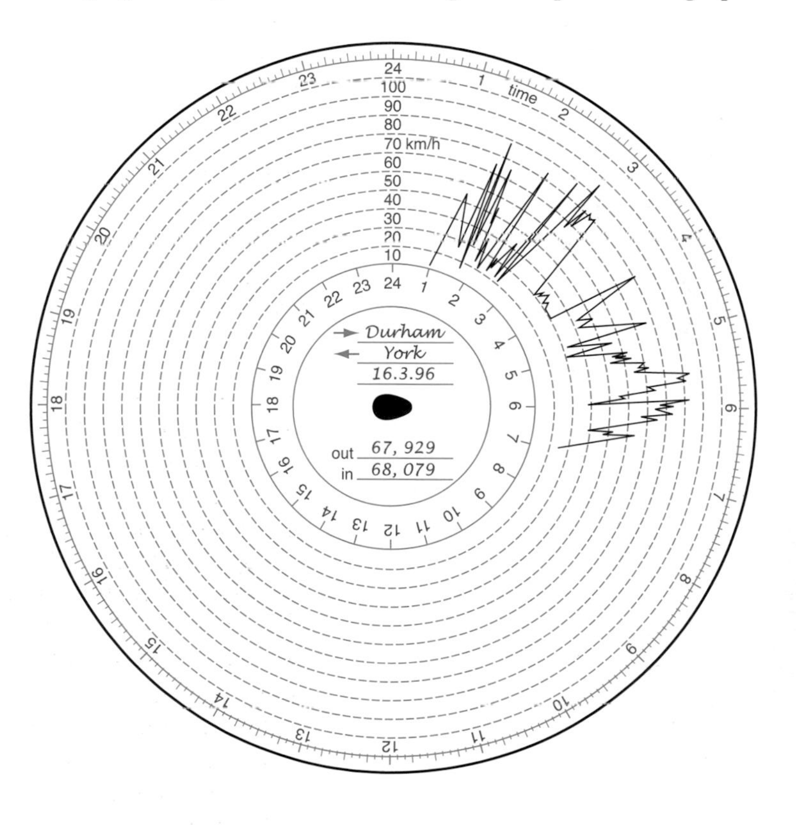

Figure 15.5 A tachometer records the vehicle's speed at any time on a tachograph.

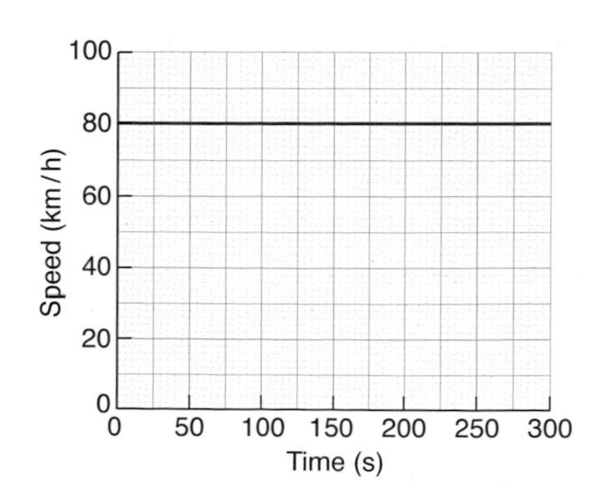

Figure 15.6 The HGV is travelling at a constant speed of 80 km/h.

The graph in Figure 15.6 is horizontal, so the HGV is travelling at a constant speed. If the speed–time graph is a horizontal line along the origin, then the HGV is at rest. It is not moving.

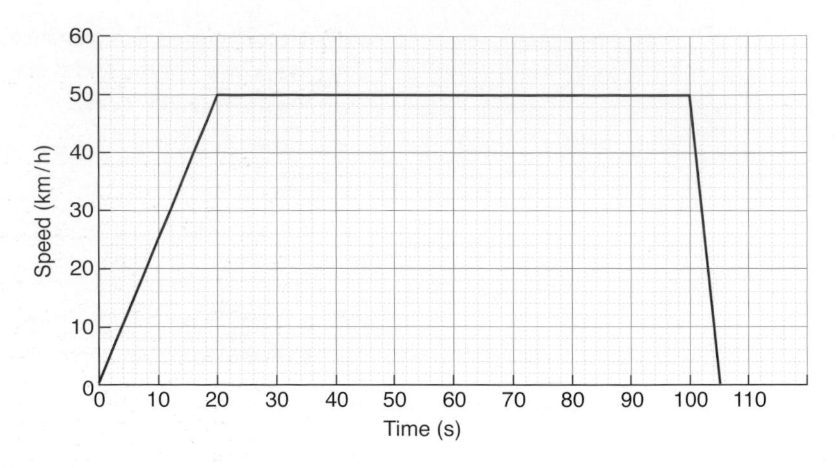

Figure 15.7 The steeper the gradient (slope) of the speed–time graph, the greater is the acceleration or deceleration.

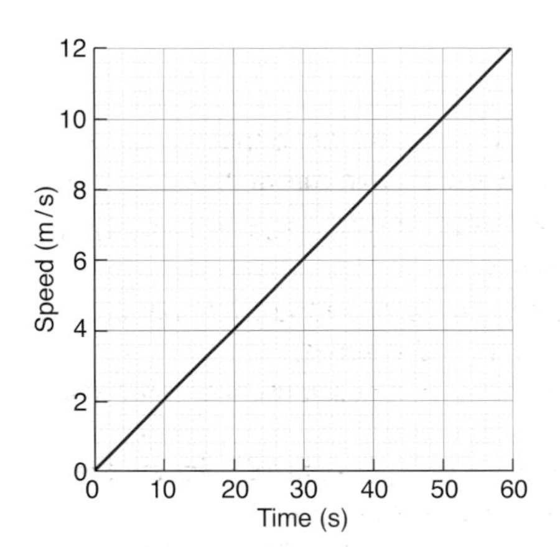

Figure 15.8 Speed–time graph for constant acceleration

The speed–time graph in Figure 15.7 shows that, at the start, the HGV is at rest. It then accelerates steadily until it reaches 50 km/h. It travels at this speed for 80 s. The driver puts the brakes on. The HGV comes to rest in 5 s.

Using speed–time graphs to find acceleration

The graph in Figure 15.8 is a straight line. This tells us that the acceleration is uniform. The slope or gradient of the line tells us what the acceleration is.

$$\text{acceleration (m/s}^2) = \frac{\text{change in speed (m/s)}}{\text{time taken (s)}}$$

$$= \frac{12\,\text{m/s}}{60\,\text{s}}$$

$$= 0.2\,\text{m/s}^2$$

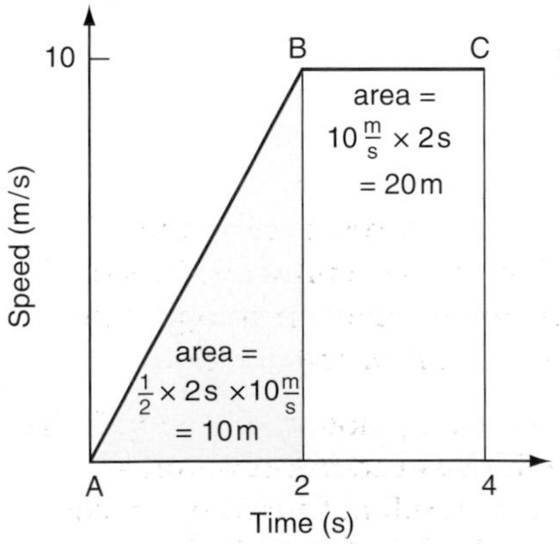

Speed (m/s)
10
B C
area = $10\frac{\text{m}}{\text{s}} \times 2\text{s}$ = 20 m
area = $\frac{1}{2} \times 2\text{s} \times 10\frac{\text{m}}{\text{s}}$ = 10 m
A 2 4
Time (s)

Figure 15.9 Speed–time graph of a sprinter

Using speed–time graphs to find distance travelled

The speed–time graph in Figure 15.9 shows how a sprinter's speed changes during the race. At first, she accelerates, and then she travels at a steady speed. We can use the area under a speed–time graph to work out the distance travelled. Remember that the distance travelled = average speed × time.

For the first part of the race, the average speed equals half the height of the yellow triangle. The time axis is the base of the triangle. Therefore:

$$\text{distance travelled} = \frac{1}{2} \times \text{height} \times \text{base}$$

$$= \frac{1}{2} \times 10\,\text{m/s} \times 2\,\text{s}$$

$$= 10\,\text{m}$$

Questions

10 The motion of two cars, **A** and **B**, is shown on the speed–time graph.

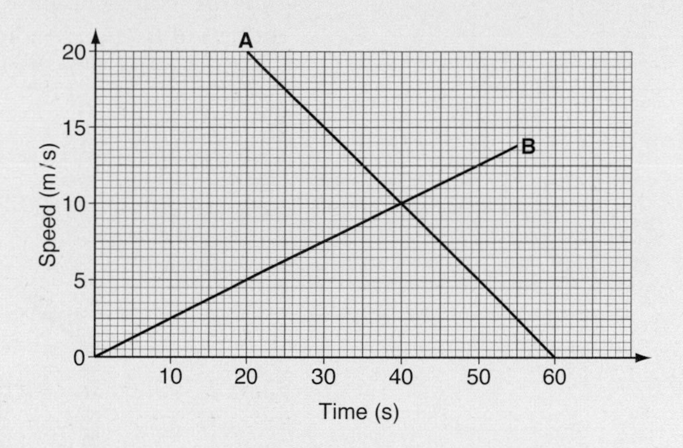

a Describe fully the motion of **A**.

b Find:

i the time when both cars travelled at the same speed

ii the difference in speed between the two cars at 20 s.

11 A car accelerates steadily from 4 m/s to 19 m/s in 10 s. It then travels at this speed (19 m/s) for another 15 s.

a On graph paper, with speed and time axes and suitable scales, draw lines to show the motion described above.

b Write down, in words, the equation connecting distance, time and speed.

c The average speed for the whole journey is 16 m/s. Calculate the distance travelled in the 25 s of the journey.

d Use the graph to find for how long the car travelled at a speed greater than 16 m/s.

e Draw a line on the graph to show the motion beyond 25 s, if the car is brought to rest in a further 5 s by a non-uniform force.

The effects of forces

Balanced and unbalanced forces

Forces have size and direction. We use an arrow to show the direction of a force. The length of the arrow represents the size of the force. If there is more than one force acting on an object the forces can be balanced (as in Figure 15.10) or unbalanced.

Balanced forces do not change the existing motion of a body. If the body is stationary, it will not move. If balanced forces act on a moving body, it will continue to move. It will not speed up or slow down. Unbalanced forces acting on a body will make it change its speed or direction of travel.

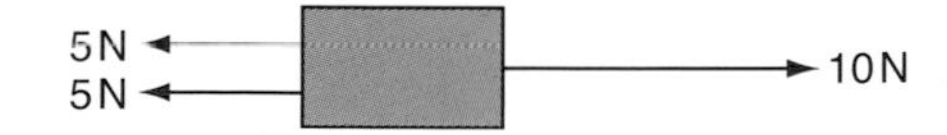

Figure 15.10 The forces on this object are balanced.

Practical work

Investigating balanced forces and constant speed

It is difficult to observe an object in motion without the everyday effects of friction. A glider on a straight, level air track rides on a cushion of air. The friction force is very low indeed.

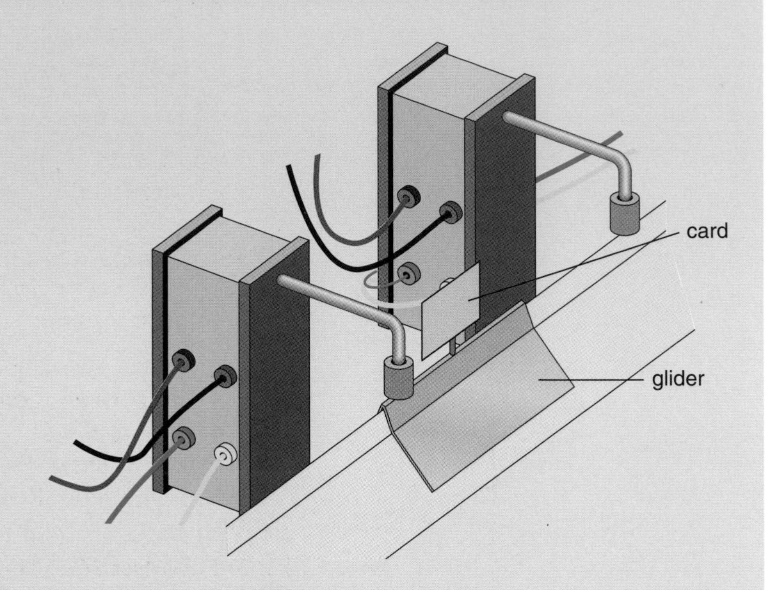

Figure 15.11 Friction forces are very small with the air track. When the card is at right angles to the direction that the glider is moving in, the air resistance is much greater and the glider comes to rest much sooner.

1. Give the glider a gentle push to start it moving.
2. It will travel backwards and forwards between the ends of the track. As there are no unbalanced forces, the glider will continue to move.
3. Fix a postcard on the glider with the card edge-on to the direction of travel.
4. The light gates, which are connected to a datalogger, record the time taken for the postcard to pass through the light gates. This means you can work out the glider's speed.
5. The speed of the glider should be constant over a period of time. The edge-on postcard has little effect on the drag force.
6. Eventually the tiny drag force will bring the glider to rest.
7. Then place the postcard at right angles to the direction of travel.
8. The drag force increases.
9. The forces are unbalanced. The glider soon comes to rest.

Did you know?

Hovercrafts float on a cushion of air, like that on a linear air track. They can travel over land and water. Their 'go anywhere' capability makes them very useful to the military.

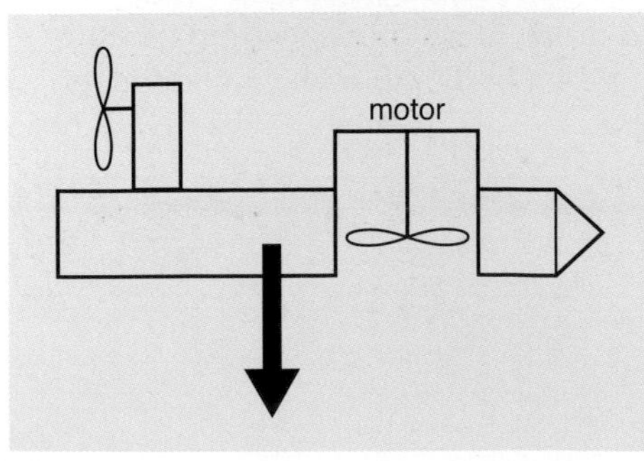

Questions

12 The diagram on the left shows a hovercraft.

a Draw all the forces which act on the hovercraft when it is moving forward at a constant speed.

b Draw all the forces which act on the hovercraft when it just begins to slow down.

13 You decide to do a bungee jump. Name the forces acting and say if they are balanced or unbalanced:
a just as you jump off
b when the bungee begins to tighten
c when the bungee is at full stretch
d when you start going up again.

The effect of unbalanced forces

To find out what happens to a body when two or more forces act on it, we imagine all the individual forces being replaced by a single force that has the same effect. This single force is called the **resultant force**. The size and direction of the resultant force determines how or even whether the body moves. If the forces are balanced, the resultant force is zero.

Push on a stationary object and you might make it move. If it doesn't move, your pushing force is balanced by a friction force. Push hard enough to overcome the force of friction and the object will move. If you keep pushing, the forces are now unbalanced and it will move faster and faster. It will accelerate.

Force and acceleration (with constant mass)

We want to find out how the size of the resultant force affects the acceleration of an object. To investigate this, we keep the mass of the object constant.

Practical work

Investigating force and acceleration with constant mass

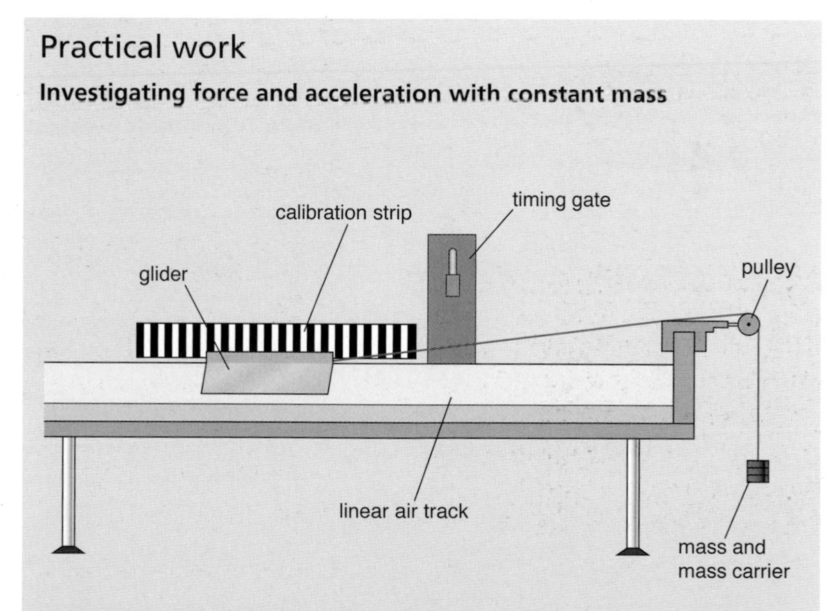

Figure 15.12 The falling weight provides an unbalanced force on the mass of the glider.

1 Level an air track and connect the light gates to the motion timer or datalogger.

2 Attach a free-running pulley to one end of the track.
3 A fine cord attached to the glider and a separate 50 g mass passes over the pulley. The falling weight provides the unbalanced force to accelerate the glider and its attachments.
4 The timer or datalogger displays information from which the acceleration may be displayed directly or calculated.
5 Adding additional masses to the cord increases the accelerating force.
6 The results show that the bigger the unbalanced force, the greater is the acceleration.

Using a trolley and ticker timer

When the distance between dots on a ticker tape changes, it means the speed of the object attached to the ticker tape is changing during that time interval. A smaller distance between dots means the object was moving more slowly during that time interval. A larger distance between dots indicates that the object was moving faster during that time interval.

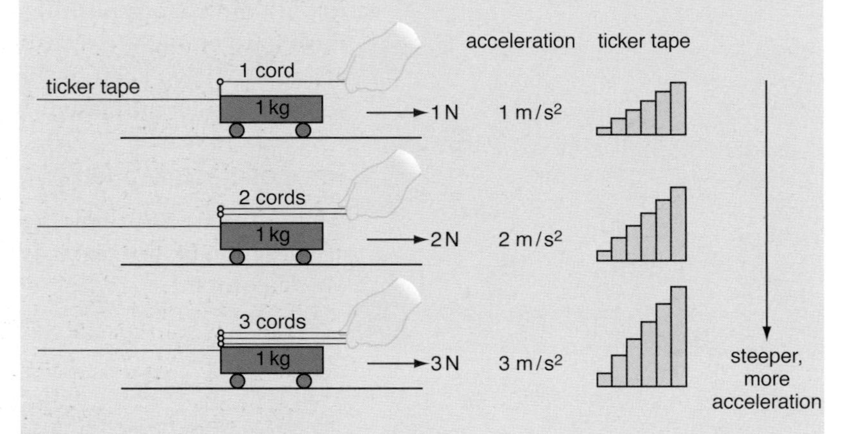

Figure 15.13 The mass is constant and the force is changed.

1 Place the trolley on the bench and connect it to a ticker tape timer.
2 Fix a piece of elastic cord to the trolley.
3 Pull the trolley along the bench, always keeping the elastic cord stretched by the same amount, to keep the force constant.
4 When you have pulled the tape through the timer, divide the tape into tick lengths so that each have ten dot-to-dot spaces.
5 Stick the ten-tick tapes side by side to make a speed–time graph.
6 Repeat the experiment with two and then three elastic cords. The graphs get steeper (greater acceleration) as the number of elastic cords (force) increases.

A barrier to prevent trolleys falling off the end of the bench is a good idea – a pile of books would do.

Mass and acceleration (with constant force)

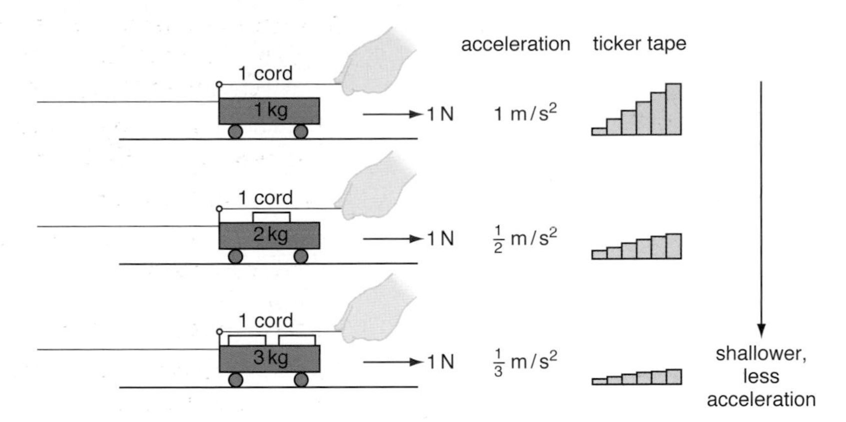

Figure 15.14 The force is kept constant and the mass is changed.

Practical work

Investigating mass and acceleration with constant force

1. Using the ticker timers, keep the force on the trolley constant by using one elastic cord.
2. Increase the mass of the trolley by 1 kg at a time.
3. The resulting acceleration graphs are less steep.
4. This shows that the acceleration is decreasing.

1. Alternatively, with the linear air track, keep the accelerating force constant by keeping the falling mass constant.
2. Take measurements of the acceleration with one glider.
3. Connect two identical gliders together to increase the mass.
4. The results from the timers or dataloggers show that doubling the mass halves the acceleration.

Analysing the results of the practical work

In the first experiment, when the pulling force is doubled, the slope of the graph doubles. The slope of a speed–time graph is the acceleration. Thus the greater the force applied, the greater is the acceleration; we say that the acceleration is proportional to the force applied:

$$\text{acceleration} \propto \text{force}$$

where $\propto$ means 'is proportional to'.

In the second experiment, the pulling force is constant. When the mass of the trolley is doubled, the acceleration is halved; the greater the mass, the lower is the acceleration. With the pulling force constant, we say that the acceleration is 'inversely proportional' to the mass moved:

$$\text{acceleration} \propto \frac{1}{\text{mass}}$$

Combining these two statements, we can say that the acceleration produced depends on the resultant force (the force moving the object) and (inversely) on the mass of the object:

$$\text{acceleration} \propto \frac{\text{force}}{\text{mass}}$$

In fact, we say that if a force is making a mass of 1 kg accelerate at 1 m/s^2, then that force is **1 newton** (1 N),

or:

$$\text{acceleration (m/s}^2\text{)} = \frac{\text{resultant force (N)}}{\text{mass (kg)}}$$

And, we can rearrange this to:

$$\text{resultant force (N)} = \text{mass (kg)} \times \text{acceleration (m/s}^2\text{)}$$

The unit of force is called the newton (N), after Isaac Newton who was the first scientist to understand the relationship between force, mass and acceleration.

Did you know?

Space probes with their fuel-efficient ion drives may replace chemical rocket thrusters.

A clear animation is available if you use a search engine like Google for: 'ESA Ion drive' and click on ESA Science Exploring space.

Example

What is the size of the force needed to give a car of mass 800 kg an acceleration of 2 m/s^2?

Formula first:

$$\text{resultant force (N)} = \text{mass (kg)} \times \text{acceleration (m/s}^2\text{)}$$

Put the numbers in:

$$\text{resultant force} = 800\,\text{kg} \times 2\,\text{m/s}^2 = 1600\,\text{N}$$

Questions

14 Why does a fully loaded car accelerate more slowly than it does with the driver alone?

15 A bike and rider have a mass of 100 kg. What is the force needed to give them an acceleration of 3 m/s^2?

16 Explain why the cyclist in Q15 could not maintain the acceleration indefinitely.

17 What happens to the speed of an object travelling at 5 m/s when balanced forces act on it?

18 Dilys tests her new car. She is happy with its 2.5 m/s^2 acceleration. The total mass of the car and driver is 750 kg. What force is needed to accelerate at this rate? She invites three friends with a combined mass of 250 kg. What is its acceleration with her friends on board?

Mass and weight

The mass of a body is the quantity of matter in a body. It does not vary. It is measured in kilograms.

Weight is a force. It is the force that attracts us to the Earth. Everything is pulled towards the centre of the Earth. The greater the mass of an object the greater is the force of attraction to the Earth. Weight depends on mass and the strength of gravity. It is measured in newtons.

Figure 15.15 The Earth's gravity attracts a 1 kg mass with a force of 10 N.

The relationship between mass and weight

As noticed earlier, the equation, resultant force (N) = mass (kg) × acceleration (m/s^2) gives us a definition of the newton. To find out what 1 N is, we write the following.

$$\text{mass} = 1\,\text{kg}$$

$$\text{acceleration} = 1\,\text{m/s}^2$$

Then:

$$\text{resultant force} = 1\,\text{kg} \times 1\,\text{m/s}^2$$
$$= 1\,\text{N}$$

Therefore a force of 1 N is the force that gives a mass of 1 kg an acceleration of 1 m/s^2.

The Earth attracts all bodies to its centre. The Earth is surrounded by a **gravitational field** which exerts a force on anything in the field.

Experiments show that this force is 9.8 N acting on 1 kg. We say that the gravitational field strength is 9.8 N/kg. Since 9.8 is almost 10, we usually say that the gravitational field strength on the Earth's surface is 10 N/kg (Figure 15.15). Therefore, if your mass is 60 kg, then your weight will be 600 N.

Figure 15.16 Skydivers jumping from an aeroplane may spread out their arms and legs and link hands when falling at terminal speed.

The effect of air resistance on falling objects

Megan is a skydiver. When she jumps from a plane, the air resistance is at first very low. The forces are unbalanced. Megan's weight accelerates her towards the ground. As she speeds up the air resistance increases, so her acceleration decreases. Her speed is still rising, but not as fast. The air resistance keeps increasing until it equals her weight. At this point the resultant force on Megan is zero and she stops accelerating. She carries on falling but at a constant speed. We call this **terminal speed**.

At a safe height, Megan pulls the ripcord to open her parachute. The canopy has a very large surface area. The upward drag force is now much greater than the downwards weight force. This unbalanced force slows her down. This reduces air resistance so that once again the two forces, weight and air resistance, become balanced so that she reaches a much slower terminal speed. She lands safely. The speed–time graph shows her motion through the air (Figure 15.17).

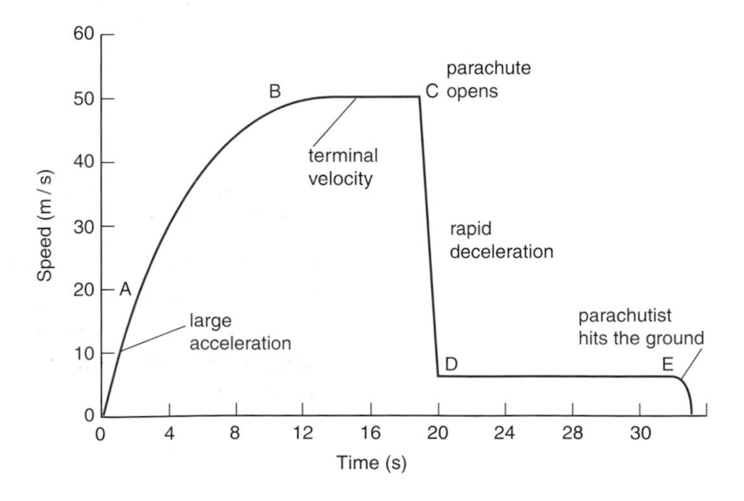

Figure 15.17 Speed–time graph for a skydiver

Did you know?

The terminal speed of a skydiver depends on altitude and surface area of the diver. It is about 50 m/s, or 190 km/h. Sometimes the people in a skydiving group link hands (Figure 15.16). The people who jump first spread out their arms and legs to increase their air resistance. This allows the others to catch up.

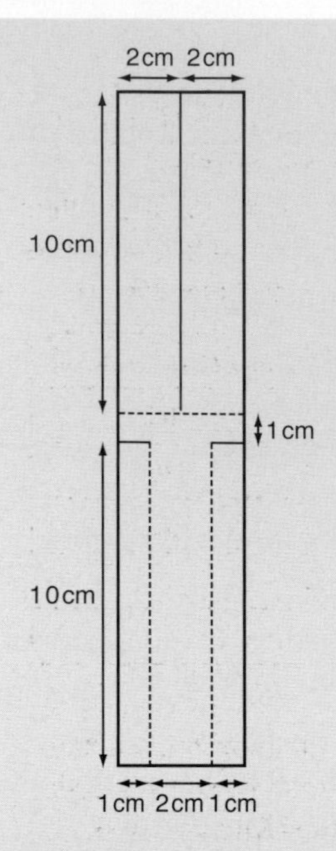

Figure 15.18 Shape of paper helicopter to be cut and folded

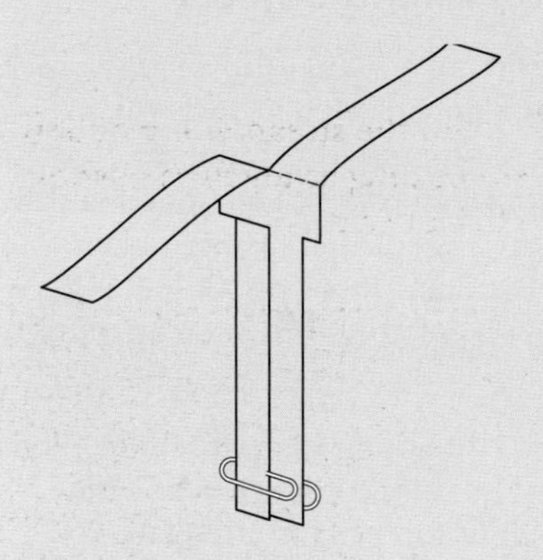

Figure 15.19 Folded helicopter with paperclip ready to be used

Practical work

Investigating forces on falling objects

Two forces act on a falling body. The weight force acts downwards. 'Lift' and air resistance forces act upwards. For convenience, we will combine the two upward forces and refer to them as a single air-resistance force. In this experiment, we use a paper helicopter to investigate how changing its weight and air resistance affects how quickly it falls through the air.

1. Draw the paper helicopter on thin card (Figure 15.18).
2. Cut along the solid lines and fold along the dotted lines (Figure 15.19).
3. Add a paperclip to the tail.
4. With care, stand on a strong stable surface.
5. Hold the model about 2 m above the floor. Let it drop and use a stopwatch to time it as it falls.
6. Record your results.
7. Add first one, then two extra paperclips.
8. Repeat step 5 each time.
9. Use scissors to shorten the wings. Cut off about 3 cm.
10. Repeat step 5.

Questions

- How did you increase the downward force?
- When you increased the downward force what happened to the time taken to fall?
- How did you change the effect of air resistance on the helicopter?
- Make a prediction. What would happen to the rate of fall if you doubled the width of the helicopter? If you have time, make the new helicopter and test your prediction.

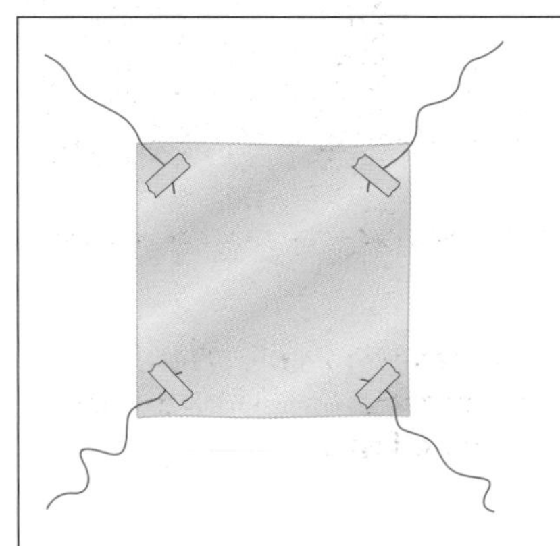

Figure 15.20 Piece of polythene with cotton attached to corners

Activity

Many parts of the world suffer from drought and famine. Quite often relief supplies need to be airlifted and dropped into remote areas. Your company has to design parachutes that will safely drop the containers without their bursting on impact. In your work group you will need to:

1. Think about what affects the rate of fall a parachute.
2. Think how to make the parachute stable so that it does not swing from side to side.
3. Think about the measurements you will make: how to record them and present them.
4. Use a polythene bin bag, cotton, sticky tape, paper clips or small weights and anything else you might need to make and test your parachute.

Questions

19 Explain how the speed of a falling sky diver changes. Use the terms 'balanced forces' and 'unbalanced forces' in your answer.

20 A skydiver falls vertically. She is acted on by two forces: gravity acting downwards, and air resistance acting upwards. The table below gives some information about the forces on the skydiver during her fall.

Speed (m/s)	Weight (N)	Air resistance (N)	Resultant force (N)
0	700	0	700
10		280	420
20		490	210
30		630	70
40	700		0

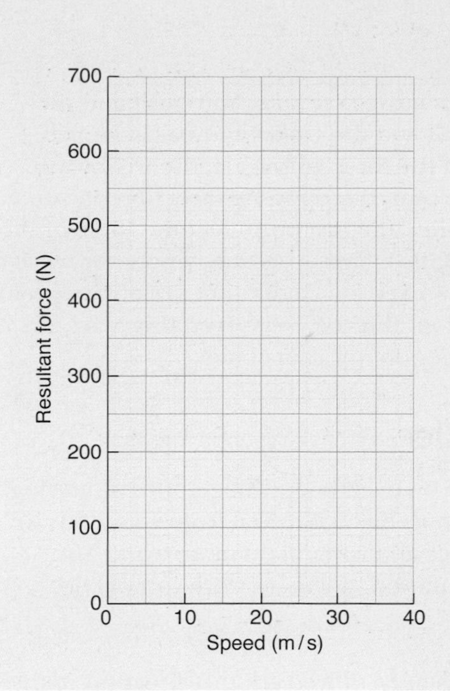

a Copy and complete the table.

b Use the information in the table to draw a graph of the resultant force against speed. Use the axes on the left and a suitable scale.

At some point in the dive, the sky diver falls at a terminal speed.

c State what is meant by terminal speed.

d State the value of the terminal speed of this skydiver.

Streamlining

Car manufacturers use wind tunnels to test the streamlining of new models (Figure 15.21). They try to reduce drag as much as possible. Lower drag means faster more fuel-efficient cars.

Formula 1 cars and wind tunnels (go to *Insight* and then *Understanding the sport*): www.formula1.com

Figure 15.21 Car manufacturers use smoke in a wind tunnel to see how air flows over the body of a car.

Athletes are keen to get the best performance possible. They wear smooth clothes made from Lycra to reduce drag. Some cyclists even shave their legs and hair so that they can go faster.

Activity

The drag coefficient is a measure of how much air resistance a car has. Use data from car makers or the internet to find the drag coefficients of modern cars. How do they compare with off-road vehicles?

Question

21 Why should a car have a low drag?

Interactions between objects

Forces come in pairs

To make a supermarket trolley move is simple. You push and the trolley moves forward. You pull and the trolley moves backwards. But, whatever the direction of the force you exert, the trolley will always exert an equal-sized force in the opposite direction on you. (The harder you push the trolley, the more you feel the force of the trolley handle against your hands.) This is because whenever two bodies interact, the forces they exert on each other are always equal in size and opposite in direction. But the two forces always act on different bodies; this is why they do not cancel out.

Work and energy transfer

If you have to push start a car you need energy to get the car moving. You have to do **work**. Energy must be transferred from your body to the car. Push with a bigger force and you will do more work. The greater the distance the car is moved, the more work you will do.

We can say that:

work done = force (N) × distance moved (in the direction of the force) (m)

If force is measured in newtons and distance is measured in metres, then work done is measured in **joules** (J). Therefore 1 J of work is done when a force of 1 N is moved through a distance of 1 m (in the direction of the force). In this case, the force overcome was the friction force. Quite often the force overcome is the force of gravity.

Example

A bag of potatoes weighs 25 N. How much work is done when the bag is lifted from the floor to a shelf 2 m higher?

Formula first:

work done (J) = force (N) × distance moved (m)

Put the numbers in:

work done (J) = 25 N × 2 m
= 50 J

Sometimes an examiner might ask a question a different way. How much energy is transferred to a 25 N bag of potatoes when it is lifted through a height of 2 m? The calculation is exactly the same. Work done is the same as energy transferred. In this case, the energy transferred is 50 J.

Energy is always **conserved** in a transfer. Whatever did the lifting lost an equivalent amount of energy, 50 J.

Example

Energy equal to 9 J is needed to raise a 3 N weight off the floor. How high is the object lifted?

Formula first:

work done (J) = force (N) × distance moved (m)

Put the numbers in:

$$9\,J = 3\,N \times \text{distance moved (m)}$$

Rearrange:

$$\text{distance moved} = \frac{9\,J}{3\,N} = 3\,m$$

Work and energy (go to *Multimedia physics studio* and then *Work and energy*):
www.physicsclassroom.com

Did you know?

The *Seawise Giant* is a very big ship. It is 460 m long, and it has a mass of 560 000 tonnes. Once it is moving, the captain needs a 10 km (over 6 miles) gap to stop in.

Example

A fast moving car has 200 000 J of energy. What force must be applied to bring it to rest in a distance of 40 m?

Formula first:

$$\text{work done (J)} = \text{force (N)} \times \text{distance moved (m)}$$

Put the numbers in:

$$200\,000\,J = \text{force (N)} \times 40\,m$$

Rearrange:

$$\text{force (N)} = \frac{200\,000\,J}{40\,m} = 5000\,N$$

Questions

22 Name the unit of work.

23 Write down the formula connecting work done, force and the distance moved in the direction of the force. State the units in which each must be measured.

24 How much work is done when a 15 N weight is lifted to a height of 4 m?

25 A moving object has 250 J of energy. A retarding force of 5 N brings it to rest. How far does it travel before stopping?

26 A moving car has 500 000 J of energy. What force will stop it in a distance of 50 m?

Kinetic energy and potential energy

H

You do work when you lift something up. You have transferred energy to it. It has energy because of its position above the Earth. We call it positional or more correctly **gravitational potential energy**.

Sometimes you do work when you change the shape of an object, for example when you pull on the elastic of a catapult or on the string of a bow ready to fire an arrow. We say the stretched elastic has stored **elastic potential energy**.

Drop the object and it falls to the ground. The stored potential energy is transferred into movement or **kinetic energy**.

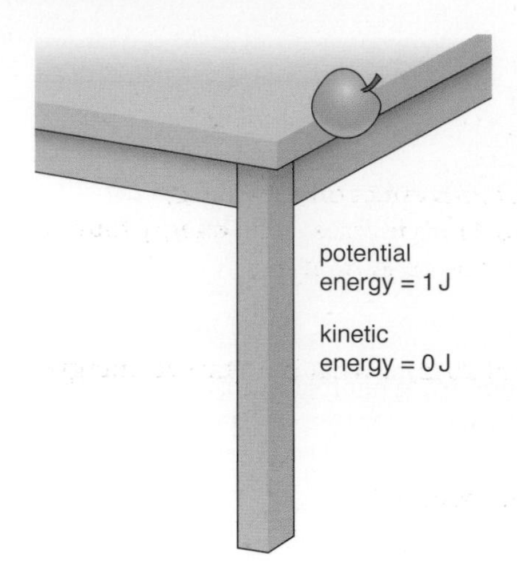

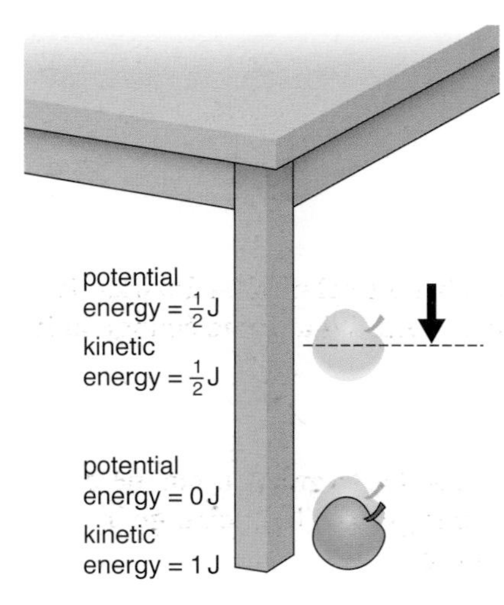

Figure 15.22 Potential energy to kinetic energy

Potential energy

We can calculate potential energy from the work we have to do to lift something up (see Figure 15.22). Find the work done in lifting the 1 N apple from the floor to a table 1 m above the floor.

Formula first:

work done (J) = force (N) × distance moved (m)

Put the numbers in:

work done (J) = 1 N × 1 m
= 1 J

Therefore the potential energy of the apple is 1 J. If the apple was dropped from the table to the floor, the potential energy would be transferred into kinetic energy. Just before it hit the ground it would have 1 J of kinetic energy.

Look at this equation again.

work done (J) = force (N) × distance moved (m)

If you drop something, the force of gravity makes it accelerate. We call it the **acceleration due to gravity** (symbol g). Its other name is **gravitational field strength**.

Remember that:

force = mass × acceleration

Therefore we can write the following:

force (N) = mass (kg) × gravitational field strength (N/kg)

If we lift the object, the distance moved is the change in height.

The equation:

work done (J) = force (N) × distance moved (m)

becomes:

work done (J) = mass (kg) × gravitational field strength (N/kg) × change in height (m)

However, if we lift the object, the work done has changed the potential energy of the object:

change in potential energy (J) = mass (kg) × gravitational field strength (N/kg or m/s^2) × change in height (m)

Kinetic energy

Kinetic or moving energy depends on the speed of the body and its mass.

kinetic energy = $\frac{1}{2}$mass × speed squared

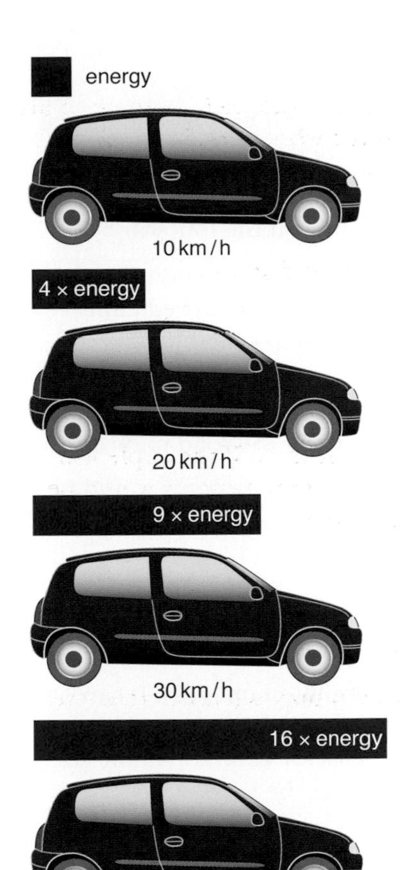

Figure 15.23 Kinetic energy is proportional to the square of the speed.

Roller-coaster (go to *Site Map*, *Exhibits*, *Amusement park physics* and then *Roller coaster*): www.learner.org

H

With the units added,

$$\text{kinetic energy (J)} = \frac{1}{2}\text{ mass (kg)} \times \text{speed (m/s) squared}$$

Increasing the speed has an enormous effect on the energy. Doubling the speed from 2 m/s to 4 m/s increases the energy four times. This has an enormous effect on road safety.

Example

A car of mass 1000 kg has a speed of 20 m/s. What is its kinetic energy?

Formula first:

$$\text{kinetic energy} = \frac{1}{2}\text{mass} \times \text{speed squared}$$

Put the numbers in:

$$\text{kinetic energy} = \frac{1}{2} \times 1000 \times 20 \times 20$$

$$= 1000 \times 10 \times 20\,\text{J}$$

$$= 200\,000\,\text{J}$$

Example

A roller coaster with all its occupants has a total mass of 1000 kg. It drops from a height of 120 m down to 20 m. What is the change in its potential energy?

Formula first:

$$\text{change in potential energy (J)} = \text{mass (kg)} \times \text{gravitational field strength (N/kg)} \times \text{change in height (m)}$$

The change in height is given by:

$$\text{change in height} = 120\,\text{m} - 20\,\text{m} = 100\,\text{m}$$

Put the numbers in:

$$\text{change in potential energy} = 1000\,\text{kg} \times 10\,\text{N/kg} \times 100\,\text{m}$$

$$= 1\,000\,000\,\text{J}$$

As 1 MJ (megajoule) = 1 000 000 J, we can write:

$$\text{change in potential energy} = 1\,\text{MJ}$$

Questions

27 The maximum kinetic energy of the roller coaster in Example 2 was only 0.8 MJ. What has happened to the 'lost' energy?

28 A roller coaster has a mass of 1000 kg and a speed of 30 m/s. What is its kinetic energy?

29 A roller coaster of mass 400 kg falls through a height of 150 m. If the gravitational field strength is 10 N/kg, what is its kinetic energy at the bottom? In this instance all the potential energy is converted to kinetic energy.

30 A car has a mass of 1000 kg and travels at a steady speed of 10 m/s. Calculate its kinetic energy. The speed of the car is doubled to 20 m/s. What is its kinetic energy? What happens to this energy when the car stops?

Stopping safely

The faster a car travels the more kinetic energy it has. To stop the car the driver has to rely on the friction in the brakes and the friction between the tyre and road surface to do work to bring it to rest.

How far a car travels before it stops depends on the following:

- How long it takes the driver to recognise a hazard and to put on the brakes. During this time the car will have moved along the road towards the hazard. This distance is called the **thinking distance.**
- The car will decelerate when the brakes are applied. During braking the car will continue to move along the road towards the hazard. This is called the **braking distance**.

The total **stopping distance** is the thinking distance plus the braking distance.

Factors which affect the thinking distance include the following:

- The speed of the car.
- The driver's **reaction time**. This is normally about 0.7 s. It can be much slower if he or she has been drinking alcohol or taking drugs; some 'over the counter' remedies for common ailments may also cause drowsiness. The driver may be tired.
- The driver may be distracted, perhaps looking for a road sign.
- The driver may have been using a mobile phone; even hands-free sets seriously affect thinking distances.

Factors which affect braking distance include the following:

- The speed of the car. The kinetic energy of the car depends on the square of its speed.
- The mass of the car.
- The condition of the brakes. Excess wear or contamination with oil and grease will seriously affect the brakes.
- The condition of the tyres. They must have 1.6 mm of tread (groove) over 75% of the width of the tyre. The grooves clear away the water on a wet road. With smooth tyres and a wet road a thin layer of water builds up between the road and the tyre. It is a dangerous condition known as aquaplaning.
- The condition of the road surface. Some are more slippery than others.
- The weather: ice and snow are obvious hazards. A light shower after a long dry spell can make the road very slippery.

Figure 15.24 shows how thinking distance and stopping distance increase with speed. Reducing speed can prevent accidents and save lives, because the stopping distance is reduced.

Did you know?

In 2001 a driver fell asleep at the wheel of his four-wheeled drive vehicle. It plunged off the M62 and onto a railway line. The driver escaped, but the car was hit by a GNER (Great North Eastern Railway) express and a coal train. Ten people were killed and 70 injured. The driver was jailed.

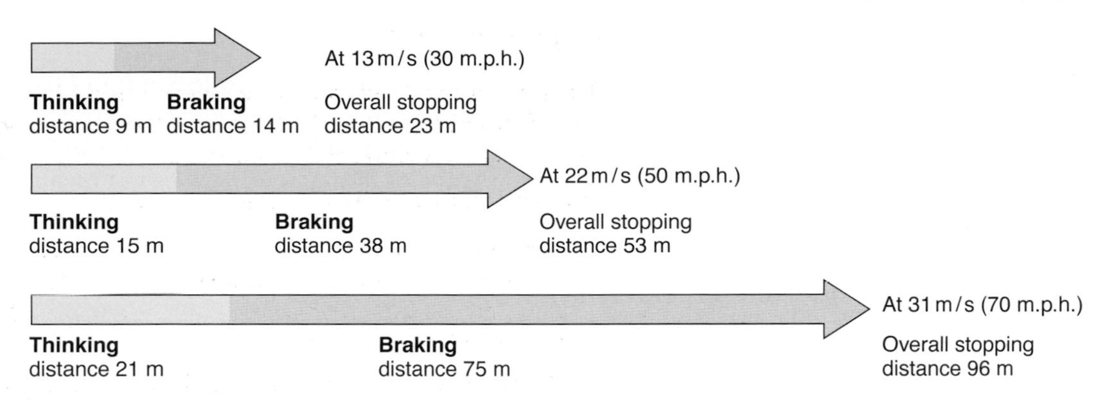

Figure 15.24 Thinking and stopping distances

Activity

Use a 50 cm rule as a reaction timer. Hold your finger and thumb either side of the lower end of the rule. Your partner will release the rule and you must catch it without moving your hand. The slower your reaction, the more of the rule will pass through your fingers.

Questions

31 Explain the terms:
  a kinetic energy
  b potential energy.

  Name the unit of kinetic energy.

32 Explain the terms, as applied to road safety:
  a thinking distance
  b braking distance.

33 Name the factors which can affect:
  a thinking distance
  b braking distance.

34 A car on a test track was brought to rest under a constant force. This was repeated using the same braking force, for cars travelling at different speeds. The graph right shows how the overall stopping distance depends on the speed of the car.
  a Use the graph to find the speed of the car, if it was stopped in 40 m.
  b Throughout the tests, a constant reaction time of 0.6 s elapsed before the braking force was applied.
  c Write down, in words, the equation connecting *distance*, *speed* and *time*.
  d For a car travelling at 30 m/s, calculate the distance travelled in the reaction time.
  e The overall stopping distance is the braking distance plus the reaction (thinking) distance. Use your answer to **d** and the graph to find the braking distance for a car travelling at 30 m/s.

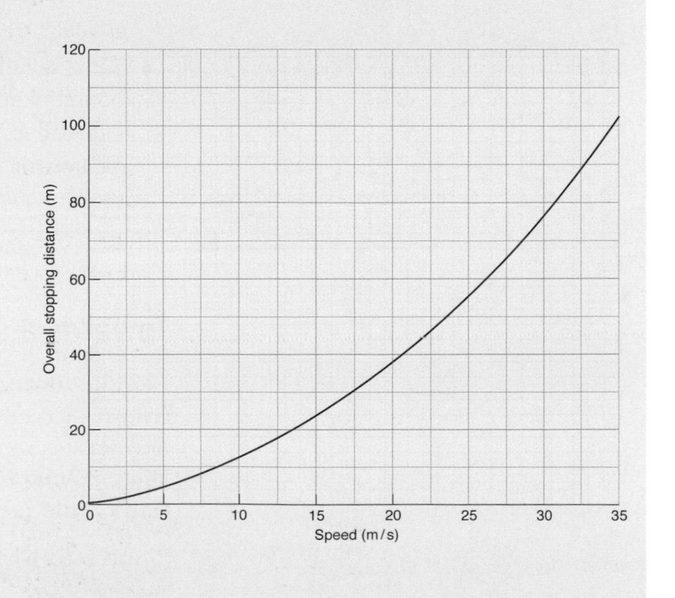

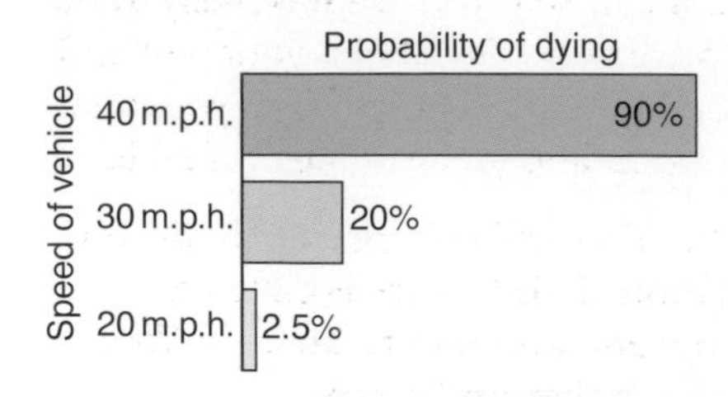

Figure 15.25 The likelihood of dying when hit by a car at speed

Safe speeds

Motorways are safe roads and traffic can travel fast on them. Town centres are busy; people crossing roads, parked cars, schools, etc. all add to the danger. In built-up areas vehicles are limited by law to a maximum speed of 30 m.p.h. This does not mean that it is safe to travel at 30 m.p.h. Drivers should always adjust their speed to the road conditions.

Not all drivers keep to the limit. To enforce the law, local authorities and the police use speed cameras. The camera takes two flash photographs of the car. White marks on the road enable police to see how far the car has travelled in the time between the flashes. They can work out the driver's speed.

Some drivers speed up after passing a camera. To stop this dangerous practice the camera is connected to a computer which records the number plate of every vehicle. A second camera that may be some miles further down the road records the number plates again. The computer uses the time interval and distance between the two to calculate speed.

Other road safety measures include the following.

- 'Sleeping policemen' are raised humps across the road. Motorists must slow down before they drive over them, or risk damage to their car.
- Road-width restrictions can be set up where one-half of the road is blocked off for a short distance.

Activity

1. Why do we have speed limits?
2. Make a list of the road safety measures used to reduce speed in your area.
3. For a selection of the measures used, discuss the effect they have had on road safety in your area. Is the road a safe place for all road users? If you can, suggest some improvements.
4. Some say speed cameras are essential for road safety. Some say they do not differentiate between those just over the limit and those driving at high speed. Others, mainly motorists, say they are a good source of income for the police and the local council. In your group discuss the case for and against speed cameras.

Towards a safer car

Car manufacturers are continually improving and adding safety features to reduce injuries in collisions. When a car crashes, it decelerates rapidly. If you are in a car that suddenly slows down, you keep moving forwards until a force acts to change your speed. That force could be between your head and the windscreen.

$$\text{force (N)} = \text{mass (kg)} \times \text{deceleration (m/s}^2\text{)}$$

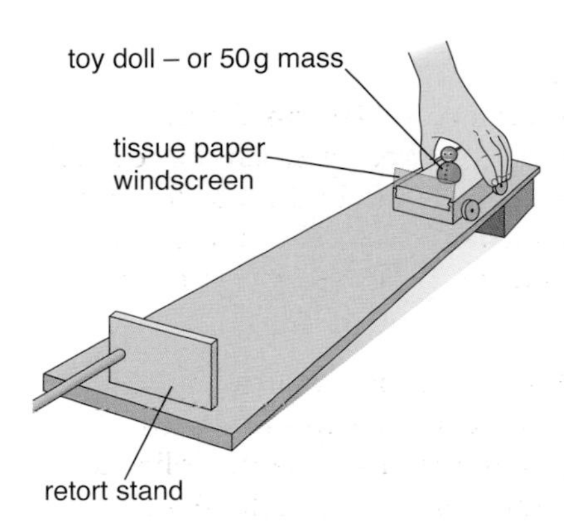

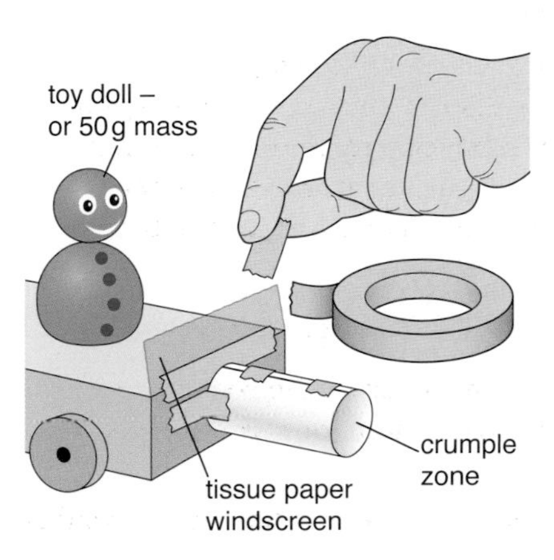

Figure 15.26 Apparatus to show the effect of a car crumple zone

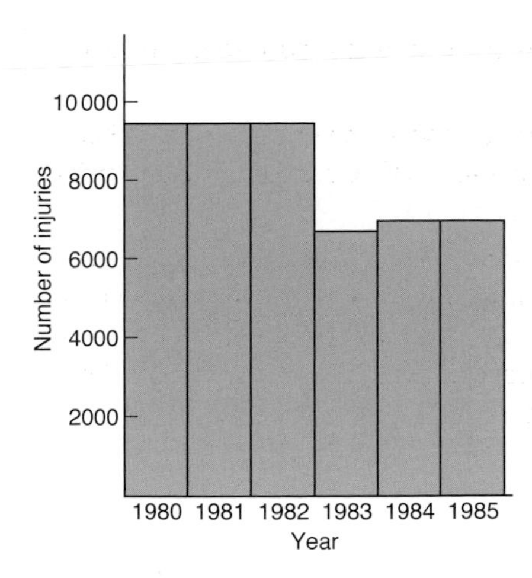

Figure 15.27 Serious injuries were reduced dramatically after the wearing of seat-belts was made compulsory in the UK in 1983.

If you decelerate quickly there is a big force on your body. Anything that reduces the deceleration will mean that you are less likely to be injured. Jump off a small step about 15 cm high. Keep your legs stiff and your body will get a jolt. If you bend your knees when you land, you will have reduced the deceleration and you will hardly feel it.

Car makers design their cars so that they will collapse gradually on impact. This reduces the deceleration. This feature is called a **crumple zone**. There is one at the front, and one at the rear. Your teacher may demonstrate this with a trolley and ramp.

The central part of the car is extremely strong and does not crumple. This **safety cage** is designed to protect all the occupants in the event of a collision. All new cars are crash-tested to check their safety. The safest cars have a five-star rating.

Practical work

Investigating the effect of a car crumple zone

1. Release the trolley and let it crash into the retort stand.
2. The dummy driver keeps on moving and smashes into the windscreen.
3. With the crumple zone in place, the trolley decelerates more slowly. The dummy driver may hit the windscreen, but with a much smaller force.
4. An elastic band makes a good seat-belt for the dummy in the trolley. Repeat the experiment to find out if it protects the driver.

Seat-belts

Safety belts have made a dramatic improvement to our chances of survival in a crash (Figure 15.27). Seat-belts provide the force that decelerates you in a crash. By stretching slightly, they increase the time taken to decelerate you, and so reduce the impact force.

Seat-belts prevent front seat passengers from hitting the windscreen, and rear seat passengers from crushing those in the front. Some cars have seat-belt pre-tensioners. They sense the impact and tighten the belt a little to reduce the effect of the collision.

Questions

35 What is the purpose of the crumple zone in a car?

36 What is the purpose of the safety cage in a car?

37 What does a seat-belt pre-tensioner do?

Airbags

Fall on the floor and it hurts. There is a big decelerating force which slows you down quickly. Fall on a soft mattress and you are slowed down much more gently. New cars are fitted with **airbags**. They inflate automatically on impact. When the driver's head hits the airbag, the bag deflates slowly as the force of impact pushes some of the gas out. The airbag increases the time it takes for the driver's head to slow down and stop. This makes the forces on the driver smaller.

Summary

1. Average speed can be calculated using the following formula.

$$\text{average speed (m/s)} = \frac{\text{distance covered (m)}}{\text{time taken (s)}}$$

2. Acceleration is the rate of change of speed.

$$\text{acceleration (m/s}^2\text{)} = \frac{\text{change in speed (m/s)}}{\text{time taken to change speed (s)}}$$

3. Speed is shown by the slope or gradient of a distance–time graph.
4. Acceleration is shown by the slope or gradient of a speed–time graph.
5. Forces can change the speed of an object, the direction in which it moves, and the shape or size of an object.
6. Balanced forces have no effect on the motion of an object.
7. Unbalanced forces can make an object accelerate, decelerate or change direction.
8. Force is measured in newtons (N).
9. Resultant force (N) = mass (kg) × acceleration (m/s^2).
10. When an object moves through the air it experiences a drag force called air resistance.
11. A body falling from a sufficient height will initially accelerate as the weight force is initially greater than the air resistance. Air resistance increases with speed until weight and air resistance forces balance; the body then falls at a terminal speed.
12. Cars and aeroplanes are streamlined to reduce air resistance.
13. Work done (J) = force (N) × distance moved (m) (in the direction of the force).
14. Kinetic energy (J) = $\frac{1}{2}$ mass (kg) × speed (m/s) squared.
15. Gravitational potential energy (J) = mass (kg) × gravitational field strength (N/kg) × height (m).
16. The faster a car is travelling, the longer it takes to stop.
17. Stopping distances depend on speed and the state of the driver (reaction time), the car (brakes and tyres) and the road surface.
18. Car makers are improving the safety of cars by fitting crumple zones, safety cages, seat-belts and airbags.
19. Local authorities impose speed restrictions in areas where speed could be a danger.
20. The police use speed cameras to enforce the law.

Periodic Table

1	2											3	4	5	6	7	0
						1 H hydrogen 1											4 He helium 2
7 Li lithium 3	9 Be beryllium 4											11 B boron 5	12 C carbon 6	14 N nitrogen 7	16 O oxygen 8	19 F fluorine 9	20 Ne neon 10
23 Na sodium 11	24 Mg magnesium 12											27 Al aluminium 13	28 Si silicon 14	31 P phosphorus 15	32 S sulphur 16	35 Cl chlorine 17	40 Ar argon 18
39 K potassium 19	40 Ca calcium 20	45 Sc scandium 21	48 Ti titanium 22	51 V vanadium 23	52 Cr chromium 24	55 Mn manganese 25	56 Fe iron 26	59 Co cobalt 27	59 Ni nickel 28	63 Cu copper 29	64 Zn zinc 30	70 Ga gallium 31	73 Ge germanium 32	75 As arsenic 33	79 Se selenium 34	80 Br bromine 35	84 Kr krypton 36
85 Rb rubidium 37	88 Sr strontium 38	89 Y yttrium 39	91 Zr zirconium 40	93 Nb niobium 41	96 Mo molybdenum 42	Tc technetium 43	101 Ru ruthenium 44	103 Rh rhodium 45	106 Pd palladium 46	108 Ag silver 47	112 Cd cadmium 48	115 In indium 49	119 Sn tin 50	122 Sb antimony 51	128 Te tellurium 52	127 I iodine 53	131 Xe xenon 54
133 Cs caesium 55	137 Ba barium 56	139 La lanthanum 57	178 Hf hafnium 72	181 Ta tantalum 73	184 W tungsten 74	186 Re rhenium 75	190 Os osmium 76	192 Ir iridium 77	195 Pt platinum 78	197 Au gold 79	201 Hg mercury 80	204 Tl thallium 81	207 Pb lead 82	209 Bi bismuth 83	210 Po polonium 84	210 At astatine 85	222 Rn radon 86
Fr francium 87	226 Ra radium 88	227 Ac actinium 89															

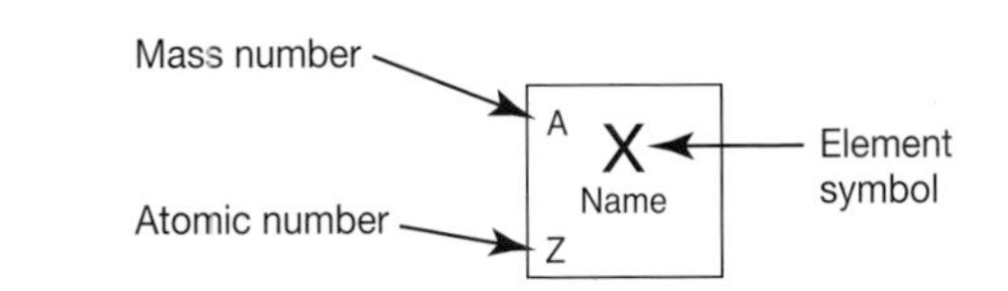

Formulae for some common ions

Positive ions		Negative ions	
Name	Formula	Name	Formula
Aluminium	Al^{3+}	Bromide	Br^{-}
Ammonium	NH_4^{+}	Carbonate	CO_3^{2-}
Barium	Ba^{2+}	Chloride	Cl^{-}
Calcium	Ca^{2+}	Fluoride	F^{-}
Copper(II)	Cu^{2+}	Hydroxide	OH^{-}
Hydrogen	H^{+}	Iodide	I^{-}
Iron(II)	Fe^{2+}	Nitrate	NO_3^{-}
Iron(III)	Fe^{3+}	Oxide	O^{2-}
Lithium	Li^{+}	Sulphate	SO_4^{2-}
Magnesium	Mg^{2+}		
Nickel	Ni^{2+}		
Potassium	K^{+}		
Silver	Ag^{+}		
Sodium	Na^{+}		

Index